Cell Engineering

Cell Engineering

Volume 2

The titles published in this series are listed at the end of this volume

CELL ENGINEERING

Vol. 2: Transient Expression

Edited by

Mohamed Al-Rubeai

School of Chemical Engineering,
The University of Birmingham,
Edgbaston, Birmingham, U.K.

KLUWER ACADEMIC PUBLISHERS
DORDRECHT / BOSTON / LONDON

Library of Congress Cataloging-in-Publication Data

ISBN 0-7923-6596-8

Published by Kluwer Academic Publishers,
P.O. Box 17, 3300 AA Dordrecht, The Netherlands.

Sold and distributed in North, Central and South America
by Kluwer Academic Publishers,
101 Philip Drive, Norwell, MA 02061, U.S.A.

In all other countries, sold and distributed
by Kluwer Academic Publishers,
P.O. Box 322, 3300 AH Dordrecht, The Netherlands.

Printed on acid-free paper

Printed in the Netherlands.

Transient Expression

Cell Engineering
Volume 2

With contributions by

E. Ailor
A. R. Bernard
M. J. Betenbaugh
J. P. Condreay
A. M. Douar
R. F. Jarman
M. Jordan
J. L. Klein
T. A. Kost
P. Liljeström
C. Mannix
B. Massie
W. Oualikene
C. Smerdou

Contents

APPLICATION OF RECOMBINANT BACULOVIRUSES IN BIOPHARMACEUTICAL RESEARCH

T. A. KOST*, J. L. KLEIN AND J. P. CONDREAY
Protein Sciences and BioMetabolism Departments, Glaxo Wellcome Research and Development, 5 Moore Drive, Research Triangle Park, North Carolina 27709 U.S.A.
**Corresponding author, fax (919)483-0585*
e-mail tak4306@glaxowellcome.com

1. Introduction

The first papers describing the use of insect cell baculoviruses as expression vectors were published in the early 1980's (Smith *et al.*, 1983; Pennock *et al.*, 1984). Interestingly, as measured by the number of published papers identified using the terms "recombinant and baculovir*" to search the PubMed database, the system had only limited use through 1990 (Fig. 1). One reason for the lack of utilization over this period may have been due to the tedious procedures initially required to identify and isolate recombinant baculoviruses (rBV). The recombination frequency of the original baculovirus expression vector system (BEVS) was very low, less than 1%. This low recombination frequency made it difficult for those unskilled in the art to identify and isolate purified recombinant viruses. In the 1990's the development of a variety of new transfer plasmids (Jarvis, 1997), linearized, gapped viral DNA's (Kitts *et al.*, 1990; Kitts and Possee, 1993) and a novel system based on site-specific transposition in *E. coli* (Luckow *et al.*, 1993) greatly simplified and rapidly expanded the use of the system. In addition, the reagents required for generating recombinant viruses became more widely available through commercial sources.

A number of attributes, highlighted in Table 1, make the BEVS system particularly attractive for the small and large scale production of recombinant proteins in insect cells (Miller, 1997; Vlak *et al.*, 1996). In order to compensate for the potential disadvantages of the system, a variety of approaches are being investigated to overcome the glycosylation deficiencies of insect cells as compared to mammalian cell lines (Jarvis *et al.*, 1998; Altmann *et al.*, 1999). Non lytic insect cell based systems have also been used successfully for the stable production of recombinant proteins (McCarroll and King, 1997; Pfeifer, 1998); however, these approaches will not be discussed in this chapter.

In addition to their application for the production of recombinant proteins in infected insect cell and larvae hosts, (rBV) have been modified to direct transient and stable expression of recombinant proteins in mammalian cells, as reviewed by Kost and Condreay (1999). rBV containing mammalian viral and cellular promoter elements have been shown to efficiently transduce a wide variety of mammalian cell types without

M. Al-Rubeai (ed.), Cell Engineering, 1–28.

discernable cytopathology. This provides a unique system, whereby one can easily generate and produce the desired recombinant viruses in the natural insect cell host and then use them to transduce mammalian cells in the absence of viral replication.

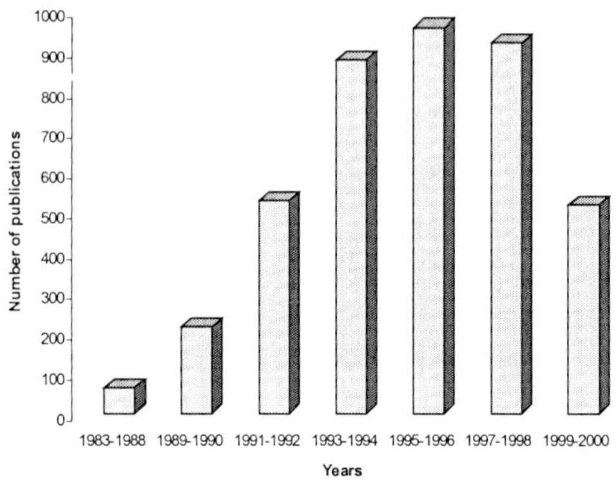

Figure 1. Publications utilizing rBV.

Another emerging use of rBV is that of virus display technology, first reported by Boublik *et al.* in 1995. A protein as large and complex as human immunodeficiency virus gp120 has been successfully displayed on the surface of baculovirus particles.

TABLE 1. Advantages and disadvantages of the BEVS for recombinant protein production in insect cells.

Advantages	**Disadvantages**
• Ease of use	• Lytic system
• Wide variety of transfer vectors	• Protein glycosylation patterns
• High expression level potential	
• Functional protein production	
• Production of multi-subunit complexes	
• Proven biosafety profile	
• Anchorage dependent or suspension culture cell growth	
• Cell growth in serum free medium	
• Ease of scale-up	

The sequence of the human genome is nearly complete. The human genome is predicted to encode upwards of 100,000 gene products. Undoubtedly rBV, as well as other viral vector systems, will serve as useful tools for the production and study of many of these proteins. The intent of this chapter is to provide a brief overview of baculovirus biology, vector technology and its application for recombinant protein production in insect and mammalian cells. Insect cell applications will focus on the production of multi-subunit protein complexes, G-protein coupled receptors, ion channels and cytochrome P450s. The more recent application of modified recombinant baculoviruses for transient and stable protein production in mammalian cells and developments in the application of baculovirus display will also be described.

2. Baculovirus Biology

Baculoviridae comprise a family of lytic viruses that are pathogens of a wide variety of arthropods. Study of the granulovirus genus of theses viruses is hampered by a lack of cell culture systems to propagate the virus (Kelly, 1982). In contrast, much is known about the biology of the nucleopolyhedrosisviruses, including the prototype of this genus, *Autographa californica* nuclear polyhedrosis virus (AcMNPV). AcMNPV can be efficiently propagated *in vivo* and *in vitro* in cultured cell lines derived from insect hosts. A collection of recent reviews of baculoviruses can be found in Miller (1997), however some relevant aspects of AcMNPV biology will be discussed as they relate to the development of this virus as an expression vector.

The genome of these viruses is a double-stranded circular DNA ranging in size from 90 to 160 kb. Currently, in addition to AcMNPV (Accession number L22858; Ayers *et al.*, 1994), the complete nucleotide sequences of four other baculoviruses (Accession numbers AF169823, L33180, AF081810, and U75930) have been determined. The AcMNPV genome is 133,894 bp and is predicted to contain approximately 150 open reading frames dispersed througout the genome on both strands. The complete pattern of all expressed genes and their functions remain to be fully elucidated.

Baculovirus genes are grouped into classes depending upon their time of expresssion after infection (Kool and Vlak, 1993). The immediate early genes do not require viral gene expression to be transcribed, and thus, transcriptional regulatory regions from these genes have found utility as promoters for the engineering of stably transfected insect cell lines (McCarroll and King, 1997; Pfeifer, 1998). However, transcription of successive groups of genes require products from the previous group for their expression (Miller, 1996). Late and very late genes contain a consensus transcriptional start site, the TAAG motif (Blissard and Rohrmann, 1990). Promoters from these highly regulated genes, that are used to direct expression of foreign genes in rBV, are not expressed in uninfected insect cells (Miller, 1996), or in other cell types in which the virus does not replicate (Carbonell *et al.*, 1985; unpublished data from our laboratory).

Baculoviruses take on one of two morphological forms that fulfill different roles in the dual life cycle of the virus (Miller, 1996). Both forms of the virus contain the same nucleocapsid structure and genetic complement. Occlusion-derived virus (ODV) consists of one or more membrane-enclosed nucleocapsids which are embedded in a protein matrix forming a polyhedron. This form of the virus is responsible for infection of the insect host in the wild. The protein matrix of the polyhedron is comprised largely of the product of the viral polyhedrin protein gene (Rohrmann, 1986), the single most highly expressed viral gene product (Jarvis, 1997). The production of polyhedra or occluded virus occurs late in infection.

Early in infection, nucleocapsids migrate out of the nucleus where the virus replicates and bud through the cytoplasmic membrane of the cell (Fraser, 1986). As it exits the cell the budded form of the virus picks up an envelope with spike structures at one end (Fig. 2) that consist of trimers of viral gp64 protein (Oomens *et al.*, 1995). The budded virus form is responsible for cell to cell transmission within an infected insect. Budded virus is the form used in the laboratory as an expression vector in cultured insect cells; it is not able to initiate an infection in a host insect (Monsma *et al.*, 1996). The gp64 protein is responsible for the fusion of the viral envelope and the cell membrane to initiate infection by AcMNPV (Blissard and Wenz, 1992; Monsma and Blissard, 1995).

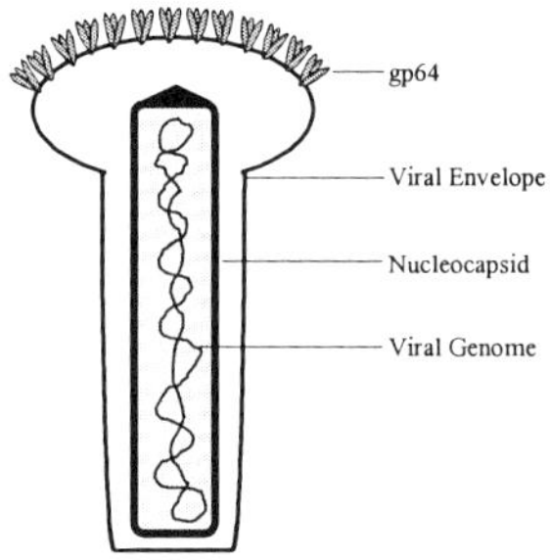

Figure 2. Schematic drawing of AcMNPV budded virion.

The polyhedrin protein gene, as mentioned previously, is involved in the formation of ODV. In the absence of polyhedrin protein budded virus is still secreted by infected cells, though polyhedra are not formed. As such, this gene is dispensible for propagation of the virus in cell culture (Miller, 1996). This, coupled with the high levels of transcription directed by the promoter for this gene provide an opportunity to exploit the biology of the system for the production of recombinant proteins. Genes placed under transcriptional control of the polyhedrin (*polh*) gene promoter can be recombined into the viral genome to produce a virus stock. The length of the viral capsid is flexible and is able to accommodate increases in the viral genome length (Fraser, 1986). Upon infection of an appropriate host cell this virus will direct high level expression of the desired gene product between 24 to 48 hours postinfection.

Some characteristics of BEVS that can limit its usefulness should be considered. Infection of cells by rBV can cause changes to the cellular processes which can lead to differences in expression and modification of proteins as infection proceeds (Kovacs *et al.*, 1991; Velardo *et al.*, 1993). Different insect cell lines that are used as hosts for rBV-mediated expression can exhibit differences in the processing of over-expressed proteins (Kost *et al.*, 1997). Furthermore, in some cases, high-level expression can exceed the cell's capacity to fully process recombinant proteins leading to heterogeneity in the product (Jarvis, 1997). However, in general, insect cell lines that are used as hosts for rBV-mediated protein overexpression carry out post-translational modifications on the expressed protein that are functionally equivalent to those found in mammalian cell systems. Insect cells are able to carry out N-linked and O-linked glycosylation of proteins at the same sites that are recognized by mammalian cells, however the composition of the carbohydrate chains added are different in the two systems (Altmann *et al.*, 1999). Other covalent modifications such as phosphorylation are achieved, proteins are secreted, inserted in the cell membrane, and can be targeted to subcellular compartments by their native signals (Jarvis, 1997). In addition, as reviewed in the chapter by Ailor and Betenbaugh, a number of strategies have been employed to enhance the production of properly processed functional proteins.

3. Vectors

3.1. CONSTRUCTION OF rBV

The baculovirus genome is too large (130 Kb) to be readily amenable to conventional cloning techniques. Therefore, "transfer plasmids" have been developed to allow cloning of the sequence of interest. The original approach used transfer plasmids designed to have the foreign gene flanked by sequences homologous to viral DNA sequences. This type of homologous recombination, allowing the introduction of a new sequence into the baculovirus genome, is termed allelic replacement. The desired rBV represents only 1% or less of progeny virus, making the identification of rBV difficult and time-consuming. Various approaches have been developed to either reduce the proportion of non-rBV or increase the frequency of recombination between the transfer plasmid and the viral genomic DNA. Currently, most of the starting materials are commercially available, thus making it relatively easy, even for a non-specialist in baculovirology to generate rBV.

3.1.1. *Allelic Replacement: The Original Approach*

The first step in constructing a rBV consists of cloning the foreign gene cDNA into a transfer plasmid. Typical transfer plasmids contain a cassette with a viral or insect cell promoter, a multiple cloning site and a termination sequence. The cassette is embedded in viral DNA sequences from the region at which it is to be inserted. The sequence identity between the transfer plasmid and viral genome allow for homologous recombination and insertion of the foreign gene into the viral genome. The target locus for insertion is often a dispensable gene for baculovirus propagation in cell culture. The transfer plasmid together with wild-type baculovirus genome DNA are then transfected into cultured insect cells. The resulting virus progeny is a mixture of wild-type parental and rBV. This method leads to the formation of 1% or less rBV, which are routinely identified by plaque morphology. This method is tedious and time-consuming due to the multiple purification rounds required to identify and isolate a clonal virus population.

3.1.2. *Linear Baculovirus DNA*

An improvement to the technique described above was the use of linearized baculovirus DNA for co-transfection. This resulted in a significant reduction in the proportion of parental virus and single-crossover recombinants present in the progeny virus population. Restriction endonuclease digestion of the baculovirus DNA at a single Bsu36 I site reduces its infectivity by 15 to 150-fold compared to the normal circular form (Kitts *et al.*, 1990). In this technique, the background of non-recombinant virus is derived in part from the non-digested viral DNA. A small fraction of uncut DNA can lead to a significant background of non-rBV, because transfected circular viral DNA is significantly more infectious than linearized DNA. The introduction of multiple Bsu36 I restriction sites, which are not normally found in wild type AcMNPV increases the efficiency of generating recombinants (Kitts and Possee, 1993; Martens *et al.*, 1995; Yang and Miller, 1998). In addition, a reduction in the background due to non-rBV can be achieved by deletion of an essential gene i.e. (ORF 1629) from the parental virus DNA. In this instance the viral DNA is rescued by homologous recombination with the transfer plasmid which transfers the sequence of interest and also restores the deleted essential gene function (Vialard and Richardson, 1993). The efficiency of rBV production with these Bsu36 I engineered viral DNAs is routinely greater than 90%.

Instead of essential genes, auxiliary viral genes have also been deleted (Bishop *et al.*, 1995). These genes encode proteins required for baculoviruses to efficiently infect insect larvae but are non-essential for the replication of the virus in cultured insect cells. The product of these auxiliary genes might interfere with the production of proteins. For example, deletion of the baculoviral cathepsin-like protease gene might avoid degradation of baculovirus-expressed recombinant proteins (Hom and Volkman, 1998).

3.1.3. *Transposon-Mediated Insertion in E.coli.*

The baculovirus genome has also been reconstituted as a replicon in *E.coli.* In this approach the modified baculovirus genome, termed a bacmid (Luckow *et al.*, 1993) is maintained in *E.coli.* Similar to wild-type baculovirus DNA, bacmid DNA is infectious when introduced into insect cells. A mini-F replicon was inserted into the polyhedrin locus of the baculovirus genome to confer autonomous replication and stable segregation of plasmids at low copy number. A selectable kanamycin resistance marker and *att*Tn7 sequence, which serves as a target site for the bacterial transposon Tn7, were also introduced into the *polh* gene locus. The *att*Tn7 sequence was introduced in frame within the *lacZα* gene which was used as a screen for recombinant bacmids. The *lacZα* sequence complements the defective β-galactosidase *lacZ*ΔM15 in a *E. coli* strain such as DH 10B giving rise to blue colonies on medium containing X-Gal. Insertions at the *att*Tn7 locus disrupt the *lacZα* gene leading to a β-galactosidase negative phenotype and the development of white colonies. In addition to the bacmid, the bacteria also contain a tetracycline resistant helper plasmid that supplies *trans*-acting functions necessary for the transposition. In the transfer plasmid the recombinant expression cassette and gentamicin acetyl transferase gene are flanked by the left and right arms of Tn7. The transposition and selection steps to give rise to a recombinant viral genome are efficiently accomplished in *E.coli.* The recombinant bacmid DNA is isolated and transfected into cultured insect cells, resulting in the isolation of essentially 100% rBV.

We have successfully applied this methodology in our laboratory to generate numerous rBV.

3.1.4. *Alternative Approaches*

Alternative approaches for generating rBV have been reported, however these techniques have not been widely adapted. To address the problem of low frequency recombination, Peakman and co-workers (1992) designed an *in vitro* system to generate recombinant baculovirus by using the Cre-loxP system of bacteriophage P1. In this system, both the transfer plasmid and baculovirus genome contain loxP sites that can mediate site specific *in vitro* recombination in the presence of the Cre recombinase enzyme. To identify recombinants, a *lacZ* gene under the control of the p10 promoter was included in the transfer plasmid. rBV are identified by the blue color of rBV derived plaques formed in medium supplemented with X-Gal. An additional approach based on homologous recombination in the yeast Saccharomyces cerevisiae has been reported (Patel *et al.*, 1992).

3.2. CHOICE OF TRANSFER PLASMIDS

There are many different transfer plasmids that can be used to insert the sequence of interest into the baculovirus genome. A variety of transfer plasmids can be obtained commercially from Clontech (Palo Alto, CA), InVitroGen (San Diego, CA), Novagen (Madison, WI), Pharmingen (San Diego, CA), Stratagene (La Jolla, CA), and Life Technologies (Rockville, MD). Both viral and cellular promoters have been used to control the expression of foreign proteins.

3.2.1. *Promoters*

Promoters of varying strengths and differential expression during the virus-life cycle can be used to control the expression of the foreign gene. The promoter of the *polh* gene has been widely used for directing the high level production of heterologous proteins. The promoter is activated very late in the baculovirus infection cycle and produces large amounts of RNA (Luckow, 1995). As discussed previously, deletion of the *polh* gene does not affect virus propagation in cell culture.

The p10 gene promoter is another very late viral promoter that has been used to regulate the expression of foreign genes. In some instances these transfer plasmids contain both the p10 promoter and p10 flanking sequences leading to the replacement of the non-essential p10 gene (Vlak *et al.*, 1990). The p10 promoter, in general, directs a lower level of foreign gene expression and has been used less frequently than the *polh* promoter. The p10 locus is frequently used to generate baculoviruses for biopesticide applications. In this case the *polh* gene cannot serve as the target for allelic replacement since baculoviruses used as pesticides must be able to produce polyhedrin to infect insect larvae (Merryweather *et al.*, 1990; McCutchen *et al.*, 1991; Stewart *et al.*, 1991).

Other promoters of either viral (ie1, gp64, or basic promoter) or cellular (OpIE2, Drosophila metallothionein, Drosophila actin 5C) origin have been used to control the expression of foreign genes (Jarvis *et al.*, 1990; Tsao *et al.*, 1990; Xie *et al.*, 1992; Arp *et al.*, 1993; Rankl *et al.*, 1994). In the transfer plasmids, these promoters are generally flanked by *polh* gene sequences for allelic replacement. Most of these promoters are not as strong as the *polh* gene promoter, and are used either to solve problems associated with the undesirable high level production of a particular protein or when high-level expression is not the goal. For example, certain proteins that are normally expressed in cells at a low level are either inactive, insoluble or both when expressed under *polh* gene promoter control (Pajot-Augy *et al.*, 1995). This may be due to saturation of the folding and secretory capacity of the insect cells (Pajot-Augy *et al.*, 1999). In addition, the *polh* gene promoter is a very late promoter and the normal physiology of the infected cells may be compromised by the infection process. (Jarvis and Summers, 1989; Murphy *et al.*, 1990). An alternative is to use either a promoter that is activated earlier during the infection cycle of the virus or is less actively transcribed than the *polh* gene promoter.

3.2.2. *Co-expression*

Co-expression of proteins in cultured cells may be achieved by co-infecting cells with viruses carrying a single gene (St. Angelo *et al.*, 1987). However, this approach can lead to a heterogeneous cell population with some cells expressing only one of the proteins. An alternative approach is to use transfer plasmids containing multiple promoters to express multiple proteins. Emery and Bishop (1987), constructed a transfer plasmid with two *polh* gene promoters resulting in a "dual" recombinant baculovirus that can express two proteins in an infected cell. Combinations of various promoters can be used to generate rBV capable of expressing multiple proteins in an infected cell (Pulford *et al.*, 1995; Belyaev *et al.*, 1995; Lee *et al.*, 1996).

Recently, a bi-cistronic transfer plasmid, containing an internal ribosome entry site (IRES) from encephalomyocarditis virus, has been described (Finkelstein *et al.*, 1999). This allows for expression of two proteins under the control of a single promoter. The expression of chloramphenicol acetyl transferase and luciferase reporter genes was detected when both genes were inserted downstream of the *polh* gene promoter. Although both gene products are expressed from the same mRNA, the gene downstream of the IRES is expressed at a lower level.

3.2.3. *Other Features*

In general, secreted proteins are easier to purify than non-secreted proteins. Vectors are available with secretion signal sequences for fusion to the N-terminal sequence of recombinant proteins to facilitate or force protein secretion (Zhou *et al.*, 2000).

The baculovirus infection cycle ends with insect cell lysis and the release of proteases into the medium. Recombinant proteins may be degraded by these proteases, especially when serum free medium is used. Protease inhibitors can be added into the medium or rBV lacking chitinase and cathepsin protease activities can be used (Pyle *et al.*, 1995; Monsma and Scott, 1997; Grosh, 1998; Hom and Volkman, 1998).

An affinity tag (6xHis, glutathione S-transferase, cellulose binding domain) followed by a proteolytic cleavage site and a multi-cloning site can also be included in the transfer vector. The recombinant protein is expressed as a tag-protein fusion and can be purified by affinity chromatography (Rudi *et al.*, 1997; Gupta *et al.*, 1997; Airenne *et al.*, 1999; Chen and Courey, 1999; Zhou *et al.*, 2000). If needed, the tag can be removed from the recombinant protein by proteolytic cleavage.

3.3. VIRUS TITRATION METHODS

Virus titers can be determined using a variety of techniques. The standard method is the plaque assay. Dilutions of culture medium containing virus are incubated with insect cells and overlaid with a semi-solid medium containing agarose. After 5-7 days plaques can be visualized and enumerated using a dissecting microscope. For plaque purification individual plaques can be picked using a Pasteur pipet. This procedure is described fully in O'Reilly *et al.* (1994). Recently, an immunological assay has been developed that allows the determination of virus titer after 48 hours (Kitts and Green, 1999). This assay takes advantage of the fact that infected cells express gp64 before plaques can be easily visualized. The assay uses a primary anti-gp64 monoclonal antibody to recognize infected cells. A secondary HRP-conjugated antibody allows the visualization of infected cells/mini-plaques which can be readily enumerated using an inverted microscope.

3.5. HOSTS

Either cultured insect cells or caterpillars can be used as hosts for foreign protein production. In the case of insect larvae infection, the *polh* gene should be expressed for the virus to be infectious. The insect cell lines most widely used for protein production are derived from either *Spodoptera frugiperda* (Sf21 or Sf9) or *Trichoplusia ni* (*T. ni*) (for a complete review of host cells used for infection see Vlak *et al.* (1996)).

4. Applications of rBV Infected Insect Cells

4.1. VIRUS-LIKE PARTICLES AND VACCINE ANTIGENS

The BEVS has proven to be valuable for studying viral assembly processes in the absence of infectious virus. Viruses encoding multiple heterologous viral proteins or panels of viruses, each expressing a single subunit, can be used to infect insect cells allowing one to follow particle assembly. The BEVS is particularly suited to this application since only the capsid proteins can be expressed and the effects of other viral gene products eliminated. In addition, the production of virus-like particles (VLP) is proving to be a useful means for the safe production of viral vaccines (Roy, 1996.).

The use of rBV expressing various herpes simplex type-1 virus (HSV-1) proteins has allowed for a detailed analysis of the viral proteins required for capsid assembly (Newcomb *et al.*, 1996, Homa and Brown, 1997). Co-infection of insect cells with six rBV, each expressing one of the six HSV-1 capsid proteins, resulted in the production of empty capsids identical in appearance and protein composition to B-capsids found in HSV-1 infected cells. These results demonstrate the utility of this approach for studying complex protein-protein interactions. More recently, Newcomb *et al.* (1999) have purified HSV-1 capsid proteins from rBV-infected cells and tested the ability of the purified proteins to assemble into procapsids *in vitro*. These studies have demonstrated that procapsids can be readily formed in the absence of additional insect cell factors.

Belyaev *et al.* (1995) described the generation of a single rBV that directed the synthesis of five Bluetongue virus (BTV) proteins in infected insect cells. Zheng *et al.* (1999) have optimized the production of BTV core-like particles (CLP) derived by co-infection of Sf9 cells with rBV encoding the BTV VP7 and VP3 proteins. This study showed that multiplicities of infection as low as 0.0001 plaque forming unit/cell could be used for the successful production of CLPs. In another report using BTV CLPs Adler *et al.* (1998) demonstrated that inclusion of a foreign T cell epitope into the CLP resulted in the production of CLPs capable of inducing a T cell response to the M1 protein of influenza virus.

Rotaviruses are the single most important etiologic agent of diarrheal disease in infants and young children worldwide. Recent studies using rotavirus VLP based vaccines produced by rBV-mediated co-expression of rotavirus capsid proteins in infected Sf9 cells have shown promise as potential vaccine candidates (Madore *et al.*, 1999, Jiang *et al.*, 1999). Jiang *et al.* (1998) have compared the kinetics of rotavirus VLP synthesis by rBV infected Sf9 and the *T. ni* derived BTI-TN-5B1-4 (High-Five) cell line. The highest yields of VLPs were obtained from High Five cells in the late phase of infection when the yield was at least 5-fold higher on a per cell basis than that from Sf9 cells.

The BEVS has also been used for the production of papillomavirus VLPs (Schiller and Roden, 1995). Immunization with purified VLPs composed of papillomavirus L1 or L1 plus L2 capsid proteins was shown to protect beagle dogs, cotton-tail rabbits and cattle against the homologous viruses (Suzich *et al.*, 1995; Breitburd *et al.*, 1995; Kirnbauer *et al.*, 1996). Since papillomaviruses cannot be readily produced in cell culture, the BEVS system has been particularly useful for the production of VLPs. In addition to VLPs containing L1 or L1 and L2 proteins, chimeric VLPs have been produced that contain a portion of the human papilloma virus E7 protein fused to L1. Vaccination of mice with these chimeric VLPs generated a human papilloma virus E7-specific cytotoxic T-lymphocyte response and induced anti-tumor immune reactivity in C57Bl/6 mice (Schäfer *et al.*, 1999). Touze *et al.* (1998) compared the production of VLPs composed of the L1 capsid proteins of six human papilloma type 16 viruses produced using the BEVS. Interestingly, even though the amino acid sequences of the L1 proteins were relatively similar the yield of VLPs varied greatly depending on the papillomavirus strain. These results are important to consider during vaccine development using rBV produced VLPs. It has also been demonstrated that BEVS produced papillomavirus VLPs are able to package unrelated plasmid DNA *in vitro* and deliver the foreign DNA to eucaryotic cells with subsequent expression of the foreign gene (Touze and Coursaget, 1998). Murine polyoma virus pseudocapsids, composed of the polyoma virus VP1 protein, have also been produced using BEVS and shown to be capable of packaging and delivering foreign DNA into mammalian cells (Forstova *et al.*, 1995; Soeda *et al.*, 1998). Thus, engineered VLPs may represent promising vehicles for foreign gene delivery.

The BEVS system has proven very useful for the production of retroviral and lentiviral VLPs (Takahashi *et al.*, 1999; Notka, *et al.*, 1999; Kang, *et al.*, 1999; Kakker *et al.*, 1999; Yao *et al.*, 2000). The choice of host cell may have a direct influence on the production of VLPs. A recent report describes a defect in the ability of High Five cells to efficiently transport and release retroviral capsids. This defect was not observed with Sf9 cells (Parker and Hunter, 2000).

The production of VLPs using BEVS has also been described for the human polyomavirus JC virus (Goldmann *et al.*, 1999), adeno-associated virus (Hoque *et al.*, 1999) and hepatitis C virus (Baumert *et al.*, 2000). In summary, the VLPs produced using the BEVS have proven very useful in studying viral assembly and have shown promise as vaccine candidates and novel gene delivery vectors.

4.2. EXPRESSION OF G- PROTEIN COUPLED RECEPTORS AND ION CHANNELS

The BEVS has been used extensively for the study and production of G-protein coupled receptors (GPCR). It has proven especially useful for studying the selectivity of coupling of receptors for distinct G proteins. Sf9 cells have a low level of endogenous G proteins, thus coupling specificity can be examined by co-expression of the receptor with different G protein subtypes. This topic has been reviewed by Bouvier *et al.* (1998). A recent example of the utility of this approach demonstrates that human 5-hydroxytryptamine$_{5A}$ receptors selectively function and couple to co-expressed mammalian G_i and G_o proteins (Francken *et al.*, 2000). In this study coupling was studied using a series of 11 mammalian G proteins. Lachance *et al.* (1999) utilized the BEVS to demonstrate that all of the properties of the β_2-adrenergic receptor observed in native receptor systems could be reconstituted in Sf9 cells by co-expression of the receptor with heterotrimeric mammalian G_s proteins. In terms of expression levels of functional GPCRs in Sf9 cells, a number of reports have indicated that in frame fusions of a heterologous signal sequence to the amino terminus of the receptor can increase the production of functional receptors (Guan *et al.*, 1992; Gimpl *et al.*, 1995; Gruenewald *et al.*, 1996; Massotte *et al.*, 1999; Hampe *et al.*, 2000; Kost *et al.*, unpublished results). It has also been reported that in the case of the μ-opioid receptor the 5'utranslated region functions to increase expression of the receptor in the BEVS (Belke-Louis *et al.*, 1999). Tate *et al.*, (1999a,b) reported that co-expression of the chaperone calnexin improves functional expression of the serotonin (5-hydroxytryptamine transporter) transporter. It has also been reported that deletion of the *chiA* gene, encoding chitinase, from AcMNPV reduced the accumulation of the nicotinic acetylcholine receptor alpha-subunit within the endoplasmic reticulum of infected cells. Approaches such as these offer promise for increasing the production of various membrane bound and secreted proteins.

The BEVS has been shown to support the assembly of both homomeric and heteromeric ion channel complexes. The ease of using rBV for the co-expression of multiple channel subunits has provided a productive approach for the expression of functional proteins, as reviewed by Radford and Buell (1998). Recent reports have extended the system for the functional expression of ATP-sensitive-K-channels (Mikhailov *et al.*, 1998, Mikhailov and Ashcroft, 2000). K_{ATP} channels are complex, made up of two components, a channel-forming subunit, Kir6.2, and a larger regulatory subunit, SUR1, which mediates the effects of sulphonylureas and dizoxide on channel activity. These studies demonstrate the capacity of the BEVS to produce functional assemblies of multi-subunit protein complexes.

4.3. PRODUCTION OF CYTOCHROME P450 ENZYMES FOR DRUG METABOLISM STUDIES

The cytochrome P450 (CYP) membrane bound monooxygenases constitute a family of enzymes that catalyze the biotransformation of a wide variety of chemically and biologically unrelated exogenous and endogenous substrates. The metabolism of drugs and xenobiotics has been shown to be one of the most important factors in determining the biological and toxicological effects of exposure. Thus, defining the role played by various CYPs in drug metabolism is an important aspect of the drug development process (Crespi and Miller, 1999).

The BEVS has been used successfully for the production of a large number of functional CYPs. In particular, the ability to infect cells with multiple viruses and viruses co-expressing both a CYP and NADPH-CYP reductase has proven very useful (Lee *et al.*, 1995; Paine *et al.*, 1996; Wang *et al.* 1996; Ong *et al.*, 1998; Ohmori *et al.*, 1998). These studies have demonstrated that high yields of functional recombinant CYPs can be expressed in insect cells in the presence of co-expressed NADPH-CYP reductase. In most of these studies either the very late *polh* gene or p10 promoter was used to regulate CYP and NADPH-CYP reductase expression. However, Paine *et al.* (1996) showed that by using the late baculovirus basic protein promoter to regulate CYP expression one could reduce the proportion of inactive protein produced.

Crespi *et al.* (1997), utilizing rBV infected insect cell CYPs, described a microtiter based plate assay for rapidly conducting higher-througput CYP inhibition studies. Palamanda *et al.* (1998), using this system for CYP2D6 inhibition studies, concluded that the system could be used for high-throughput assays, but human liver microsomes should subsequently be used to confirm the assay results on the compounds selected for further development. An *in situ* toxicity assay system using the BEVS has also been described (Grant *et al.*, 1996). Although baculovirus infection is a lytic process, the authors found that after infection a 24-48 hour timeframe exists during which *in situ* toxicity could be measured.

5. Baculovirus-Mediated Expression in Mammalian Cells

Interest in interactions between baculoviruses and non-target cells was fueled in the 1980s by the desire to use these pathogens as biological pesticides. Volkman and Goldsmith (1983) incubated a panel of vetebrate cell lines with both budded virus and occluded virus forms of AcMNPV. Virus could be detected in several of the cell lines both by immunostaining and electron microscopy. No evidence of viral replication was detectable in the non-insect cells. Similar results have been reported after treatment of HeLa, primary human embryonic kidney, BHK21, or CV-1 cells with AcMNPV (Tjia *et al.*, 1983). Viral DNA could be detected in the nuclei of virus-treated cells out to 24 hours post-inoculation, but several passages later, no viral sequences were detected in the cell cultures. Viral DNA was neither replicated nor transcribed in these cells, and once again the virus failed to propagate in cultures of mammalian cells. The inability of AcMNPV to replicate or persist in mammalian cells has also been demonstrated in human WI-38 cells and foreskin fibroblasts (Hartig *et al.*, 1991).

Until recently the ability of recombinant baculoviruses carrying reporter genes under the transcriptional control of highly active mammalian promoters had only been investigated in a few mammalian cell lines. A virus carrying the chloramphenicol acetyltransferase (CAT) gene controlled by the Rous sarcoma virus long terminal repeat was originally reported to give rise to a very low level of CAT expression upon prolonged exposure of mouse L929 cells to the virus (Carbonell *et al.*, 1985). Subsequently it was demonstrated (Carbonell and Miller, 1987) that the low level of CAT expression seen in both L929 cell and the human lung cell line A549 was also detected in the presence of protein synthesis inhibitors. Low-level CAT activity was detected in purified virus particles, thus it was concluded that the CAT activity was carried into the mammalian cells along with the virus inoculum and was not due to *de novo* gene expression.

5.1. TRANSIENT EXPRESSION IN MAMMALIAN CELLS

Within a few months of each other two groups reported success using recombinant baculoviruses carrying mammalian transcription cassettes to effect gene expression in cells primarily of hepatic origin. Viruses containing either a cytomegalovirus (CMV) promoter luciferase cassette (Hofmann *et al.*, 1995), or a Rous sarcoma virus promoter β-galactosidase cassette (Boyce and Bucher, 1996) were tested for gene delivery on a panel of cell lines. High levels of reporter gene expression were seen in primary hepatocytes and hepatoma cell lines within 24 to 36 hours after virus treatment. Lower levels of reporter gene expression were seen in a few other cell lines (COS-1, T-47D, and 293). In these studies the level of gene expression was dose-dependent with respect to the virus inoculum.

Further reports have expanded the list of cell lines that are susceptible to rBV mediated gene delivery. High level gene expression was demonstrated in several cell lines, including HeLa, COS-7, and CPK, using a baculovirus containing transgenes under the control of a hybrid CMV/β-actin promoter (Shoji *et al.*, 1997; Yap *et al.*, 1998). Furthermore, the study by Yap and co-workers used a two component expression system with one virus expressing the T7 RNA polymerase and a second virus carrying a reporter gene controlled by a T7 promoter to show that cells could be simultaneously transduced with with two different viruses. Efficient, high-level baculovirus-mediated expression of green fluorescent protein (GFP) driven by the CMV promoter was demonstrated in a large panel of cell lines that included primary cultures of human keratinocytes and bone marrow fibroblasts (Condreay *et al.*, 1999). The observed transduction efficiency and expression level was low in cells of hematopoetic origin and PC12 cells.

A.

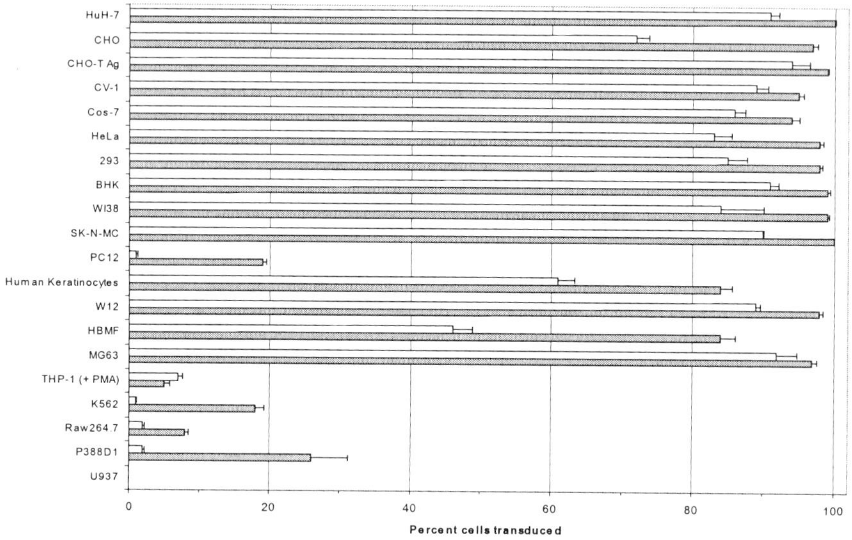

B.

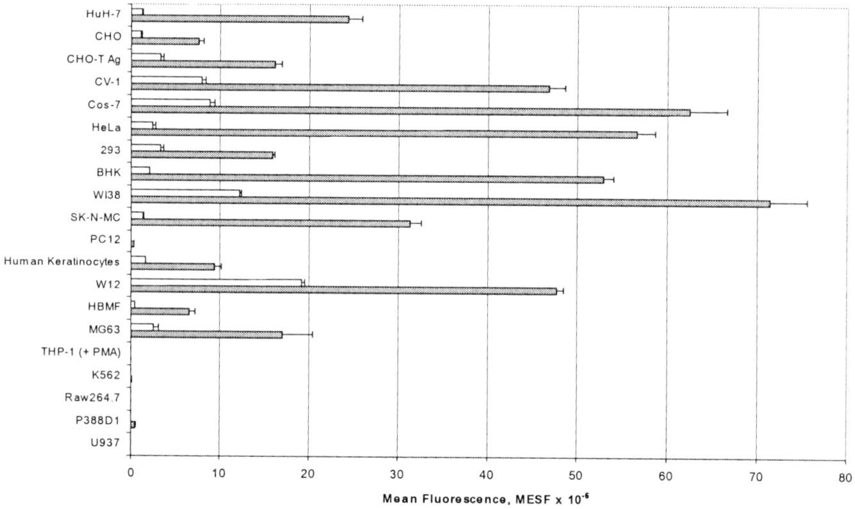

Figure 3. Transient transduction of cell lines with rBV. A. Transduction efficiency. B. GFP expression levels. Experiment and quantitation are as described in Condreay *et al.* (1999). Open bars, virus-treated; shaded bars, virus-treated with addition of 10 m*M* butyrate.

5.2. MECHANISM OF ENTRY AND TRANSDUCTION

The mammalian cell surface molecule that baculoviruses interact with has not been identified. In one study competition for virus binding to HuH7 cells was observed at high ratios of virus without a mammalian expression cassette when compared to virus with such a cassette. Cells treated with trypsin exhibited this competition at lower ratios of the two viruses (Hofmann *et al.*, 1995). Baculovirus-mediated gene expression is also inhibited when cells are treated with chloriquine (Hofmann *et al.*, 1995; Boyce and Bucher, 1996). Collectively, these data seem to suggest that baculoviruses bind to a protein receptor on the cell surface and this complex is endocytosed for subsequent DNA delivery into the nucleus. Electrostatic interactions between the virus and heparan sulfate residues on the surface of cells have also been implicated in virus attachment (Duisit *et al.*, 1999). Polycationic molecules such as polybrene inhibits the uptake of virus by 293 cells. Virus binds to heparin columns *in vitro* and virus is not taken up by cells in the presence of heparin. A role for another receptor can not be completely ruled out, as enzymatically stripping cells of heparan sulfate residues resulted in only a 50% reduction in gene expression levels.

As mentioned above, baculovirus appears to be taken up into endosomes where it is processed and eventually the viral DNA is delivered into the nucleus for transcription. The ability of the viral DNA to escape these intracellular vesicles might explain why some cells are more susceptible to baculovirus transduction (Barsoum *et al.*, 1997). Pseudotyping baculovirus with the G glycoprotein of vesicular stomatitis virus (VSV G) markedly enhances its ability to transduce several cell types, including those that were inefficiently transduced by the unmodified baculovirus. The effect of the VSV G protein in the pseudotyped virion apppears to be an enhancement of viral DNA delivery to the nucleus. Once delivered to the nucleus viral DNA can also be transcriptionally silenced by being bound by histones and condensed into chromatin. This type of silencing can be reversed by inhibitors of histone deacetylase (Chen *et al.*, 1997). This phenomenon, of at least partial silencing, occurs in baculovirus transduced cells (Condreay *et al.*, 1999) and could be another explanation for differences in transduction efficiences among different cell types.

5.3. PRACTICAL APPLICATIONS OF THE TECHNOLOGY

5.3.1. *Engineering of Cell Lines*

The transduction efficiencies and broad range of cell lines that are reported for transient expression of recombinant genes mediated by baculovirus (Fig. 3; Shoji *et al.*, 1997; Condreay *et al.*, 1999) make it an attractive choice as a gene delivery system with few disadvantages (Table 2). One trend that our laboratory has noted is that cells of hematopoietic origin are not efficiently transduced by baculovirus (Condreay *et al.*, 1999). A common feature of these cell lines is that they all were grown in suspension, whereas the cell types that were efficiently transduced were all attachment-dependent. However, as shown in Fig. 4, we have used rBV to transduce CHO cells adapted to growth in suspension, demonstrating that culture conditions are not to blame for the difficulty in transducing hematopoietic cells. Our laboratory has used rBV to routinely express a variety of proteins in different mammalian cells (unpublished data). These include cytoplasmic proteins (annexin V), secreted proteins (hepatocyte growth factor), nuclear receptors (estrogen receptors-α and $-\beta$), and membrane-bound proteins (human ileal bile acid transporter and K_{ATP} ion channel). A recent report by Alessi *et al.*, presented at the 3rd Annual Meeting on Process Development and Production Issues for Baculovirus and Insect Cell Culture held in San Antonio, Texas, February 2000, described the use of rBV to deliver ion channel genes to mammalian cells for the development of functional screening assays.

TABLE 2. Advantages and disadvantages of rBV as a vector for mammalian cells.

Advantages	**Disadvantages**
• Broad cell type susceptibility	• Low level expression (without butyrate)
• Transduces primary cell cultures	• Hematopoietic lines not transduced
• High transduction efficiency	• System relatively uncharacterized
• Lack of toxicity	
• Transient and stable transductants	
• Ease of construction and propagation	
• Ease of manipulation	
• Non-replicative in mammalian cells	

One method commonly employed to engineer cell lines for particular uses is to transfect a cell line with two recombinant cassettes. One cassette expresses the gene of interest, and another expresses a gene product that will confer resistence to a cytotoxic compound, thus allowing the selection of derivatives that maintain the recombinant genes in their genome and stably express the gene of interest. Work from our laboratory has demonstrated that rBV can deliver their DNA to the nucleus of a cell with sufficient integrity that stable maintenance of expression cassettes in transduced cells can be selected (Condreay *et al.*, 1999). In our initial findings the state of the viral DNA that was maintained in stable derivatives was unclear, however recent data from our laboratory has revealed that the viral-derived cassettes are integrated, apparently randomly, into the genome of the host cell (Merrihew *et al.*, submitted for publication).

A list of cell types that have been successfully transduced by rBV for stable expression of recombinant genes is given in Table 3. Most notable in this collection are primary cell types derived from different tissues. Efficient, non-toxic transient expression of primary cells is one hallmark of the rBV system (Condreay *et al.*, 1999; Fig 5). In our laboratory we have successfully used rBV that direct expression of establishment genes, such as the large T antigen of SV40 (Shay *et al.*, 1991), to establish continuous cell lines from primary cultures of human and rodent cells. Gene delivery to primary cells is a major limitation to the derivation of cell lines from normal (i.e. non-transformed) tissues. Delivery techniques must be efficient and non-delterious to cells due to difficulties in obtaining human tissues and isolating large quantities of cells (Katakura *et al.*, 1998). The use of rBV expressing SV40 T antigen to transduce primary cells, such as hepatocytes, gives rise to multiple colonies of cells that proliferate long-term in culture (unpublished results).

A. B.

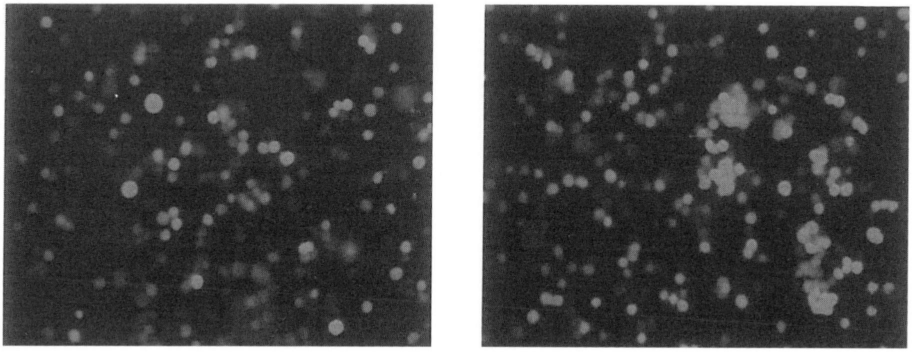

Figure 4. Transient transduction of CHO cells grown in suspension culture. Cells transduced with CMV-GFP virus were visualized by fluorescence microscopy 24 hours post-transduction. A, virus-treated; B, virus-treated with addition of 2 m*M* butyrate.

Another baculovirus-based vector that is capable of efficient stable delivery of recombinant genes into cells is described by Palombo and co-workers (1998). They constructed a baculovirus in which the expression cassettes (β-galactosidase and resistance to hygromycin) are flanked by the inverted terminal repeats (ITR) of adeno-associated virus (AAV). The *rep* gene products responsible for site-specific integration of AAV DNA were expressed in a derivative of the chimera from their native promoters, but were not enclosed within the ITRs. In the absence of *rep* gene functions a low level of hygromycin resistant colonies are derived from cultures transduced with the virus. However, when *rep* functions are provided during transduction of mammalian cells, the frequency of stable clones is enhanced 10 to 50-fold and approximately 40% of the clones surveyed are integrated into the site on chromosome 19 associated with AAV integration. This hybrid vector has the potential to help overcome size limitations on recombinant inserts placed into AAV, one of the major disadvantages of these vectors.

TABLE 3. Cell types that have yielded G418 resistant stable transductants.

• CHO	• P388D1
• HuH7	• K562
• HepG2	• CV-1
• Hep3B	• COS-7
• HeLa	• 293
• Mouse L6	
• Human hepatocytes*	• Murine smooth muscle*
• Human chondrocytes*	• Guinea pig smooth muscle*
• Human bone marrow stromal cells*	• Rat bone marrow stromal cells*

* Primary cell cultures transduced with CMV-*ts* SV40 T Ag virus; cultures propagate at least 10 passages.

5.3.2. *Chimeric Viruses*

As exemplified above with the baculovirus-AAV chimera, the advantages of rBVs as delivery vehicles for mammalian cells make it an excellent platform to launch infections by viruses that are hampered by the lack of accessible cell culture systems for their propagation. Baculovirus has the capacity to accept large inserts into its genome and is able to deliver DNA to a wide variety of cells for transcription. Chimeric rBV are suitable as vectors for viruses whose genome is comprised of positive polarity RNA or that utilize a positive strand RNA as an intermediate in the replication of their genome. Examples of this application will be discussed below.

Poliovirus. A two baculovirus system was used to derive infectious poliovirus from cloned cDNA (Yap *et al.*, 1997). The DNA copy of the positive strand RNA poliovirus genome was placed under transcriptional control of a T7 RNA polymerase promoter in a rBV. Co-transduction of this virus and a second virus that expresses T7 RNA polymerase from a strong mammalian promoter into mammalian cells gives rise to high titers of infectious poliovirus at 2 to 3 days post-transduction.

Hepatitis C Virus. HCV is a positive strand RNA hepatotropic virus the study of which is hampered by a lack of cell culture systems to elucidate its basic molecular biology (Moradpour *et al.*, 1998). A rBV was constructed that placed a cDNA of the HCV genome under control of the CMV promoter (Fipaldini *et al.*, 1999). This virus directs expression of the HCV polyprotein in the human hepatoma cell line Huh7. Furthermore, the polyprotein appears to be correctly processed and HCV proteins can be detected in cultures out to 49 days post-transduction. No evidence of HCV nucleic acid replication was obtained, possibly because the HCV clone used may not have been infectious. However, this report demonstates the feasibility that a productive HCV infection could be launched using a baculovirus hybrid.

Hepatitis B Virus. Replication of the DNA genome of HBV proceeds through an RNA intermediate and a productive infection can be lauched by the production of a greater-than-unit length, or pregenomic, RNA (Kann and Gerlich, 1997). Cell culture systems have been developed to study the replication of HBV that generate the pregenomic RNA from either the natural HBV promoter (Sells *et al.*, 1987), or from strong external promoters (Fallows and Goff, 1995). However, these systems suffer from a variety of limitations such as inefficient transfection techniques or low virus production. Delaney and Isom (1998) constructed a rBV in which the pregenomic RNA is transcribed from endogenous HBV promoters. HBV gene expression, nucleic acid replication and high levels of HBV viral particles are detected in human hepatoma cells transduced with the HBV-baculovirus for up to 11 days post-transduction. Furthermore, this system can be utilized to examine the effects of anti-viral drugs on various aspects of the HBV replication cycle (Delaney *et al.*, 1999). We have observed similar results using a baculovirus-HBV chimera in which the pregenomic RNA is driven from the CMV promoter. This system allows for the production of very high levels of HBV polymerase active core particles from transduced HepG2 cells (unpublished data).

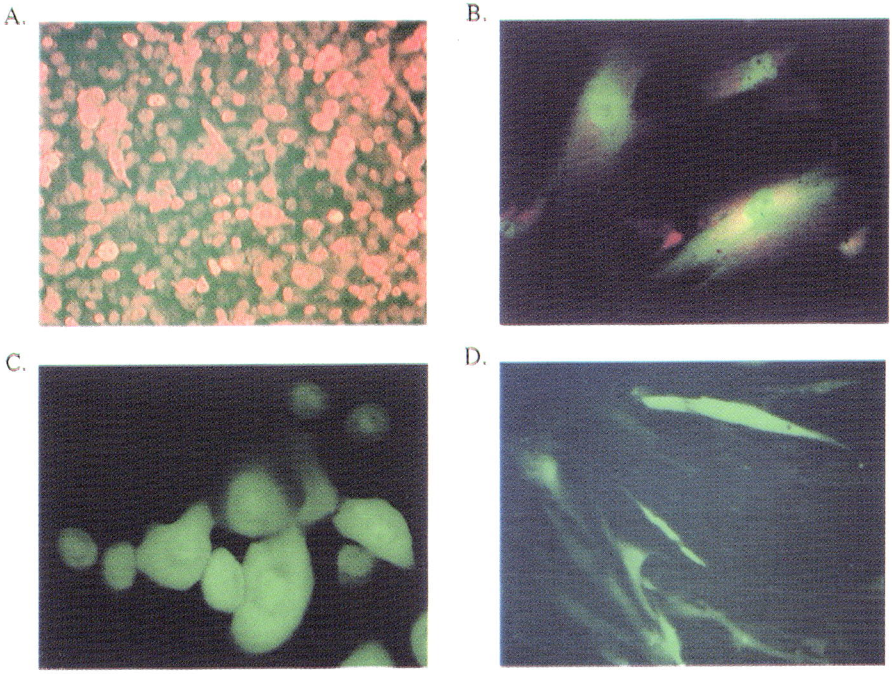

Figure 5. Transient transduction of primary human cells. A. Transduced with CMV-SV40 *ts* TAg virus and visualized with immuncytochemistry using anti-TAg antibody; B.-D. Transduced with CMV-GFP virus and visualized by fluorescence microscopy. A. hepatocytes; B. bone marrow fibroblasts; C. keratinocytes; D. bladder smooth muscle.

5.3.3. *Delivery of Genes In Vivo*

The inherent inability of AcMNPV to replicate in mammalian cells (discussed above) make it an attractive candidate for development as a vector for *in vivo* gene delivery since this would negate the concerns about replication competent breakthroughs that exist for the vectors more commonly used for this application. Unfortunately, this is presently an unrealized dream. Virus is inactivated by heat-sensititve serum components thus preventing gene delivery when administered to rats and mice *in vivo* by a variety of routes (Sandig *et al.*, 1996). Further investigation has implicated the classical complement pathway in the inactivation of the virus (Hofmann and Strauss, 1998). However, there are potential strategies that show promise for overcoming this limitation for the *in vivo* administration of baculovirus vectors (Hofmann and Strauss, 1998; Hofmann *et al.*, 1999). An additional problem in attempting to use these vectors *in vivo* is in purification and concentration of virus. This can be achieved by ultracentrifugation (O'Reilly *et al.*, 1994), though it is difficult to scale-up this procedure. Baculovirus particles can be concentrated on cation exchange resins by binding in the low pH of insect cell media followed by subsequent elution using physiological buffer solutions (Barsoum, 1999). One additional limitation on this use is that the abortive infection of baculovirus in mammalian cells is still relatively uncharacterized, especially in terms of the extent of viral gene expression in non-conventional host cells. Using GFP as a reporter protein, we have demonstrated that the promoter for the *ie-1* gene of the virus is active in mammalian cells (unpublished observations). Furthermore, the product of the *ie-1* gene is able to activate certain viral promoters in mammalian cells (Murges *et al.*, 1997) raising the possibility of some immediate early and early viral gene expression during rBV transduction of mammalian hosts.

6. Display of Proteins on the Surface of Baculovirus Virions

In recent years several groups have demonstrated that foreign proteins can be displayed on the surface of baculovirus virions by construction of a genetic fusion between the protein and a second copy of the gp64 major surface glycoprotein of AcMNPV. The foreign proteins that have been displayed include not only normally extracellular protiens (Boublik *et al.*, 1995; Grabherr *et al.*, 1997; Mottershead *et al.*, 1997), but also proteins that are normally found intracellularly (Mottershead *et al.*, 1997; Lindley *et al.*, 2000). A transfer vector with restriction endonuclease sites engineered between the signal sequence and mature domain of a gp64 gene under *polh* gene promoter control is available from Novagen to construct these recombinants.

This technology has been applied to library generation by Ernst and co-workers (1998). A library of epitopes centered around the HIV-1 gp41 epitope "ELDKWA" was constructed in a site on the influenza virus hemagglutinin protein expressed on the surface of baculovirus particles. The library was screened with a neutralizing antibody that recognizes the epitope and a clone was identified that is bound by the antibody better than the non-modified epitope. This result serves as proof-of-principle that baculovirus display is a viable system for library generation and screening. Another report that demonstrates a valuable application of this technology uses baculovirus particles displaying gp64 fusions to regions of nuclear receptors (LXR-β and FXR) as an antigen delivery vehicle to immunize mice and derive monoclonal antibodies that recognize the nuclear receptor (Lindley *et al.*, 2000). This provides a novel method of generating and screening for monoclonal antibodies to a target protein without the need for purified protein, one need only purify virions by ultracentrifugation.

Baculovirus particles can also be used to display recombinant proteins without the need to construct fusions with gp64. Budded virions derive their envelope from the cytoplasmic membrane of the infected cell and presumably are able to pick up other membrane proteins in addition to gp64. A rBV directing the expression of the β_2-adrenergic receptor under the control of the *polh* gene promoter has been shown to display the receptor on the surface of isolated virions. The receptor is glycosylated and exhibits ligand binding activity (Loisel *et al.*, 1997). We have obtained similar results in our laboratory with NPY5 receptor, CCR5 receptor, and human ileal bile acid transporter (unpublished observations). A recent report by Aoki *et al.* (1999) used a rBV with a mammalian expression cassette to induce an antibody response against pseudorabies virus glycoprotein B (gB). These investigators constructed a rBV with the gB gene under the control of a strong mammalian promoter and demonstrated gB expression in a variety of mammalian cell lines. Mice that were inoculated with this virus exhibited an antibody response against gB as measured by ELSIA assay. It is possible that the injected virus enters an appropriate cell in the animal and expresses the antigen *de novo*. However, it is also possible that the virion particles acquire gB on their surface due to expression from the mammalian promoter in the infected insect cells during virus propagation. These virion particles may then display the antigen and induce an antibody response similar to the results described by Lindley *et al.* (2000).

7. Concluding Remarks

The BEVS has been used extensively for producing recombinant proteins in insect cells. Since its inception in 1983 numerous improvements and modifications have been made to transfer vectors. Simpler methods for generating recombinant viruses have been devised and new host cell lines have been developed making the system increasingly accessible to investigators in a variety of research fields. In addition, although not discussed in this chapter, significant improvements have been made in the scale-up of rBV infected insect cell cultures. The ability to co-express multiple proteins by co-infection has proven extremely useful for studying protein-protein interactions, producing VLPs and CYPs. As discussed in the accompanying chapter in this book by Ailor and Betenbaugh, the application of novel engineering approaches to enhance the production and secretion of functional proteins will continue to expand the use of the BEVS for recombinant protein expression.

The application of recombinant baculoviruses for gene delivery and expression in mammalian cells provides a unique opportunity for cellular engineering. One can generate and produce the rBV in insect cells for transduction into mammalian cells in the absence of viral replication. In our laboratory we have found this to be one of the most non-cytotoxic approaches for gene delivery that is available, providing the opportunity for a variety of novel gene delivery applications. This is clearly shown by the studies demonstrating that rBVs can be used to launch polio and hepatitis B virus infections in transduced mammalian cells. Undoubtedly, in the future one can expect to see the development of additional chimeric viruses and other novel applications of this unique gene delivery system.

8. Acknowledgements

The authors wish to acknowledge their colleagues at Glaxo Wellcome Research and Development for their contributions to the unpublished work mentioned in this manuscript.

9. References

Adler, S., Reay, P., Roy, P., and Klenk, H.-D. (1998) Induction of T cell response by bluetongue virus core-like particles expressing a T cell epitope of the M1 protein of influenza A virus. *Med. Microbiol. Immunol. (Berl.)* **187**, 91-96.

Airenne, K. J., Laitinen, O. H., Alenius, H., Mikkola, J., Kalkkinen, N., Arif, S. A. M., Yeang, H. Y., Palosuo, T., and Kulomaa, M. (1999) Avidin is a promising tag for fusion proteins produced in baculovirus-infected insect cells. *Protein Expr. Purif.* **17**, 139-145.

Altmann, F., Staudacher, E., Wilson, I.B., and Marz, L. (1999) Insect cells as hosts for the expression of recombinant glycoproteins, *Glyconj. J.* **16**, 109-123.

Aoki, H., Sakoda, Y., Jukuroki, K., Takada, A., Kida, H., and Fukusho, A. (1999) Induction of antibodies in mice by a recombinant baculovirus expressing pseudorabies virus glycoprotein B in mammalian cells, *Vet. Microbiol.* **68**, 197-207.

Arp, J., Ford, C. M., Palker, T. J., King, E. E., and Dekaban, G. A. (1993) Expression and immunogenicity of the entire human T cell leukemia virus type I envelope protein produced in a baculovirus system. *J. Gen. Virol.* **74**, 211-222.

Ayers, M.D., Howard, S.C., Kuzio, J., Lopez-Ferber, M., and Possee, R. D. (1994) The complete DNA sequence of *Autographa californica* nuclear polyhedrosis virus. *Virology* **202**, 586-605.

Barsoum, J. (1999) Concentration of recombinant baculovirus by cation-exchange chromatography. *BioTechniques* **26**, 834-840.

Barsoum, J., Brown, R., McKee, M., and Boyce, F.M. (1997) Efficient transduction of mammalian cells by a recombinant baculovirus having the vesicular stomatitis virus G glycoprotein, *Hum. Gene Ther.* **8**, 2011-2018.

Baumert, T. F., Vergalla, J., Satoi, J., Thomson, M., Lechmann, M.,Herion, D. Greenberg, H.B., Ito, S., and Liang, T. J. (1999) Hepatitis C virus-like particles synthesized in insect cells as a potential vaccine candidate. *Gastroenterology* **117**, 1397-1407.

Belke-Louis, G., F., Wehmeyer, A., and Schulz, R. (1999) μ-opioid receptor expression in High Five^{TM} insect cells is regulated by 5' untranslated region (5'UTR). *Life Sci.*, **64**: 913-921.

Belyaev, A. S., Hails, R. S., and Roy, P. (1995) High-level expression of five foreign genes by a single recombinant baculovirus. *Gene* **156**, 229-233.

Bishop, D. H. L., Novy, R., and Mierendorf, R. (1995) The BacVector system: Simplified cloning and protein expression using novel baculovirus vectors. *Innovations* **4**, 1.

Blissard, G.W. and Rohrmann, G.F. (1990) Baculovirus diversity and molecular biology. *Annu. Rev. Entomol.* **35**, 127-155.

Blissard, G.W. and Wenz, J.R. (1992) Baculovirus gp64 envelope glycoprotein is sufficient to mediate pH-dependent membrane fusion. *J. Virol.* **66**, 6829-6835.

Boublik, Y., DiBonito, P., and Jones, I. M. (1995) Eukaryotic virus display: engineering the major surface glycoprotein of the *Autographa californica* nuclear polyhedrosis virus (AcNPV) for the presentation of foreign proteins on the virus surface. *Bio/Technol*ogy **13**, 1079-1084.

Bouvier, M., Menard, L., Dennis, M., and Marullo, S. (1998) Expression and recovery of functional G-protein coupled receptors using baculovirus expression systems. *Curr. Opin. Biotechnol.* **9**, 522-527.

Boyce, F.M. and Bucher, N.L.R. (1996) Baculovirus-mediated gene transfer into mammalian cells, *Proc. Natl. Acad. Sci USA* **93**, 2348-2352.

Breitburd, F., Kirnbaurer, R., Hubbert, N. L., Nonnenmacher, B., Trin-Dinh-Desmarquet, C., Orth, G., Schiller, J. T., and Lowy, D. R. (1995) Immunization with virus-like particles from cottentail rabbit papilloma virus (CRPV) can protect against experimental CRPV infection. *J. Virol.* **69**, 3959-3963.

Carbonell, L.F., Klowden, M.J., and Miller, L. K. (1985) Baculovirus-mediated expression of bacterial genes in dipteran and mammalian cells, *J. Virol.* **56**, 153-160.

Carbonell, L.F. and Miller, L.K. (1987) Baculovirus interaction with nontarget organisms: a virus-borne reporter gene is not expressed in two mammalian cell lines. *Appl. Environ. Microbiol.* **53**, 1412-1417.

Chen, G. and Courey, A. J. (1999) Baculovirus-transfer vector for eukaryotic expression and immuno-affinity purification of Gal4-fusion proteins. *BioTechniques* **26**, 808-814.

Chen, W.Y., Bailey, E.C., McCune, S.L., Dong, J.Y., and Townes, T.M. (1997) Reactivation of silenced, virally transduced genes by inhibitors of histone deacetylase. *Proc. Natl. Acad. Sci. USA* **94**, 5798-5803.

Condreay, J.P., Witherspoon, S.M., Clay, W.C., and Kost, T.A. (1999) Transient and stable expression in mammalian cells transduced with a recombinant baculovirus vector, *Proc. Natl. Acad. Sci. USA* **96**, 127-132.

Crespi, C. L. and Miller, V. P. (1999) The use of heterologously expressed drug metabolizing enzymes-state of the art and prospects for the future. *Pharmacol. Ther.* **84**, 121-131.

Crespi, C. L., Miller, V. P., and Penman, B. W. (1997) Microtiter plate asays for inhibition of human, drug-metabolizing cytochromes P450. *Anal. Biochem.* **248**, 188-190.

Delaney, W.E. and Isom, H.C. (1998) Hepatitis B virus replication in human HepG2 cells mediated by hepatitis B virus recombinant baculovirus. *Hepatology* **28**, 1134-1146.

Delaney, W.E., Miller, T.G., and Isom, H.C. (1999) Use of the hepatitis B virus recombinant baculovirus-HepG2 system to study the effects of (-)-β-2',3'-dideoxy-3'-thiacytidine on replication of hepatitis B virus and accumulation of covalently closed circular DNA, *Antimicrob. Agents Chemother.* **43**, 2017-2026.

Duisit, G., Saleun, S., Douthe, S., Barsoum, J., Chadeuf, G., and Moullier, P. (1999) Baculovirus vector requires electrostatic interactions including heparan sulfate for efficient gene transfer in mammalian cells. *J. Gene Med.* **1**, 93-102.

Emery, V. C. and Bishop, D. H. L. (1987) The development of multiple expression vectors for high level synthesis of eukaryotic proteins: Expression of LCMV-N and AcNPV polyhedrin protein by a recombinant baculovirus. *Protein Eng.* **1**, 359-366.

Fallows, D.A. and Goff, S.P. (1995) Mutations in the ε sequences of human hepatitis B virus affect both RNA encapsidation and reverse transcription. *J. Virol.* **69**, 3067-3073.

Finkelstein, Y., Faktor, O., Elroy-Stein, O., and Ben-Zion, L. (1999) The use of a bi-cistronic transfer vector for the baculovirus expression system. *J. Biotechnol.* **75**, 33-44.

Fipaldini, C., Bellei, B., and LaMonica, N. (1999) Expression of hepatitis C virus cDNA in human hepatoma cell line mediated by a hybrid baculovirus-HCV vector. *Virology* **255**, 302-311.

Forstova, J., Krauzewicz, N., Sandig, V., Elliott, J., Palkova, Z., Strauss, M., and Griffin, B. E. (1995) Polyoma virus pseudocapsids as efficient carriers of heterologous DNA into mammalian cells. *Hum. Gene Ther.* **6**, 297-306.

Francken, B. J. B., Josson, K., Lijnen, P., Jurzak, M., Luyten, W. H. M. L., and Leysen, J. E. (2000) Human 5-hydroxytryptamine$_{5A}$ receptors activate coexpressed G_i and G_o proteins in *Spodoptera frugiperda* 9 cells. *Mol. Pharmacol.* **57**, 1034-1044.

Fraser, M.J. (1986) Ultrastructural observations of virion maturation *in Autographa californica* nuclear polyhedrosis virus infected *Spodoptera frugiperda* cell cultures. *J. Ultrastruct. Mol. Struct. Res.* **95**, 189-195.

Gimpl, G., Klein, U., Reiländer, H., and Fahrenholz, F. (1995) Expression of the human oxytocin receptor in baculovirus-infected insect cells: high-affinity binding is induced by a cholesterol-cyclodextrin complex. *Biochemistry* **34**, 13794-13801.

Goldmann, C., Petry, H., Frye, S., Ast, O.,, Ebitsch, S., Klaus-Dieter, J., Kaup, F.J., Weber, F.,. Trebst, C., Nisslein, T., Hunsmann, G., Weber, T., and Luke, W. (1999) Molecular cloning and expression of major structural protein VP1 of the human polyomavirus JC virus: formation of virus-like particles useful for immunological and therapeutic studies. *J. Virol.* **73**, 4465-4469.

Grabherr, R., Ernst, W., Doblhoff-Dier, O., Sara, M., and Katinger, H. (1997) Expression of foreign proteins on the surface of *Autographa californica* nuclear polyhedrosis virus. *BioTechniques* **22**, 730-735.

Grant, D. F., Greene, J. F., Pinot, F., Borhan, B., Moghaddam, M. F., Hammock, B. D., McCutchen, B., Ohkawa, H., Luo, G., and Guenthner, T. M. (1996) Development of an *in situ* toxicity system using recombinant baculoviruses. *Biochem. Pharmacol.* **51**, 503-515.

Grosh, H.-W. (1998) Protection of proteolysis-prone recombinant proteins in baculovirus expression systems. *BioTechniques* **24**, 930-931.

Grünewald, S., Reiländer, H., and Michel, H. (1996) *In* vivo reconstitution of of dopamine D_{2S} receptor-mediated G protein activation in baculovirus-infected insect cells: preferred coupling to G_{i1} versus G_{i2} *Biochemistry* **35**, 15162-15173.

Guan, X. M., Kolbilka, T. S., and Kobilka, B. K. (1992) Enhancement of membrane insertion and function in a type IIIb membrane protein following introduction of a cleavable signal peptide. *J. Biol. Chem.* **267**, 21995-21998.

Gupta, G., Kim, J., Yang, L., Sturley, S. L., and Danziger, R. S. (1997) Expression and purification of soluble, active heterodimeric guanylyl cyclase from baculovirus. *Prot. Expr. Purif.* **10**, 325-330.

Hampe, W., Voss, R.-H., Haase, W., Boege, F., Michel, H., and Reiländer, H. (2000) Engineering of a proteolytically stable human β_2-adrenergic receptor/maltose-binding protein fusion and production of the chimeric protein in *Escherichia coli* and baculovirus-infected insect cells. *J. Biotechnol.* **77**, 219-234.

Hartig, P.C., Cardon, M.C., and Kawanishi, C.Y. (1991) Insect virus: assays for viral replication and persistence in mammalian cells. *J. Virol. Meth.* **31**, 335-344.

Hofmann, C., Huser, A., Lehnert, W., and Strauss, M. (1999) Protection of baculovirus-vectors against complement-mediated inactivation by recombinant soluble complement receptor type 1. *Biol. Chem.* **380**, 393-395.

Hofmann, C, Sandig, V., Jennings, G., Rudolph, M., Schlag, P., and Strauss, M. (1995) Efficient gene transfer into human hepatocytes by baculovirus vectors. *Proc. Natl. Acad. Sci. USA* **92**, 10099-10103.

Hofmann, C. and Strauss, M. (1998) Baculovirus-mediated gene transfer in the presence of human serum or blood facilitated by inhibition of the complement system. *Gene Ther.* **5**, 531-536.

Hom, L. G., and Volkman, L. E. (1998) Preventing proteolytic artifacts in the baculovirus expression system. *BioTechniques* **25**, 18-20.

Homa, F. L. and Brown, J. C. (1997) Capsid assembly and DNA packaging in herpes simplex virus. *Rev. Med. Virol.* **7**, 107-122.

Hoque, M., Shimizu, N., Ishizu, K.-I., Yajima, H., Arisaka, F., Suzuki, K., Watanabe, H., and Handa, H. (1999) Chimeric virus-like particle formation of adeno-associated virus. *Biochem. Biophys. Res. Commun.* **266**, 371-376.

Jarvis, D.L. (1997) Baculovirus expression vectors, in L.K. Miller (ed.) *The Baculoviruses*, Plenum Press, New York, pp. 389-431.

Jarvis, D. L., Fleming, J., G., W., Kovacs, G., R., Summers, M., D., and Guarino, L., A. (1990) Use of early baculovirus promoters for continuous expression and efficient processing of foreign gene products in stably-transformed lepidopteran insect cells. *Bio/Technology* **8**:50-55.

Jarvis, D. L., Kawar, Z. S., and Hollister, J. R. (1998) Engineering N-glycosylation pathways in the baculovirus-insect cell system. *Curr. Opin. Biotechnol.* **9**, 528-533.

Jarvis, D. L., and Summers, M. D. (1989) Glycosylation and secretion of human tissue plasminogen activator in recombinant baculovirus-infected cells. *Mol. Cell. Biol.* **9**, 214-223.

Jiang, B., Barniak, V., Smith, R. P. Sharma, R. Corsaro, B. Hu, B., and Madore, H. P. (1998) Synthesis of rotavirus-like particles in insect cells: comparative and quantitative analysis. *Biotechnol. Bioeng.* **60**, 369-374.

Jiang, B., Estes, M. K., Barone, C., Barniak, V., O' Neal, C. M., Ottaiano, A., Madore, H. P., and Conner, M. E. (1999) Heterotypic protection from rotavirus infection in mice vaccinated with virus-like particles. *Vaccine* **17**, 1005-1013.

Kakker, N. K., Mikhailov, M. V., Nermut, M. V., Burny, A., and Roy, P. (1999) Bovine leukemia virus gag particle assembly in insect cells: formation of chimeric particles by domain-switched leukemia/lentivirus gag polyprotein. *Virology* **265**, 308-318.

Kang, C.Y., Luo, L., Wainberg, M.A., and Li, Y. (1999) Development of HIV/AIDS vaccine using chimeric gag-env virus-like particles. *Biol. Chem.* **380**, 353-364.

Kann, M. and Gerlich, W.H. (1997) Replication of hepatitis B virus, in T.J. Harrison and A.J. Zuckerman (eds.) *The Molecular Medicine of Viral Hepatitis*, John Wiley & Sons, Chichester, pp. 63-87.

Katakura, Y., Alam, S., and Shirahata, S. (1998) Immortalization by gene transfection, in J.P. Mather and D. Barnes (eds.) *Animal Cell Culture Methods*, Academic Press, San Diego, pp. 69-91.

Kelly, D.C. (1982) Baculovirus replication. *J. Gen. Virol.* **63**, 1-13.

Kirnbauer, R., Chrandrachud, L. M., O'Neil, B. W., Wagner, E. R., Grindlay, G. J., Armstrong, A. McGarvie, G. M., Schiller, J. T., Lowy, D. R., and Campo, M. S. (1996) Virus-like particles of bovine papilloma virus type 4 in prophylactic and therapeutic immunization. *Virology* **219**, 37-44.

Kitts, P. A., Ayers, M. D., and Possee, R. D. (1990) Linearization of baculovirus DNA enhances the recovery of recombinant baculovirus expression vectors. *Nucl. Acids Res.* **18**, 5667-5672.

Kitts. P. A. and Green,G. (1999) An immunological assay for determination of baculovirus titers in 48 hours. *Anal. Biochem.* **268**, 173-178.

Kitts, P. A., and Possee, R. D. (1993) A method for producing recombinant baculovirus expression vectors at high frequency. *BioTechniques* **14**, 810-817.

Kool, M. and Vlak, J.M. (1993) The structural and functional organization of the *Autographa californica* nuclear polyhedrosis virus genome. *Arch. Virol.* **130**, 1-16.
Kost, T. A. and Condreay, J. P. (1999) Recombinant baculoviruses as expression vectors for insect and mammalian cells. *Curr. Opin. Biotechnol.* **10**, 428-433.
Kost, T.A., Ignar D.M., Clay, W.C., Andrews, J., Leray, J.D., Overton, L., Hoffman, C.R., Kilpatrick, K.E., Ellis, B., and Emerson, D.L. (1997) Production of a urokinase plaminogen activator-IgG fusion protein (uPA-IgG) in the baculovirus expression system. *Gene* **190**, 139-144.
Kovacs, G.R., Guarino, L.A., Graham, B.L., and Summers, M.D. (1991) Identification of spliced baculovirus RNAs expressed late in infection. *Virology* **185**, 633-643.
Lachance, M., Ethier, N., Wolbring, G., Schnetkamp, P., P., M., and Hébert, T., E. (1999) Stable association of G proteins with β_2AR is independent of the state of receptor activation. *Cell Signal.* **11**, 523-533.
Lee, C. A., Kadwell, S. H., Kost, T. A., and Serabjit-Singh, C. J. (1995) CYP3A4 expressed by insect cells infected with a recombinant baculovirus containing both CYP3A4 and human NADPH-cytochrome P450 reductase is catalytically similar to human liver microsomal CYP3A4. *Arch. Biochem. Biophys.* **319**, 157-167.
Lee, C. A., Kost T. A., and Serabjit-Singh, C. J. (1996) Recombinant baculovirus strategy for coexpression of functional human cytochrome P450 and P450 reductase. *Methods Enzymol.* **72**, 86-94.
Lindley, K.M., Su, J.-L., Hodges. P.K., Wisely, G.B., Bledsoe, R.K., Condreay, J.P., Winegar, D.A., Hutchins, J.T., and Kost, T.A. (2000) Production of monoclonal antibodies using recombinant baculovirus displaying gp64-fusion proteins. *J. Immunol. Meth.* **234**, 123-135.
Loisel, T. P., Ansanay, H., St-Onge, S., Gay, B., Boulanger, P., Strosberg, A. D., Marullo, S., and Bouvier, M. (1997) Recovery of homogeneous and functional β_2-adrenergic receptors from extracellular baculovirus particles. *Nat. Biotechnol.* **15**, 1300-1304.
Luckow, V. A. (1995) Protein production and processing from baculovirus expression vectors, M. L. Shuler, W. H. Alan, R. R. Granados, and D. A. Hammer, eds., in *Baculovirus Expression Systems and Biopesticides* Wiley-Liss, New York, pp. 51-90.
Luckow, V. A., Lee, S. C., Barry, G. F., and Olis, P. O. (1993) Efficient generation of infectious recombinant baculoviruses by site-specific transposon-mediated insertion of foreign genes into a baculovirus genome propagated in *Escherichia coli*. *J. Virol.* **67**, 4566-4579.
McCarroll, L. and King, L. A. (1997) Stable insect cell cultures for recombinant protein production. *Curr. Opin. Biotechnol.* 8, 590-594.
McCutchen, B. F., Choudary, P. V., Crenshaw, R., Maddox, D., Kamita, S. G., Palekar, N., Volrath, S., Fowler, E., Hammock, B. D., and Maeda, S. (1991) Development of a recombinant baculovirus expressing an insect-selective neurotoxin: Potential for pest control. *Bio/Technology* **9**, 848-852.
Madore, H. P., Estes, M. K., Zarley, C. D., Hu, B., Parsons, S., Digravio, D., Greiner, S., Smith, R. Jiang, B., Corsaro, B. Barniak, V., Crawford, S., and Conner, M. E. (1999) Biochemical and immunologic comparison of virus-like particles for a rotavirus subunit vaccine. *Vaccine* **17**, 2461-2471.
Martens, J. W., Van Oers, M. M., Van de Bilt, B. D., Oudshoorn, P., and Vlak, J. M. (1995) Development of a baculovirus vector that facilitates the generation of p10-based recombinants. *J. Virol. Meth.* **52**, 15-19.
Massotte, D., Pereira, C., A., Pouliquen, Y., and Pattus, F. (1999) Parameters influencing human μ opioid receptor over-expression in baculovirus-infected insect cells. *J. Biotechnol.* **69**, 39-45.
Merryweather, A. T., Weyer, U., Harris, M. P., Hirst, M., Booth, T., and Possee, R. D. (1990) Construction of genetically engineered baculovirus insecticides containing the *Bacillus thuringiensis* subsp. kurstaki HD-73 delta encotoxin. *J. Gen. Virol.* **71**, 1535-1544.
Mikhailov, M. V., and Ashcroft, S. J. (2000) Interactions of the sulfonylurea subunit in the molecular assembly of beta-cell K(ATP) channels. *J. Biol. Chem.* **275**, 3360-3364.
Mikhailov, M. V., Proks, P., Ashcroft, F. M., and Ashcroft, S. J. H. (1998) Expression of functionally active ATP-sensitive K-channels in insect cells using baculovirus. *FEBS Lett.* 429, 390-394.
Miller, L.K. (1996) Insect viruses, in B.N. Fields, D.M. Knipe, and P.M. Howley (eds.), *Fields Virology, 3rd Edition*, Lippincott-Raven Publishers, Philadelphia, pp. 533-556.
Miller, L.K. (1997) *The Baculoviruses*, Plenum Press, New York.
Monsma, S.A. and Blissard, G.W. (1995) Identification of a membrane fusion domain and an oligomerization domain in the baculovirus gp64 envelope fusion protein. *J. Virol.* **69**, 2583-2595.
Monsma, S. A., and Scott, M. (1997) Bacvector-3000: An engineered baculovirus for foreign protein expression, lacking viral chitinase and cathepsin protease activities. *FASEB J.* **11**, A1399.
Monsma, S.A., Oomens, A.G., and Blissard, G.W. (1996) The gp64 envelope fusion protein is an essential baculovirus protein required for cell-to-cell transmission of infection. *J. Virol.* **70**, 4607-4616.

Moradpour, D., Kary, P., Rice, C.M., and Blum, H.E. (1998) Continuous human cell lines inducibly expressing hepatitis C virus structural and nonstructural proteins. *Hepatology* **28**, 192-201.

Mottershead, D., van der Linden, I., Von Bonsdorff, C.-H., Keinanen, K., and Oker-Blom, C. (1997) Baculoviral display of the green fluorescent protein and rubella virus envelope proteins. *Biochem. Biophys. Res. Commun.* **238**, 717-722.

Murges, D., Kremer, K., and Knebel-Morsdorf, D. (1997) Baculovirus transactivator IE1 is functional in mammalian cells. *J. Gen. Virol.* **78**, 1507-110.

Murphy, C. I., Lennick, M., Lehar, S. M., Beltz, G. A., and Young, E. (1990) Temporal expression of HIV-1 envelope proteins in baculovirus-infected cells: implication for glycosylation and CD4 binding. *Genet. Anal. Tech. Appl.* **7**, 160-171.

Newcomb, W.W., Homa, F.L., Thomsen, D.R., Booy, F.P., Trus, B.L., Steven, A.C., Spencer, J.V., and Brown, J.C. (1996) Assembly of the herpes simplex virus capsid: characterization of intermediates observed during cell-free capsid assembly. *J. Mol. Biol.* **263**, 432-446.

Newcomb, W.W., Homa, F.L., Thomsen, D.R., Trus, B.L., Cheng, N., Stevens, A., Booy, F., Brown, J.C. (1999) Assembly of the herpes simplex virus procapsid from purified components and identification of small complexes containing the major capsid and scaffolding proteins. *J. Virol.* **73**, 4239-4250.

Notka, F., Stahl-Hennig, C., Dittmer, U., Wolf, H., and Wagner, R.(1999) Construction and characterization of recombinant VLPs and Semliki-Forest Virus live vectors for comparative evaluation in the SHIV monkey model. *Biol. Chem.* **380**, 341-352.

Ohmori, S., Fujiki, N., Nakasa, H., Nakamura, H., Ishii, I., Itahashi, K., and Kitada, M. (1998) Steroid hydroxylation by human fetal CYP3A7 and human NADPH-cytochrome P450 reductase coexpressed in insect cells using baculovirus. *Res. Commun. Mol. Path. Pharmacol.* **100**, 15-28.

Ong, C., E., Miners, J. O., Birkett, D. J., and Bhrasker, C. R. (1998) Baculovirus-mediated expression of cytochrome P4502C8 and human NADPH-cytochrome P450 reductase: optimization of protein expression. *Xenobiotica* **28**, 137-152.

Oomens, A.G.P., Monsma, S.A., and Blissard, G.W. (1995) The baculovirus gp64 envelope fusion protein: synthesis, oligomerization, and processing. *Virology* **209**, 592-603.

Paine, J., I., M., Gilham, D., Roberts, G., C., K., and Wolf, C. R. (1996) Functional high level expression of cytochrome P450 CYP2D6 using baculovirus expression systems. *Arch. Biochem. Biophys.* **328**, 143-150.

Pajot-Augy, E., Bozon, V., Remy J.-J., Couture L., and Salesse, R. (1999) Critical relationship between glycosylation of recombinant lutropin receptor ectodomain and its secretion from baculovirus-infected insect cells. *Eur. J. Biochem.* **260**, 635-648.

Pajot-Augy, E., Couture, L., Bozon, V., Remy, J.-J., Biache, G., Severini, M., Huet, J. C., Pernollet, J. C., and Salesse, R. (1995) High-level expression of recombinant porcine LH receptor in baculovirus-infected insect cells or caterpillars. *J. Mol. End.* **14**, 51-66.

Palamanda, J. R., Favreau, F., Lin, C.-C., and Nomeir, A.A. (1998) Validation of a rapid microtiter plate assay to conduct cytochrome P450 2D6 enzyme inhibition studies. *Drug Discov. Today* **3**, 466-470.

Palombo, F., Monciotti, A., Recchia, A., Cortese, R., Ciliberto, G., and La Monica, N. (1998) Site-specific integration in mammalian cells mediated by a new hybrid baculovirus-adeno-associated virus vector. *J. Virol.* **72**, 5025-5034.

Parker, S. D. and Hunter, E. (2000) A cell line specific defect in the intracellular transport and release of assembled retroviral capsids. *J. Virol.* **74**, 784-795.

Patel, G., Nasmyth, K. and Jones, N. (1992) A new method for the isolation of recombinant baculovirus. *Nucleic Acids Res.* **20**, 97-104.

Peakman, T. C., Harris, R. A., and Gewert, D. R. (1992) Highly efficient generation of recombinant baculoviruses by enzymatically mediated site-specific *in vitro* recombination. *Nucl. Acids Res.* **20**, 495-500.

Pennock, G. D., Shoemaker, C., and Miller, L. K. (1984) Strong and regulated expression of *Escherichia coli* β-galactosidase in insect cells with a baculovirus vector. *Mol. Cell. Biol.* **4**, 399-406.

Pfeifer, T. A. (1998) Expression of heterologous proteins in stable insect cell culture. *Curr. Opin. Biotechnol.* **9**, 518-521.

Possee, R. D., Thomas, C. J., and King, L. A. (1999) The use of baculovirus vectors for the production of membrane proteins. *Biochem. Soc. Trans.* **27**, 928-932.

Pulford, D. J., Lowrey, P., and Morgan, A. J. (1995) Co-expression of the Epstein-Barr virus BXLF2 and BKRF2 genes with a recombinant baculovirus produces gp85 on the cell surface with antigenic similarity to the native protein. *J. Gen. Virol.* **76**, 3145-3152.

Pyle, L. E., Barton, P., Fujiwara, Y., Mitchell, A., and Fidge, N. (1995) Secretion of biologically active human proapolipoprotein A-I in a baculovirus-insect cell system: protection from degradation by protease inhibitors. *J. Lipid Res.* **36**, 2355-2361.

Radford, K., and Buell, G. (1998) Expression of ligand-gated ion channels using Semliki Forest virus and baculovirus. *Methods Enzymol.,* **293**: 459-483.

Rankl, N. B., Rice, J. W., Gurganus, T. M., Barbee, J. L., and Burns, D. J. (1994) The production of an active protein kinase C-delta in insect cells is greatly enhanced by the use of the basic protein promoter. *Prot. Expr. Purif.* **5**, 346-356.

O'Reilly, D. R., Miller, L. K. and Luckow, V. A. (1994) *Baculovirus Expression Vectors—A Laboratory Manual.* Oxford University Press, New York.

Rohrmann, G.F. (1986) Polyhedrin structure..*J. Gen. Virol.* **67**, 1499-1513.

Roy, P. (1996) Genetically engineered particulate virus-like structures and their use as vaccine delivery systems. *Intervirology* **39**, 62-71.

Rudi, H., Doan, D. N. P., and Olsen, O.-A. (1997) A $(His)_6$-tagged recombinant barley (*Hordeum vulgare L.*) endosperm ADP-glucose pyrophosphorylase expressed in the baculovirus-insect cell system is insentive to allosteric regulation by 3-phosphoglycerate and inorganic phosphate. *FEBS Lett.* **419**, 124-130.

St. Angelo, C., Smith, G. E., Summers, M. D., and Krug, R. M. (1987) Two of the three influenza viral polymerase proteins expressed by using baculovirus vectors form a complex in insect cells. *J. Virol.* **61**, 361-365.

Sandig, V., Hofmann, C., Steinert, S., Jennings, G., Schlag, P., and Strauss, M. (1996) Gene transfer into hepatocytes and human liver tissue by baculovirus vectors. *Hum. Gene Ther.* **7**, 1937-1945.

Schäfer, K., Müller, M., Faath, S., Henn, A., Osen, W., Zentgraf, H., Benner, A., Gissman, l., and Jochmus, I. (1999) Immune response to human papilloma virus 16 L1E7 chimeric virus-like particles: induction of cytotoxic T-cells and specific tumor protection. *Int. J. Cancer.* **81**, 881-888.

Schiller, J. T. and Roden, R. B. S. (1995) Papilloma-like particles. *Papillomavirus Rep.* **6**, 121-128.

Sells, M., Chen, M.-L., Acs, G. (1987) Production of hepatitis B particles in HepG2 cells transfected with cloned hepatitis B virus DNA. *Proc. Natl. Acad. Sci. USA* **84**, 1005-1009.

Shay, J.W., Wright, W.E., and Werbin, H. (1991) Defining the molecular mechanisms of human cell immortalization. *Biochim. Biophys. Acta* **1072**, 1-7.

Shoji, I., Aizaki, H., Tani, H., Ishii, K., Chiba, T., Saito, I., Miyamura, T., and Matsuura, Y. (1997) Efficient gene transfer into various mammalian cells, including non-hepatic cells, by baculovirus vectors. *J. Gen. Virol.* **78**, 2657-2664.

Smith, G. E., Summers, M. D., and Fraser, M. J. (1983) Production of human β-interferon in insect cells infected with a baculovirus expression vector. *Mol. Cell. Biol.* **3**, 2156-2165.

Soeda, E., Krauzewicz, N. Cox, C., Stokrova, J., Forstova, J., and Griffin, B. E. (1998) Enhancement by polylysine of transient, but not stable, expression of genes carried into cells by polyoma VP1 pseudocapsids. *Gene Ther.* **5**, 1410-1419.

Stewart, L. M., Hirst, M., Lopez Ferber, M., Merryweather, A. T., Cayley, P. J., and Possee, R. D. (1991) Construction of an improved baculovirus insecticide containing an insect-specific toxin gene. *Nature* **352**, 85-88.

Suzich, J. A., Ghim, S. J., Palmer-Hill, F. J., White, W. I., Tamura, J. K., Bell, J. A., Newsome, J. A., Jenson, A. B., and Schlegel, R. S. (1995) Systemic immunization with papillomavirus l1 protein prevents the development of viral mucosal papillomas. *Proc. Natl. Acad. Sci. USA* **92**, 11553-11557.

Takahashi, R.H., Nagashima, K., Kurata, T., and Takahashi, H. (1999) Analysis of human lymphotropic T-cell virus type II-like particle production by recombinant baculovirus-infected insect cells. *Virology* **256**, 371-380.

Tate, C. G., Whiteley, E., and Betenbaugh, M. J. (1999a) Molecular chaperones improve functional expression of the serotonin (5-hydroxytryptamine) transporter in insect cells. *Biochem. Soc.* **27**, 932-936.

Tate, C., G., Whiteley, E., and Betenbaugh, M., J. (1999b) Molecular chaperones stimulate the functional expression of the cocaine-sensitive serotonin receptor. *J. Biol. Chem.* **274**, 17551-17558.

Tjia, S.T., zu Altenschildesche, G.M., and Doerfler, W. (1983) *Autographa californica* nuclear polyhedrosis virus (AcNPV) DNA does not persist in mass cultures of mammalian cells. *Virology* **125**, 107-117.

Touze, A. and Coursaget, P. (1998) *In vitro* gene transfer using human papillomavirus-like particles. *Nucleic Acids Res.* **26**, 1317-1323.

Touze, A., Mehdaoui, S., Sizaret, P. Y., Mougin, C., Munoz, N., and Coursaget, P. (1998) The L1 major capsid protein of human papillomavirus type 16 variants affects yield of virus-like particles produced in an insect cell expression cell system. *J. Clin, Microbiol.* **36**, 2046-2051.

Tsao, T., Hsieh, J. C., Durkin, M. E., Wu, C. Y., Chakravarti, S., Dong, L. J., Lewis, M., and Chung, A. E. (1990) Characterization of the basement membrane glycoprotein entactin synthesized in a baculovirus expression system. *J. Biol. Chem.* **265**, 5188-5191.

Velardo, M.A., Bretthauer, R.K., Boutaud, A., Reinhold, B., Reinhold, V.N., and Castellino, F.J. (1993) The presence of UDP-N-acetylglucosamine:alpha-3-D-mannoside beta 1,2-N-acetylglucosamineaminyl-transferase I activity in *Spodoptera frugiperda* cells (IPLB-SF-21AE) and its enhancement as a result of baculovirus infection. *J. Biol. Chem.* **268**, 17902-17907.

Vialard, J. E. and Richardson, C. D. (1993) The 1,629-nucleotide open reading frame located downstream of the *Autographa californica* nuclear polyhedrosis virus polyhedrin gene encodes a nucleocapsid-associated phosphoprotein. *J. Virol.* **67**, 5859-5866.

Vlak, J. M., deGooijer, C. D., Tramper, J., and Miltenberger, H. G. (eds.) (1996) *Insect Cell Cultures: Fundamental and Applied Aspects*, Kluwer Academic Publishers, Dordrecht.

Vlak, J. M., Schouten, A., Usmany, M., Belsham, G. J., Klinge-Roode, E. C., Maule, A. J., Van Lent, J. W., and Zuidema, D. (1990). Expression of cauliflower mosaic virus gene I using a baculovirus vector based upon the p10 gene and a novel selection method. *Virology* **179**, 312-320.

Volkman, L.E. and Goldsmith, P.A. (1983) *In vitro* survey of *Autographa californica* nuclear polyhedrosis virus interaction with nontarget vertebrate host cells. *Appl. Environ. Microbiol.* **45**, 1085-1093.

Wang, M.-H., Patten, C. J., Yang, G.-Y., Paranawithana, S., R., Tan, Y., and Yang, C., S. (1996) Expression and coupling of human cytochrome P450 2E1 and NADPH-cytochrome 450 oxidoreductase in dual expression and co-infection systems with baculovirus in insect cells. *Arch. Bochem. Biophys.* **334**, 380-388.

Xie, Y. B., Sui, Y. P., Shan, L. X., Palvimo, J. J., Phillips, D. M., and Janne, O. A. (1992) Expression of androgen receptor in insect cells. Purification of the receptor and renaturation of its steroid- and DNA-binding functions. *J. Biol. Chem.* **267**, 4939-4948.

Yang, S., and Miller, L. K. (1998) Expression and mutational analysis of the baculovirus very late factor 1 (*vlf-1*) gene. *Virology* **245**, 99-109.

Yao, Q., Kuhlmann, F. M., Eller, R., Compans, R. W., and Chen, C. (2000) Production and characterization of simian-human immunodeficiency virus-like particles. *AIDS Res. Hum. Retroviruses* **16**, 227-236.

Yap, C.C., Ishii, K., Aizaki, H., Tani, H., Aoki, Y., Ueda, Y., Matsuura, Y., and Miyamura, T. (1998) Expression of target genes by co-infection with replication-deficient viral vectors. *J. Gen. Virol.* **79**, 1879-1888.

Zheng, Y. Z., Greenfield, P. F., and Reid, S. (1999) Optimized production of recombinant bluetongue core-like particles produced by the baculovirus expression system. *Biotechnol. Bioeng.* **65**, 600-604.

Zhou, D., Malissard, M., Berger, E. G., and Hennet, T. (2000) Secretion and purification of recombinant β1-4 galactosyltransferase from insect cells using pFmel-protA, a novel transposition-based baculovirus transfer vector. *Arch. Biochem. Biophys.* **374**, 3-7.

ENGINEERING POST-TRANSLATIONAL PROCESSING OF RECOMBINANT PROTEINS PRODUCED IN INSECT CELL CULTURE

E. AILOR AND M.J. BETENBAUGH*

The Johns Hopkins University
Department of Chemical Engineering
3400 North Charles Street
Baltimore, Maryland 21218 U.S.A.
**Corresponding Author, Tel.: 410-516-5461, Fax: 410-516-5510, e-mail: beten@jhu.edu*

1. Introduction

A number of host organisms are used today to facilitate the production of recombinant proteins. Recombinant proteins are produced in cultures derived from both prokaryotic and eukaryotic sources. The proteins obtained from these sources are used in a wide range of applications including diagnostics, structure-function studies, therapeutics, and vaccines. The choice of which host to employ for production of a particular protein depends upon a number of factors including, but not limited to, the type of protein, the quantity and quality of protein desired, the specific application, and the available resources. Each host includes properties that will dictate its suitability for a particular product of interest.

Due to several biological properties, insect cells are used to produce recombinant proteins for both research and industrial purposes (Luckow and Summers, 1988; Possee, 1997). Insect cells can be cultured under conditions without the use of serum supplements or regulated CO_2 levels. When using a baculovirus to drive expression of a heterologous gene, strong baculovirus promoters often result in the production of high levels of recombinant protein. Furthermore, the eukaryotic protein-processing environment of insect cells is more similar to that of mammalian cells than bacteria and yeast. Many of the post-translational modifications of proteins (e.g. proteolytic processing, disulfide bond formation, phosphorylation, and glycosylation) which normally occur in mammalian cells also take place in insect cells (O'Reilly *et al.*, 1993; Klenk, 1996; Jarvis, 1997). These features are especially important when the recombinant protein of interest is a secreted or membrane-bound protein. Though these processing events occur in insect cells, the modifications are not always identical to those that occur in mammalian cells. For certain applications these differences may prove to be extremely important.

M. Al-Rubeai (ed.), Cell Engineering, 29–42.

Many post-translational processing events occur within the secretory pathway. The secretory pathway is a multi-organelle system that transports polypeptides through multiple compartments where they are subsequently cleaved, folded, modified, assembled, and secreted or displayed at the membrane surface. Within these compartments are a number of cellular proteins which interact with and promote the folding, assembly, chemical modification, and transport of newly synthesized polypeptides (Whiteley and Betenbaugh, 2000). These cellular factors must be present in the proper balance in order to ensure successful secretion of a target protein. Consequently, engineering of the secretory pathway may be necessary if these factors are absent or limited in insect cells during production of heterologous proteins.

2. Polypeptide Cleavage

A number of proteins require cleavage at specific sites along the polypeptide chain in order to attain their final functional form. Enzymes located in various cellular compartments process these cleavage sites, delineated by recognizable amino acid motifs. As hosts, insect cells appear to contain some of the apparatus necessary to accurately process pre- and prosequences often associated with secreted mammalian proteins. However, in certain situations, polypeptide processing may be limited in insect cells over-expressing large amounts of heterologous polypeptides.

2.1. SIGNAL PEPTIDE PROCESSING

Secreted and many membrane polypeptides are targeted to the endoplasmic reticulum (ER) membrane through the use of a signal peptide located at the N-terminus of a polypeptide (von Heijne, 1994). As part of the translocation process, a membrane resident enzyme, signal peptidase, removes the signal peptide. In some cases, polypeptides expressed in insect cells accumulate as both a precursor and a final processed form. Proteins such as tissue plasminogen activator, insulin receptor binding domain, papain, and immunoglobulin single chain antibody fragment (scFv) are processed improperly and retained within the cell following over-expression in insect cells (Jarvis and Summers, 1989; Sissom and Ellis, 1991; Tessier, *et a*l., 1991; Ailor *et al.*, 1999). These polypeptides accumulate in the insoluble fraction so aggregation may be due to the presence of unprocessed signal peptide. When the native prepropapain signal sequence was replaced with one from the insect protein honeybee melittin, the accumulation of prepropapain was reduced and a five-fold increase in propapain secretion was observed (Vernet *et al.*, 1990). However, other attempts to enhance processing and secretion of recombinant proteins by replacing the native signal sequence with insect-derived signal peptides such as bee melittin, cecropin B and baculovirus 64K glycoprotein were not as successful (Jarvis *et al.*, 1993). A large fraction of an over-expressed scFv was sequestered within insect cells as pre-processed polypeptide chains even though the honeybee melittin signal sequence was used (Ailor *et al.*, 1999). In another study, Golden *et al.* (1998) demonstrated that a human signal peptide derived from tissue plasminogen activator was more efficient in secreting the HIV-1 gp120 glycoprotein from insect cells than signal sequences from either human placental alkaline phosphatase or baculovirus gp67. Effective signal sequences are likely protein dependent so selection of a signal sequence that will promote efficient secretion is likely to remain empirical until more is known about this process.

An alternative method to improve signal peptide processing would be to enhance signal peptidase (SP) activity. Increasing SP activity facilitated an improvement in signal peptide processing in bacteria (van Dijl *et al.*, 1991). SP has been identified in a number of pro- and eukaryotes. While bacterial signal peptidases are monomeric, eukaryotic signal peptidase complexes are oligomeric, having two to five components. Although varying in structure, signal peptidases from bacteria and eukaryotes have similar substrate specificity and are capable of processing signal sequences from either source (Dalbey and von Heijne, 1992). In order to reduce a scFv precursor accumulating as insoluble protein within insect cells a SP from *Bacillus subtilis* was co-expressed (Ailor *et al.*, 1999). The accumulation of precursor scFv was nearly completely eliminated even though a bacterial SP was used to cleave a eukaryotic signal peptde in a eukaryotic host. In addition to reducing the pre-processed form of scFv, extracellular levels of scFv were increased over 300%. Thus, over-expression of signal peptidase in insect cells may represent an effective means for improving the efficiency of signal peptide removal from secreted polypeptide precursors.

2.2. PROSEQUENCE PROCESSING

Other proteolytic processing steps are necessary for the production of functional proteins and these may or may not be performed correctly and/or efficiently in insect cells. Some viral polyproteins (Sindbis and rubella) have been processed to release correct individual subunits while others such as HIV gp160 were not efficiently cleaved to gp120 and gp41 (Luckow, 1995). In cases where protein synthesis is extremely high, the cleavage of protein precursors may be substantially slower in insect cells with processing half-times that are more than triple those observed in the native host. These differences are likely due to a deterioration of cell activity following viral infection or a lack of sufficient levels of a particular processing enzyme. In some instances, cleavage of protein precursors will occur in insect cells at positions that are different from those found in the native species, while in others, cleavage is minimal or completely absent. In the latter case, co-expression of processing proteases may accomplish the desired cleavage goals. Co-expression of heterologous furin, the first member of the mammalian proprotein convertase family to be characterized, has been used to enhance proprotein processing of recombinant proteins in insect cells. Cleavage efficiencies of HIV-1 gp160, transforming growth factor beta 1, and E-cadherin were greatly enhanced when co-expressed with furin derived from mammalian sources (Morikawa *et al.*, 1993; Laprise *et al.*, 1998; Posthaus *et al.*, 1998). Although furin has been identified and characterized in *Spodoptera frugiperda* cells (Sf) (Cieplik *et al.*, 1998) and can correctly process proteins such as human fibrillin-1 (Ritty *et al.*, 1999), a lack of proprotein processing capacity in insect cells limits processing of other heterologous proteins. This deficiency may be due to an absence of other proprotein convertase enzymes in insect cells that are typically found in mammalian hosts.

3. Folding and Assembly

Although the amino acid sequence of a nascent polypeptide contains all the information necessary to provide for its correct folding *in vitro* (Anfinsen, 1973), a newly synthesized polypeptide chain is exposed to an environment *in vivo* that is very different from *in vitro* conditions. Interactions with neighboring cellular components may lead to improper folding and incorrect associations. It is within this environment that a number of cellular factors are present that specifically associate with the polypeptides to help bring about their correct conformational structure.

3.1. CYTOSOL TO ENDOPLASMIC RETICULUM TRANSPORT

Following the start of translation from the ribosome-message RNA complex, polypeptides destined for secretion are directed to the ER membrane. Polypeptides are transported across the ER membrane either post-translationally or co-translationally (Rapoport, 1996). Co-translational translocation is the predominant mechanism used in mammalian cells to import polypeptides into the ER, however, both mechanisms are used extensively in yeast. Co-translational processing is initiated by the binding of the signal recognition particle (SRP) to the signal sequence as it emerges from the ribosome in the cytosol (Wilkinson *et al.*, 1997). Translation is halted until the SRP-ribosome-polypeptide complex interacts at the ER membrane with the signal receptor (also called docking protein or SRP receptor). The SRP-ribosome complex also binds to a heterotrimeric complex, sec61p, which forms the polypeptide-conducting channel from the cytosol to the ER. The post-translational processing apparatus is best characterized in yeast. Cytosolic hsp70 maintains the polypeptide precursor in a translocation-competent unfolded, or loosely folded, conformation until transport into the ER and translocation is completed (Hartl and Martin, 1995). Hsp70 is also believed to help prevent aggregation during this process by screening exposed hydrophobic regions on the nascent polypeptides.

Hsp70 proteins involved in translocation include ATPase and peptide binding domains (Cyr and Neupert, 1996). These folding assistant proteins have been found to operate synergistically with other co-factors including hsp40, Hip, and Hop. The hsp40 protein, which is also required for post-translational translocation, facilitates the hydrolysis of ATP and conversion of hsp70 into a tight binding conformation (Sugito *et al.*, 1995). Stabilization of the tight binding form is facilitated by the co-chaperone Hip (Hohfeld *et al.*, 1995). Eventual release of the ADP and bound polypeptide is enhanced by another co-chaperone, Hop (Johnson *et al.*, 1998). As a collective, these chaperones enhance hsp70's ability to bind and release unfolded polypeptides. Although the primary mechanism for polypeptide import into the ER in insect cells is unknown at this time, the accumulation of aggregated non-glycosylated precursor polypeptides (containing the signal sequence) suggests that at least one of these translocation mechanisms may be saturated during the expression of recombinant proteins.

To investigate this possibility, hsp70 was co-expressed in *Trichoplusia ni* (*T. ni*) cells with an immunoglobulin G (IgG) to determine if an increase in the chaperone concentration could help prevent aggregation of immunoglobulin chains prior to their import into the ER (Ailor and Betenbaugh, 1997). These initial studies revealed the binding of hsp70 to IgG chains in soluble cell fractions. When the cells were co-infected with recombinant IgG and hsp70 baculoviruses, pulse-chase analysis revealed increased levels of the light chain precursor as well as enhanced processing of the light chain precursor into its mature form. As a result, enhancement of IgG secretion was obtained.

One hypothesis for the improved processing is that hsp70 maintains immunoglobulin light chain precursors or segment of the precursors in the cytosol in a conformation conducive to post-translational translocation into the ER. Due to the strong baculovirus promoter, a recombinant protein may be produced at such a high level that a fraction of the protein may undergo partial or complete polypeptide translation in the cytosol of infected cells. Although secreted polypeptides are thought to enter the ER by a co-translational translocation mechanism, the post-translational translocation mechanism may serve as an alternate route to the ER for secreted proteins expressed using the baculovirus system. Another alternative is that hsp70 increases the efficiency of the co-translational pathway by preserving ribosome bound polypeptides, which may also be in excess, in a form conducive to insertion into the membrane. An indirect mechanism is also possible in which hsp70 increases the functioning of other polypeptide assistance factors that enhance immunoglobulin processing.

3.2. ENDOPLASMIC RETICULUM

Once the secreted polypeptide begins the translocation process, whether post- or co-translationally, the emerging polypeptide undergoes additional processing in the ER lumen. In the ER newly imported polypeptides are cleaved and begin to fold and assemble into multimers (Helenius *et al.*, 1992). Due to the high concentrations of unfolded and partially folded polypeptides, folding assistance factors are necessary to minimize aggregation and accelerate folding (Gething and Sambrook, 1992). Jarvis and co-workers (1990) observed the accumulation of misfolded polypeptides within baculovirus-infected insect cells and suggested that folding and disulfide formation were limited within the infected insect cells.

3.2.1. Chaperones

During or immediately following translocation polypeptides may interact with several ER resident chaperones prior to proper folding and assembly. One such chaperone, the Immunoglobulin Heavy Chain Binding Protein (BiP) or GRP78, is a hsp70 family member that binds to polypeptides along segments of 7 to 9 amino acids of significant hydrophobic content (Blond-Elguindi *et al.*, 1993). BiP is known to complex with immunoglobulin chains and other polypeptides and may be involved in assisting translocation of polypeptides into the ER (Vogel *et al.*, 1990). As mentioned previously, immunoglobulins were observed to undergo significant aggregation following their expression in insect cells. In an attempt to reduce aggregation, BiP was co-expressed with IgG in *T. ni* cells (Hsu and Betenbaugh, 1997). Co-expressed BiP was shown to bind to immunoglobulin chains and significantly increase soluble and secreted IgG levels. Functional IgG, as determined by antigen binding studies, was also improved in the presence of BiP. Unfortunately, even with BiP co-expression, significant levels of insoluble immunoglobulin remained suggesting that other bottlenecks to folding exist in insect cells.

As with cytosolic hsp70, a number of co-chaperones have been identified that enhance the binding and release function of BiP. Several co-chaperones of the hsp40 protein family including the lumenal protein SCJ1 and the membrane homologue Sec63p have been observed in the ER. Modeling studies suggest that BiP's efficiency as a chaperone may be improved by the presence of additional partner proteins. Lowering the kinetics of BiP release from a bound polypeptide by expressing a co-chaperone could reduce polypeptide aggregation and further increase secreted protein yields (Whiteley *et al.*, 1997a).

In addition to BiP, several other chaperones have been identified in the ER that aid in the correct folding and assembly of membrane and secreted proteins. For many membrane proteins and some secreted glycoproteins the assistance of the membrane-bound chaperone calnexin is essential to ensure proper folding and assembly. Calnexin selectively binds to the structure $Glc_1Man_9GlcNAc_2$ of glycoproteins in a lectin-like binding fashion (Vassilakos *et al.*, 1998). Calnexin was first identified as a chaperone in association with a number of membrane proteins including MHC antigen, VSV G protein, influenza trimer, cystic fibrosis transmembrane conductance regulator, and serotonin transporter (Anderson and Cresswell, 1994; Hammond *et al.*, 1994; Pind *et al.*, 1994; Tate and Blakely, 1994). More recently, calnexin has also been observed to associate with secreted glycoprotein folding intermediates including transferrin, α1-antitrypsin, α-fetoprotein, gp80, and thyroglobulin prior to folding and transport from the ER (Ou *et al.*, 1993; Wada *et al.*, 1994; Kim and Arvan, 1995).

Calnexin also serves as part of the ER proofreading mechanism by inhibiting unfolded polypeptides from leaving the ER in a processing cycle with UDP-glucose:glycoprotein glucosyltransferase and glucosidase II. Calnexin's role in the folding of proteins in insect cells was revealed by co-expressing the chaperone with recombinant sertonin transporter (Myc-SERT), a membrane glycoprotein. Calnexin was shown to increase active Myc-SERT levels by nearly three fold over levels obtained in Sf-9 cells in the absence of any co-expressed chaperone (Tate *et al.*, 1999). Immunoblot analysis revealed a specific interaction between calnexin and Myc-SERT suggesting that calnexin association with Myc-SERT was responsible for facilitating the improvement in transporter activity.

Calnexin is known to cooperate with other folding assistance proteins including calreticulin and ERP57 to facilitate folding (Tatu and Helenius, 1997; Oliver *et al.*, 1997). While calnexin binds preferentially to sugars in the stem domain of a protein, calreticulin binds to glycans in the upper reaches of membrane proteins. Although not as effective as calnexin, co-expressed calreticulin was able to increase functional levels of Myc-SERT by approximately 30% (Tate *et al.*, 1999).

Calnexin is also known to facilitate protein folding in concert with other chaperones including BiP. Calnexin and BiP recognize different structural motifs sequentially during polypeptide folding (Hammond and Helenius, 1994; Kim and Arvan, 1995). Another ER chaperone, GRP94, a member of the hsp90 chaperone family, has been observed to interact with newly synthesized immunoglobulins in the ER following BiP association and prior to assembly (Melnick *et al.*, 1992). In fact, GRP94 binding to immunoglobulin light chains is prevented by BiP mutants that do not release the polypeptide, this suggests a coordinated and sequential binding process of multiple chaperones (Melnick *et al.*, 1994). Thus, future studies in insect cells should consider coordinated expression of several ER chaperones in order to optimize the folding of heterologous proteins.

3.2.2. Foldases

In addition to chaperones, the ER contains catalytic enzymes that accelerate rate-limiting reactions along the folding pathway. Without the presence of these enzymes polypeptides may form improper intrachain and/or interchain associations with other polypeptides due to slow folding at a rate-limiting step. One important folding event along the polypeptide processing pathway is the formation of disulfide bonds which occurs as polypeptides leave the cytosol and enter the oxidizing environment of the ER. As a catalyst in the oxidization, reduction, and isomerization of disulfide bonds, Protein Disulfide Isomerase (PDI) can improve recombinant protein folding and secretion. PDI over-expression was shown to increase immunoglobulin solubility, secretion, and activity in *T. ni* cells (Hsu et al., 1996). Since PDI is known to contain two distinct folding assistant domains, C- and N-terminal thioredoxin active sites, further studies were performed to examine the importance of these regions on polypeptide folding in insect cells. When several PDI mutants were co-expressed, the C-terminus thioredoxin domain was found to be effective in enhancing IgG solubility and secretion (Whiteley *et al.*, 1997b). However, mutations in the thioredoxin domain at the N-terminus led to a reduction in IgG solubility. The mutations at the N-terminus may have destabilized PDI and facilitated aggregation of other polypeptides in an anti-chaperone role (Puig and Gilbert, 1994; Puig *et al.*, 1994).

Another enzyme suspected to be involved in the folding process is *cis-trans* peptidyl-prolyl isomerase (PPI). PPI catalyzes the *cis-trans* isomerization of the Xaa-Pro peptide bond in the polypeptide chain (Schmid, 1993). The presence of PPI has accelerated the *in vitro* refolding of murine IgG light chain and monoclonal antibody F_{ab} fragments by 7 and 3.5 fold, respectively, to increase product yield (Lang *et al.*, 1987; Lilie *et al.*, 1993). The importance of PPI to protein folding *in vivo* has been suggested by the retardation of procollagen and transferrin maturation in the presence of cyclosporin A, an inhibitor of PPI activity in mammalian cells (Steinmann *et al.*, 1991; Lodish and Kong, 1991). PPI activity may be limiting in insect cells as well. Sf-9 cells stably transformed to produce the *Drosophila melanogaster* protein ninA, a membrane bound cyclophilin with PPIase activity, imported dopamine at a rate over five times faster than that of wild type Sf-9 cells (Lenhard and Reilander, 1997). The increase in dopamine uptake was suggested to be a result of improved dopamine transporter folding facilitated by over-expression of PPI.

4. Polypeptide Modifications

In many cases a secreted or membrane protein will require additional modifications to the polypeptide chain to confer authentic protein structure and function. These modifications typically involve the covalent linkage of a donor molecule to a specific site on the polypeptide. The donor is transferred to the polypeptide at a recognition site, designated by one or more amino acid groups, with the aid of a catalytic enzyme specific for the reaction.

4.1. OLIGOSACCHARIDE PROCESSING

One post-translational process in which insect and mammalian cells differ is glycosylation. Both N-linked and O-linked glycans have been observed on recombinant glycoproteins produced in insect cells. N-linked glycans are particularly significant to glycoprotein structure and function (Jenkins *et al.*, 1996). In mammalian cells, N-linked oligosaccharides are complex and often contain terminal sialic acid moieties. In contrast, N-glycans produced in insect cells are comprised primarily of high ($Man_{9-5}GlcNAc_2$) or truncated low (paucimannose) ($Man_{3-2}GlcNAc_2$) mannose (Man) structures. A few hybrid and complex oligosaccharide structures have been observed but these generally do not include galactose (Gal) or sialic acid. As in mammalian cells, N-glycosylation in insect cells begins in the ER with the transfer of Glc3Man9GlcNAc2 from a P-P-dolichol-Glc3Man9GlcNAc2 complex to specific asparagine sites along the polypeptide chain (Altmann, 1996; Marchal *et al.*, 1999). The asparagines targeted for N-glycosylation are found as part of an Asn-X-Ser/Thr sequence within the polypeptide. However, not all of these acceptor sites may be occupied on a particular protein. In addition, N-glycosylation sites that are normally occupied on native proteins may not be occupied when produced as recombinant proteins in insect cells (Tate and Blakely, 1994; Hooker *et al.*, 1999).

Following its addition to the polypeptide chain, the core oligosaccharide is processed by a number of enzymes that remove and add carbohydrate groups to the core during the glycoprotein's passage through the ER and Golgi compartments. The initial carbohydrate trimming of glucose and outer mannose sugars is similar in mammalian and insect cells but the processing events appear to diverge during the carbohydrate addition steps.

Initial studies, performed primarily on Sf-9 or Sf-21 cells from *Spodoptera frugiperda*, suggested that insect cells could trim N-linked oligosaccharides but lacked the capacity to elongate these trimmed structures to produce complex carbohydrates (Marz *et al.*, 1995). Studies using other insect cell lines, including *T. ni* and *Estigmena acrea* (Ea-4), have demonstrated that insect cells can produce small amounts of biantennary hybrid (structures with one terminal Man branch and one branch with terminal Gal, N-acetylglucosamine [GlcNAc], or another sugar) and complex (structures with two non-Man termini) N-glycans (Altmann *et al.*, 1999). The idea that insect cells could produce at least limited amounts of these hybrid and complex N-linked oligosaccharides was further supported by the finding that these cells had low levels of GlcNAc transferase I, GlcNAc transferase, fucosyltransferase, mannosidase I, mannosidase II, and galactosyltransferase (Velardo *et al.*, 1993; Altmann *et al.*, 1993; van Die *et al.*, 1996).

The presence of a processing N-acetylglucosaminidase which cleaves GlcNAc residues attached to the α(1,3) Man arm may explain the appearance of the paucimannosidic structures observed on glycoproteins produced in insect cells (Licari *et al.*, 1993; Altmann *et al.*, 1995). Attempts at limiting the activity of this enzyme with chemicals has met with some success, but significant amounts of paucimannosidic N-glycans are still produced even in the presence of these inhibitors (Wagner *et al.*, 1996).

Genetic engineering of the oligosaccharide processing pathway is another approach used to produce more complex N-glycans in insect cells. Over-expression of human β(1,2)-N-acetylglucosamine transferase I was shown to provide a greater than fourfold increase of terminal N-acetylglucosamine found on the N-glycans of fowl plague virus hemagglutinin expressed in Sf-9 cells (Wagner *et al.*, 1996). Although this method was effective in elongating N-glycans, no complex structures were detected. In studies in Sf-9 cells, expression of a mammalian β(1,4)-galactosyltransferase with tissue plasminogen activator by either a recombinant baculovirus (Jarvis and Finn, 1996) or a stably transformed cell line (Hollister *et al.*, 1998) successfully elongated N-glycans to include galactose as detected by lectin blot analysis. Using 2D HPLC techniques, the specific N-glycan structures attached to recombinant human transferrin from *T. ni* cells were determined following co-expression of galactosyltransferase. Although no complex structures were detected, the majority of the hybrid structures were elongated to include a β-linked terminal galactose on the α(1,6) branch (Ailor *et al.*, 2000). Recently, The over-expression of mammalian α(2,6)-sialyltransferase in Sf-9 cells stably transformed to express mammalian β(1,4)-galactosyltransferase was reported to produce N-glycans from baculovirus gp64 that contain galactose and sialic acid (Jarvis *et al.*, 1998).

Whereas knowledge of N-linked glycosylation in insect cells has increased in recent years, less is known about O-linked glycosylation. O-glycosylation was observed to occur at the same site on interferon-α2 expressed in Sf-9 cells as its native counterpart (Sugyiama *et al.*, 1993). However, insect cells, unlike mammalian cells, appear to lack the capacity to sialylate O-linked oligosaccharides. In fact, several studies have shown that O-glycans from recombinant glycoproteins produced in insect cells are comprised of a single N-acetylgalactosamine (GalNAc) moiety or the disaccharide Gal-GalNAc (Marz *et al.*, 1995).

4.2. OTHER MODIFICATIONS

In addition to glycosylation, insect cells are capable of performing other covalent chemical modifications as well. In some cases these modifications are identical to those occurring in mammalian cells, however, other processing events may be different or not occur at all in insect cells.

Phosphorylation of proteins at serine, threonine, or tyrosine residues by protein kinases and the removal of phosphate residues by phosphotases are key regulatory events in a cell. Many phosphorylated proteins have been correctly expressed and modified in insect cells though different phosphorylation patterns have been observed in select cases

(O'Reilly *et al.*, 1993). Serine and threonine phosphorylation appears to be common in insect cells but tyrosine phosphorylation is often less effective. Insect cell-derived proteins can be underphosphorylated, especially those proteins expressed near the end of the viral infection cycle. In one study, enhanced DNA binding activity and phosphorylation of STAT3 was facilitated by the co-expression of the src-family kinase Lck in insect cells (Lund *et al.*, 1999).

Structural proteins of the budded form of AcMNPV undergo fatty acid acylation so it is not surprising that insect cells can also perform this post-translational processing event on recombinant proteins (O'Reilly *et al.*, 1993). Using radioactive labeled acid precursors, baculovirus expressed proteins were observed to undergo myristoylation, palmitylation, and isoprenylation. These processing events are especially prominent for viral membrane proteins and membrane-associated proteins involved in signal transduction, such as the ras-family of proteins. In some cases, variant forms that are insufficiently modified are also observed, although these structures may not be processed to the membrane.

Some proteins, usually small peptides, may undergo α-amidation. α-Amidation is a sequential process that starts with the cleavage and processing of a protein precursor at the C-terminus to produce a terminal glycine residue. The glycine is then converted to an α-amide group by a series of enzymatic reactions (Jarvis, 1997). This process has been carried out successfully for proteins expressed in insect larvae (Hellers *et al.*, 1991). However, the successful execution in insect cells may require co-expression of an α-hydroxylating monooxygenase that performs an initial conversion of glycine to the hydroxylated form.

The N-terminal methionine residue of some intracellular mammalian proteins is removed followed by the acetylation of the following residue. In baculovirus-infected insect cells, this same process has been observed as well as the acetylation of the N-terminal methionine itself (Jarvis, 1997). Thus, the cleavage and acetylation processes may be different and less efficient in insect cells as compared to those observed in native mammalian environments.

The addition of sulfate to tyrosine and oligosaccharides is another common modification found on proteins from eukaryotic sources. Sulfation requires the generation, via a two-step process, of the donor substrate 3'-phosphoadenosine-5'-phosphosulfate (PAPS). Sulfation of heterologous proteins expressed in *Spodoptera frugiperda* cells indicates that insect cells possess the machinery to synthesize PAPS and transfer sulfate groups to tyrosine residues (Zhao and Sane, 1993; Brown-Augsburger *et al.*, 1994).

5. Surface Transport Apparatus

Once the polypeptide is folded and assembled in the ER, the protein is transported from the ER to the Golgi and into the extracellular space. This process occurs through a sequence of steps involving repeated cycles of vesicle budding, transport, and fusion through the Golgi compartments. The protein transport process is likely to include a number of cellular proteins such as vesicle coating proteins, ADP ribosylation factor

(ARF), v-snares and t-snares. Kunz and co-workers (1993) observed a 30% drop in the intracellular concentration of endogenous ARF in *Spodoptera frugiperda* following baculovirus infection. Furthermore, co-expression of a recombinant ARF2 was able to increase the intracellular ARF level 5 fold. Other transport factors in the secretory pathway may be similarly affected. However, the components and processes of the post-assembly protein transport pathway are still being elucidated. These transport steps, not yet characterized in insect cells, may also deteriorate following baculovirus infection.

6. Concluding Remarks

Due to several key features, including the ability to perform post-translational modifications, the baculovirus expression vector system is highly popular for the rapid production of eukaryotic secreted and membrane proteins. Post-translational processing in insect cells can be similar to that in mammalian cells for some proteins, while the modifications may be inefficient, different, or absent altogether for others. In many cases the post-translational processing apparatus naturally present in insect cells will suffice. For other cases, inefficient protein modifications can lead to lower product yields and inferior product quality. In these instances, additional co-factors can be expressed to increase and/or make insect cell-derived proteins more similar to those generated in the native host. By developing cell lines and vectors with additional processing functions, the capacity to generate complex mammalian-like modifications is likely to improve in the future.

However, the direct expression of co-factors and enzymes is only one of several ways to manipulate post-translational processing of proteins in cultured insect cells. These events may also be influenced by a number of other factors including the insect cell line, the infection strategy, and the cell culture conditions. In fact, each of these contributing factors must be considered in order to optimize the large scale production of functional secreted or membrane proteins needed by the biotechnology and pharmaceutical industries.

7. References

Ailor, E., Takahashi, N., Tsukamoto, Y., Masuda, K., Abdul-Rahman, B., Jarvis, D.L., Lee, Y.C., and Betenbaugh, M.J. (2000) *N* glycan patterns of human transferrin produced in *Trichoplusia ni* insect cells: effects of mammalian galactosyltransferase. *Glycobiology*, in press.

Ailor, E., Pathmanathan, J., Jongbloed, J.D.H., and Betenbaugh, M.J. (1999) A bacterial signal peptidase enhances processing of a recombinant single chain antibody fragment in insect cells. *Biochem. Biophys. Res. Commun.* **255**, 444 - 450.

Ailor, E. and Betenbaugh, M.J. (1997) Overexpression of a cytosolic chaperone to improve solubility and secretion of a recombinant IgG protein in insect cells. *Biotech. Bioeng.* **58**, 196 - 203.

Altmann, F., Staudacher, E., Wilson, I.B.H., and Marz, L. (1999) Insect cells as hosts for the expression of recombinant glycoproteins. *Glycoconjugate J.* **16**, 109 - 123.

Altmann, F. (1996) N-glycosylation in insects revisited. *Trends Glycosci. Glycotech.* **8**, 101 - 114.

Altmann, F., Schwihla, H., Staudacher, E., Glossl, J., and Marz, L. (1995) Insect cells contain an unusual, membrane-bound β-N-acetylglucosaminidase propably involved in the processing of protein N-glycans. *J. Biol. Chem.* **270**, 17344 - 17349.

Altmann, F., Kornfeld, G., Dalik, T., Staudacher, E., and Glossl, J. (1993) Processing of asparagine-linked oligosaccharides in insect cells. N-acetylglusoaminyltransferase I and II activities in cultured lepidopteran cells. *Glycobiology* **3,** 619 - 625.

Anderson, K.S. and Cresswell, P. (1994) A role for calnexin (IP90) in the assembly of class II MHC molecules. *EMBO J.* **13,** 675 - 682.

Anfinsen, C.B. (1973) Principles that govern the folding of protein chains. *Science* **181,** 223 - 230.

Blond-Elguindi, S., Fourie, A.M., Sambrook, J.F., and Gething, M.J. (1993) Affinity panning of a library of peptides displayed on bacteriophages reveals the binding specificity of BiP. *Cell* **75,** 717 - 728.

Brown-Augsburger, P., Broekelmann, T., Mecham, L., Mercer, R., Gibson, M.A., Cleary, E.G., Abrams, W.R., Rosenbloom, J., and Mecham, R.P. (1994) Microfibril-associated glycoprotein binds to the carboxyl-terminal domain of tropoelastin and is a substratefor transglutaminase. *J. Biol. Chem.* **269,** 28443 - 28449.

Cieplik, M., Klenk, H.D., and Garten, W. (1998) Identification and characterization of *Spodoptera frugiperda* furin: a thermostable subtilisin-like endopeptidase. *Biol. Chem.* **379,** 1433 - 1440.

Cyr, D.M. and Neupert, W. (1996) Roles for hsp70 in protein translocation across membranes of organelles. *EXS.* **77,** 25 - 40.

Dalbey, R.E. and von Heijne, G. (1992) Signal peptidases in prokaryotes and eukaryotes - a new protease family. *TIBS* **17,** 474 - 478.

Gething, M.J. and Sambrook, J. (1992) Protein folding in the cell. *Nature* **355,** 33 - 45.

Golden, A., Austen, D.A., van Schravendijk, M.R., Sullivan, B.J., Kawasaki, E.S., and Osburne, M.S. (1998) Effect of promoters and signal sequences on the production of secreted HIV-1 gp120 protein in the baculovirus system. *Prot. Expr. Purif.* **14,** 8 - 12.

Hammond, C., Braakman, I., and Helenius, A. (1994) Folding of VSV G protein: sequential interaction with BiP and calnexin. *Science* **266,** 456 - 458.

Hammond, C. and Helenius, A. (1994) Role of N-linked oligosaccharide recognition, glucose trimming, and calnexin in glycoprotein folding and quality control. *Proc. Nat. Acad. Sci. USA* **91,** 913 - 917.

Hartl, F. U. and Martin, J. (1995) Molecular chaperones in cellular protein folding. *Cur. Opin. Struc. Biol.* **5,** 92 - 102

Helenius, A., Marquardt, T., and Braakman, I. (1992) The endoplasmic reticulum as a protein-folding compartment. *Trends Cell Biol.* **2,** 227 - 231.

Hellers, M., Gunne, H., Steiner, H. (1991) Expression of post-translational processing of preprocecropin A using a baculovirus vector. *Eur. J. Biochem.* **199,** 435 - 439.

Hohfeld, J., Minami, Y., and Hartl, F.U. (1995) Hip, a novel cochaperone involved in the eukaryotic hsc70/hsp40 reaction cycle. *Cell* **83,** 589 - 598.

Hollister, J.R., Shaper, J.H., and Jarvis, D.L. (1998) Stable expression of mammalian β1-4-galactosyltransferase extends the N-glycosylation pathway in insect cells. *Glycobiology* **8,** 473 - 480.

Hooker, A.D., Green, N.H., Baines, A.J., Bull, A.T., Jenkins, N., Strange, P.G., and James, D.C. (1999) Constraints on the transport and glycosylation of recombinant IFN-γ in chinese hamster ovary and insect cells. *Biotech. Bioeng.* **63,** 559 - 572.

Hsu, T.-A. and Betenbaugh, M. J. (1997) Coexpression of molecular chaperone BiP improves immunoglobulin solubility and IgG secretion from *Trichoplusia ni* insect cells. *Biotech. Prog.* **13,** 96 - 104.

Hsu, T-A., Watson, S., Eiden, J.J., and Betenbaugh, M.J. (1996) Rescue of immunoglobulins from insolubility is facilitated by PDI in the baculovirus expression system. *Prot. Expr. Purif.* **7,** 281 - 288.

Jarvis, D.L., Kawar, Z.S., and Hollister, J.R. (1998) Engineering N-glycosylation pathways in the baculovirus-insect cell system. *Curr. Opin. Biotech.* **9,** 528 - 533.

Jarvis, D.L. (1997) Baculovirus Expression Vectors, in: *The Baculoviruses*, Miller, L.K., ed., Plenum Publishing Corporation, New York, pp. 389 - 431.

Jarvis, D.L and Finn, E.E. (1996) Modifying the insect cell N-glycosylation pathway with immediate early baculovirus expression vectors. *Nature Biotech.* **14,** 1288 - 1292.

Jarvis, D.L., Summers, M.D., Garcia, A., and Bohlmeyer, D.A. (1993) Influence of different signal peptides and prosequences on expression and secretion of human tissue plasminogen activator in the baculovirus system. *J. Biol. Chem.* **268,** 16754 - 16762.

Jarvis, D.L., Fleming, J.G.W., Kovacs, G.R., Summers, M.D., and Guarino, L.A. (1990) Use of early baculovirus promoters for continuous expression and efficient processing of foreign gene products in stably transformed lepidopteran cells. *Bio/Tech.* **8,** 950 - 955.

Jarvis, D.L. and Summers, M.D. (1989) Glycosylation and secretion of human tissue plasminogen activator in recombinant baculovirus-infected insect cells. *Mol. Cell. Biochem.* **9,** 214 - 223.

Jenkins, N., Parekh, R.B., and James, D.C. (1996) Getting the glycosylation right: implications for the biotechnology industry. *Nature Biotech.* **14,** 975 - 980.

Johnson, B.D., Schumacher, R.J., Ross, E.D., and Toft, D.O. (1998) Hop modulates hsp70/hsp90 interactions in protein folding. *J. Biol. Chem.* **273,** 3679 - 3686.

Kim, P.S. and Arvan, P. (1995) Calnexin and BiP act as sequential molecular chaperones during thyroglobulin folding in the endoplasmic reticulum. *J. Cell. Biol.* **128,** 29 - 38.

Klenk, H-D. (1996) Post-translational modifications in insect cells. *Cytotechnology* **20,** 139 - 144.

Kunz, B.C., Muczynski, K.A., Welsh, C.F., Stanley, S.J., Tsai, S.C., Adamik, R., Chang, P.P., Moss, J., and Vaughn, M. (1993) Characterization of recombinant and endogenous ADP-ribosylation factors synthesized in Sf9 cells. *Biochemistry* **32,** 6643 - 6648.

Lang, K., Schmid, F., and Fischer, G. (1987) Catalysis of protein folding by prolyl isomerase. *Nature* **329,** 268 - 270.

Laprise, M-H., Grondin, F., Dubois, C.M. (1998) Enhanced TGFb1 maturation in High Five cells coinfected with recombinant baculovirus encoding the convertase furin/pace: improved technology for the production of recombinant proproteins in insect cells. *Biotech. Bioeng.* **58,** 85 - 91.

Lenhard, T. and Reilander H. (1997) Engineering the folding pathway of insect cells: generation of a stably transformed insect cell line showing improved folding of a recombinant membrane protein. *Biochem. Biophys. Res. Commun.* **238,** 823 - 830.

Licari, P.J., Jarvis, D.L., and Bailey, J.E. (1993) Insect cell hosts for baculovirus expression vectors contain endogenous exoglycosidase activity. *Biotech. Prog.* **9,** 146 - 152.

Lilie, H., Lang, K., Rudolph, R., and Buchner, J. (1993) Prolyl isomerases catalyze antibody folding in vitro. *Prot. Science* **2,** 1490 - 1496.

Lodish, H.F. and Kong, N. (1991) Cyclosporin A inhibits an initial step in folding of transferrin within the endoplasmic reticulum. *J. Biol. Chem.* **266,** 14835 - 14838.

Luckow, V.A. (1995) Protein production and processing from baculovirus expression vectors, in: *Bacuolvirus expression systems and biopesticides*, Shuler, M. L., Wood, H. A., Granados, R. R., and Hammer, D. A., eds., John Wiley and Sons, New York, pp. 51 - 90.

Luckow, V.A. and Summers, M.D. (1988) Trends in the development of baculovirus expression vector. *Bio/Tech.* **6,** 47 - 55.

Lund, T.C., Coleman, C., Horvath, E., Sefton, B.M., Jove, R., Medveczky, M.M., and Medveczky, P.G. (1999) The Src-family kinase Lck can induce STAT3 phosphorylation and DNA binding activity. *Cell Signal.* **11,** 789 - 796.

Marchal, I., Mir, A-M., Kmiecik, D., Verbert, A., and Cacan, R. (1999) Use of inhibitors to characterize intermediates in the processing of N-glycans synthesized by insect cells: a metabolic study with Sf9 cell line. *Glycobiology* **9,** 645 - 654.

Marz, L., Altmann, F., Staudacher, E., and Kubelka, V. (1995) Protein glycosylation in insects, in: *Glycoproteins*, Montreuil, J., Schachter ,H., and Vliegenthart, J.F.G., eds., Elsevier Science, pp. 543 - 563.

Melnick, J., Dul, J.L., and Argon, Y. (1994) Sequential interaction of the chaperone BiP and GRP94 with immunoglobulin chains in the endoplasmic reticulum. *Nature* **370,** 373 - 375.

Melnick, J., Aviel, S., and Argon, Y. (1992) The endoplasmic reticulum stress protein GRP94, in addition to BiP, associates with unassembled immunoglobulin chains. *J. Biol. Chem.* **267,** 21303 - 21306.

Morikawa, Y., Barsov, E., and Jones, I. (1993) Legitimate and illegitimate cleavage of human immunodeficiency virus glycoproteins by furin. *J. Virol.* **67,** 3601 - 3604.

O'Reilly, D.R., Miller, L.K., and Luckow, V. A. (1992) Baculovirus expression vectors. W. H. Freeman and Company, New York.

Oliver, J.D., van der Wal, F.J., Bullied, N.J., and High, S. (1997) Interaction of the thiol-dependent reductase Erp57 with nascent glycoproteins. *Science* **275,** 86-88.

Ou, W., Cameron, P.H., Thomas, D.Y., and Bergeron, J.J.M. (1993) Association of folding intermediates of glycoproteins with calnexin during protein maturation. *Nature* **364,** 771 - 776.

Pind, S., Riordan, J.R., and Williams, D.B. (1994) Participation of the endoplasmic reticulum chaperone calnexin (p88, IP90) in the biogenesis of the cystic fibrosis transmembrane conductance regulator. *J. Biol. Chem.* **269,** 12784 - 12788.

Possee, R.D. (1997) Baculoviruses as expression vectors. *Curr. Opin. Biotech.* **8,** 569 - 572

Posthaus, H., Dubois, C.M., Laprise, M.H., Grondin, F., Suter, M.M., and Muller, E. (1998) Proprotein cleavage of E-cadherin by furin in baculovirus over-expression system: potential role of other convertases in mammalian cells. *FEBS Lett.* **38,** 306 - 310.

Puig, A. and Gilbert, H.F. (1994) Protein disulfide isomerase exhibits chaperone and anti-chaperone activity in the oxidative refolding of lysozyme. *J. Biol. Chem.* **269,** 7764 - 7771

Puig, A., Lyles, M.M., Noiva, R., and Gilbert, H.F. (1994) The role of the thiol/disulfide centers and peptide binding site in the chaperone and anti-chaperone activities of protein disulfide isomerase. *J. Biol. Chem.* **269,** 19128 - 19135

Rapoport, T. A. (1996) Transport of proteins across the enoplasmic reticulum membrane. *Science* **258,** 931 - 936.

Ritty, T.M., Broekelmann, T., Tisdale, C., Milewicz, D.M., and Mecham, R.P. (1999) Processing of the fibrillin-1 carboxyl-terminal domain. *J. Biol. Chem.* **274,** 8933 - 8940.

Schmid, F. (1993) Prolyl isomerase: enzymatic catalysis of slow protein-folding reactions. *Annu. Rev. Biophys. Biomol. Struct.* **22,** 123 - 143.

Sissom, J.F. and Ellis, L. (1991) Biosynthesis of the precursor of a soluble human insulin ectodomain in insect Sf9 cells infected with a recombinant baculovirus. *Biochem. Biophys. Res. Comm.* **177,** 764 - 770.

Steinmann, B., Bruckner, P., and Superti-furga, A. (1991) Cyclosporin A slows collagen triple-helix formation *in vivo*: indirect evidence for a physiologic role of peptidyl-prolyl *cis-trans* isomerase. *J. Biol. Chem.* **266,** 1299 - 1303.

Sugyiama, K., Ahorn, H., Maurer-Fogy, I., and Voss, T. (1993) Expression of human interferon-alpha 2 in Sf9 cells. Characterization of O-linked glycosylation and protein heterogeneities. *Eur. J. Biochem.* **217,** 921-927.

Sugito K., Yamane, M., Hattori, H., Hayashi, Y., Tohnai, I., Ueda, M., Tsuchida, N., and Ohtsuka, K. (1995) Interaction between hsp70 and hsp40, eukaryotic homologues of dnaK and dnaJ, in human cells expressing mutant-type p53. *FEBS Lett.* **358,** 161 - 164.

Tate, C.G., Whiteley, E., and Betenbaugh, M.J. (1999) Molecular chaperones stimulate the functional expression of the cocaine-sensitive serotonin transporter. *J. Biol. Chem.* **274,** 17551 - 17558.

Tate, C.G. and Blakely, R.D. (1994) The effect of N-linked glycosylation on activity of the Na(+)- and Cl(-)-dependent serotonin transporter expressed using recombinant baculovirus in insect cells. *J. Biol. Chem.* **269,** 26303 - 26310.

Tatu, U. and Helenius, A. (1997) Interactions between newly synthesized glycoproteins, calnexin and a network of resident chaperones in the endoplasmic reticulum. *J. Cell. Biol.* **136,** 555 - 565.

Tessier, D.C., Thomas, D.Y., Khouri, H. E., Laliberte, F., and Vernet, T. (1991) Enhanced secretion from insect cells of a foreign protein fused to the honeybee melittin signal peptide. *Gene* **98,** 177 - 183.

van Die, I., van Tetering, A., Bakker, H., van den Eijnden, D.H., and Joziasse, D.H. (1996) Glycosylation in lepidopteran insect cells: identification of a β1-4-N-acetylgalactosaminyltransferase involved in the synthesis of complex-type oligosaccharide chains. *Glycobiology* **6,** 157 - 164.

van Dijl, J.M., de Jong, A., Smith, H., Bron, S., and Venema, G. (1991) Signal peptidase I overproduction results in increased efficiencies of export and maturation of hybrid secretory proteins in *Escherichia coli. Mol. Gen. Genet.* **227,** 40 - 48.

Vassilakos, A., Michalak, M., Lehrman, M.A., and Williams, D.B. (1998) Oligosaccharide binding characteristics of the molecular chaperones calnexin and calreticulin. *Biochemistry* **37,** 3480 - 3490.

Velardo, M.A., Bretthauer, R.K., Boutaud, A., Reinhold, B., Reinhold, V.N., and Castillino, F.J. (1993) The presence of UDP-N-acetylglucosamine: alpha-3-d-mannoside beta 1,2-N-acetylglucosaminyltransferase I activity in *Spodoptera frugiperda* cells (IPLB-Sf-21AE) and its enhancement as a result of baculovirus infection. *J. Biol. Chem.* **268,** 17902 - 17907.

Vernet, T., Tessier, D.C., Richardson, C., Laliber, F., Khouri, H.E., Bell, A.W., Storer, A.C., and Thomas, D.Y. (1990) Secretion of funtional papain precursor from insect cells. *J. Biol. Chem.* **265,** 16661 - 16666.

Vogel, J.P., Misra, L.M., and Rose, M.D. (1990) Loss of BiP/GRP78 functions blocks translocation of secretory proteins in yeast. *J. Cell. Biol.* **110,** 1885 - 1895.

von Heijne, G. (1994) Signal peptidases and protein trafficking, in: *Signal peptidases,* von Heijne, G., ed., R.G. Landes Company, Austin, pp. 1 - 3.

Wada, I., Ou, W., Liu, M., and Scheele, G. (1994) Chaperone function of calnexin for the folding intermediate of gp80, the major secretory protein in MDCK cells. *J. Biol. Chem.* **269,** 7464 - 7472.

Wagner, R., Liedke, S., Kretzschmar, E., Geyer, H., Geyer, R., and Klenk, H-D. (1996) Elongation of the *N*-glycans of fowl plague hemagglutinin expressed in *Spodoptera frugiperda* (Sf9) cells by coexpression of human β1,2-N-acetylglucosaminyltransferase I. *Glycobiology* **6,** 165 - 175.

Whiteley, E.M. and Betenbaugh, M.J. (2000) Protein processing in the endoplasmic reticulum and golgi network, in: *Encyclopedia of Cell Technology*, Spier, R.E., ed., John Wiley & Sons, New York, in press.

Whiteley, E.M., Hsu, T-A., and Betenbaugh, M.J. (1997a) Modeling assembly, aggregation, and chaperoning of immunoglobulin G production in insect cells. *Biotech. Bioeng.* **56,** 106 - 116.

Whiteley, E.M., Hsu, T-A., and Betenbaugh, M.J. (1997b) Thioredoxin domain non-equivalence and anti-chaperone activity of protein disulfide isomerase *in vivo. J. Biol. Chem.* **272,** 22556 - 22563.

Wilkinson, B.M., Regnacq, M., and Stirling, C.J. (1997) Protein translocation across the membrane of the endoplasmic reticulum. *J. Membr. Biol.* **155,** 189 - 197.

Zhao, Y. and Sane, D.C. (1993) Expression of a recombinant baculovirus for vitronectin in insect cells: purification, characterization ofpost-translational modifications and functional studies of the recombinant protein. *Arch. Biochem. Biophys.* **304,** 434 - 442.

A GUIDE TO SUCCESSFUL SCALE-UP OF THE BACULOVIRUS EXPRESSION SYSTEM

C. MANNIX [a]* AND R. F. JARMAN [b]
[a] *Fermentation and Cell Sciences, New Frontiers Science Park (North), Third avenue, Harlow, Essex, CM19 5AW, UK.*
[b] *School of Chemical Engineering, University, Birmingham, Edgbaston, Birmingham, B15 2TT, UK.*
* *Corresponding author, fax +1279-622100,*
e-mail Christopher_J_Mannix@sbphrd.com

1. Introduction

The baculovirus expression system has come to be accepted as one of the most efficient methods for production of many recombinant proteins. The system is based on the insect nuclear polyhedrosis viruses (NPV), a group of large rod-shaped DNA viruses, into which the gene or genes of interest are cloned. The recombinant virus can then be used to infect insect cells in culture, with subsequent expression of recombinant proteins (O'Reilley *et al.*, 1994). The virus *Autographa californica* (AcNPV), which infects a limited number of insect species including lepidoptera, is the most widely used baculovirus. The baculovirus system has many advantages, including short time-scales, the safety and relative simplicity of the insect cell culture systems, and the high levels of production that are frequently achieved. In addition, many recombinant proteins are produced in soluble form and with authentic post-translational modifications. Other expression systems (bacterial, yeast, mammalian) each offer some of these benefits, and there are many situations in which such systems may be preferred. However, the advantages of baculovirus make it the system of first choice for many products. This chapter is intended primarily as a guide to those researchers who have little previous experience of the baculovirus expression system, but it is hoped that it will also prove of interest to those familiar with baculoviruses. An overview of the baculovirus system is followed by a more detailed discussion of the factors important to the successful scale-up of recombinant protein production, and the possibilities opened up by recent developments.

M. Al-Rubeai (ed.), Cell Engineering, 43–55.

2. Background

The advantages of the baculovirus system are rooted in the properties of the virus and the host (insect, or cell lines derived from it). During the normal infection cycle, two forms of the virus are produced: an early budded virus (BV) form (Kost *et al.*, 2000), in which the viral DNA and structural proteins are surrounded by membrane derived from the infected cell; and a late occluded form (occlusion-derived virus, ODV), consisting of enveloped viral cores which are embedded in a crystal matrix of viral proteins. The principal component of the matrix is the abundantly expressed protein polyhedrin. The budded virus rapidly spreads the infection from cell to cell within the insect host, resulting ultimately in the complete liquefaction of the host, and release of occluded virus into the environment. The occluded form protects the released virus, allowing it to survive for long periods in the environment until ingested by another host. In the alkaline environment of the insect gut, the protective protein matrix is removed, and the life cycle is repeated. In insect cell cultures, only the BV form of baculovirus is required, and the polyhedrin gene may be replaced with the gene for the recombinant protein. An additional benefit of replacing or deleting polyhedrin is that it effectively makes the virus unable to survive outside the laboratory, an advantage in terms of environmental safety. The system is intrinsically safe to animals, being unable to replicate in species other than a limited range of insects.

Early baculovirus expression systems were constructed by using recombination between the AcNPV circular double stranded DNA (dsDNA) genome and a vector carrying the gene of interest. The recombinant viruses (which lack polyhedrin) were identified by the absence of occluded virus particles (polyhedra). However, the recombination frequency of this system was very low (0.1-0.2%), and the isolation of recombinant baculoviruses was extremely tedious. A major advance was the introduction of specific restriction endonuclease sites in the baculovirus genome, allowing the linearisation of the DNA and the removal of viral sequences essential to replication (Kitts and Possee, 1993). The linearised DNA was then co-transfected into insect cells with a plasmid containing both the essential viral sequences and the cloned gene under control of the polyhedrin promoter. Recombination gave progeny virus expressing the desired product, and allowed high levels of recombination (approaching 100%) with low background. The method is simple and rapid, and is widely available in kit form for generation of recombinant baculoviruses (e.g. BaculoGold from Pharmingen; BacPAK6 from Clontech). For a short practical introduction to the preparation of recombinant baculoviruses see Griffiths and Page, 1997. For additional detail, refer to O'Reilley *et al.*, 1994; Kost *et al.*, 2000 and King and Possee, 1992.

In addition to polyhedrin, many other gene products (p10, p35, p39) are expressed at high levels, but are dispensable for virus replication in cell culture, and may be deleted or replaced. The promoters of such genes may be used to drive expression of heterologous proteins. Other non-essential products, such as proteases and chitinase, may be detrimental to the production of recombinant proteins, and their deletion may improve product yield or quality.

3. Advantages and Limitations of the Baculovirus Expression System

Until the early 1990s, the majority of recombinant proteins were expressed in either microbial (predominantly *E. coli*) or mammalian cell culture systems. While the microbial systems offer a relatively fast route to expression, and can be readily scaled-up, many eukaryotic proteins produced in such systems are incorrectly folded, forming insoluble products, or are not appropriately processed. Post-translational processing of eukaryotic proteins to give authentic product may involve such processes as phosphorylation, acylation, glycosylation, secretion, and proteolytic cleavage. Recombinant mammalian cells have been utilised to give appropriately processed product, with considerable success. However, deriving productive cell lines is a lengthy process (3-6 months), and the levels of expression are frequently low (1-10 mg/L). In addition, the culture systems are relatively complex and expensive to scale-up.

Over the last decade, and particularly since the introduction of the rapid methods using restricted and linearised DNA, the baculoviruses have offered significant advantages over microbial and mammalian systems for expression of many proteins. It is now routinely possible to go from gene to expression of recombinant protein in less than a month, including cloning (plaque purification) of the virus. Amplification of virus stocks, and expression at scales of up to 100 L, can readily be achieved in an additional month. Expression levels (using the polyhedrin promoter) are usually high, in the range 10-100 mg/L. The most widely used insect cell lines, Sf-21, Sf-9 (a clonal line from Sf-21), and TN-5B1-4 ("High-5^{TM}") (Davis *et al.*, 1992; Davis *et al.*, 1993) grow well in suspension in a range of culture media, including serum-free media, and scale-up is relatively straightforward. A powerful advantage of the baculovirus expression system is that the two components, virus and cells for production, can be progressed independently prior to the required expression, thus adding to the speed and flexibility of the system. Stocks of cells can be maintained continuously at an appropriate scale, and used as needed for evaluation or production; and baculovirus stocks can be prepared, the titre of the suspension determined and stored until required. The product can be changed at a very late stage by the choice of a different virus, without the need to grow and scale-up different cell lines. Two or more proteins can be expressed simultaneously by co-infection with different viruses (Kost *et al.*, 2000), facilitating the production of multi-subunit products, and enabling the study of different protein-protein interactions. Multiple subunits can also be encoded into a single baculovirus, under the control of the same, or of different, promoters.

Insect cells can carry out many of the post-translational modifications of higher eukaryotes, including acylation, acetylation, α-amidation, and phosphorylation. In addition, proteins with appropriate signal sequences may be secreted, and glycosylated at the correct sites. However, insect cells appear to lack the enzymes required for trimming and extension of the basic core polysaccharide, and N-glycosylation in insect cell systems results primarily in high-mannose structures, rather than the complex glycosylation patterns normally observed in mammalian systems (Hooker *et al.*, 1999; Altmann *et al.*, 1999). Therefore, proteins that require glycosylation for biological function (or stability) may not be suited to production using the baculovirus system.

Other components of the secretion apparatus in insect cells, such as signal peptidases (Ailor *et al.,* 1999a; Ailor and Betenbaugh 2000) and proprotein cleavage enzymes (Laprise *et al.,* 1998; Cieplik *et al.,* 1998; Ailor and Betenbaugh 2000), may be rate-limiting for the processing of heterologous proteins. As a result secreted proteins are generally expressed at much lower levels than intracellular products in the baculovirus system. Secretion may be improved by use of particular cell lines such as the *Trichoplusia ni* cell line TN-5B1-4 ("High-5™"), by utilisation of insect signal sequences (Tessier *et al.,* 1991), or by genetic modification of cell lines to overcome rate-limiting steps (Ailor *et al.,* 1999a; Cieplik *et al.,* 1998). Although work is in progress to engineer insect cell lines for the expression of secreted and glycosylated products (Ailor and Betenbaugh 1999; Ailor and Betenbaugh 2000; Altmann *et al.,* 1999), this poses an enormous challenge, requiring the manipulation of multiple biosynthetic pathways. For the present, baculovirus is the first-choice system for rapid production of intracellular (non-secreted) eukaryotic proteins, and for many prokaryotic proteins. However, it may not be appropriate for production of secreted proteins, and is unlikely to be the most suitable system for proteins that require glycosylation to function appropriately.

4. Overview of the Production System

Autographa californica and other baculoviruses adsorb to insect cells by interaction of the major viral envelope protein gp64 with an unknown receptor protein on the cell surface, and are rapidly (within 30 min) taken up into the cell (Hefferon *et al.,*1999). The viral dsDNA is transported to the nucleus, where replication takes place. Cell division ceases within a few hours of infection, infected cells begin to increase in volume, and host DNA synthesis is halted (Ikeda and Kobayashi 1999). Viral DNA synthesis is detectable from about 8h post-infection (pi), and reaches a maximal rate at 10-12h pi (Ikeda and Kobayashi 1999): by 16h pi, the amount of viral DNA may exceed the amount of cellular DNA by 2-fold or more, representing up to approximately 10^5 copies of the viral genome per cell. The viral coat glycoprotein gp64 is detectable on the surface of infected cells from about 12h pi, and shortly after this time viral nucleocapsids begin to bud through the cell membrane to form new virus particles. Gp64 is required to drive the budding process, and without it very few (and non-infectious) virus particles are formed (Oomens and Blissard, 1999).

For recombinant proteins under control of the common late promoters such as polyhedrin and p10, product synthesis begins about one day post-infection, and continues to 3-4 days pi: generally, there is relatively little additional product produced after 3 days pi. Over the same period (1-4 days pi), viral proteases and other degradative enzymes are produced, and may be detrimental to some products. It is likely that many of these enzymes are irrelevant to recombinant baculovirus replication and product synthesis, and could be deleted from the virus genome to give improved expression systems. Cell viability (measured by dye exclusion) is likely to start to decrease from 3 days post-infection, and by day 6 the majority of the cells in the culture will have lysed.

The timescales outlined above are those observed for populations of cells in culture, and it is important to note that within these populations there is a significant degree of heterogeneity, with a non-uniform progression of the infection. This is the case even for populations infected at high multiplicities of infection (MOI >10, more than ten infectious particles per cell), in which essentially every cell is infected simultaneously. For example, some cells show high level expression of gp64 at 14h pi, while others may not start to express gp64 until 18h or more post-infection (C.M. results not shown). Similar results have been obtained for virus structural protein markers (Braunagel *et al.*, 1998). In the same way, synthesis of product in some cells will be advanced, and in others retarded, compared to the average of the population. The phase of the cell cycle may affect the progression of infection within individual cells, though there is as yet no definitive evidence for this (Brown and Faulkner 1975; Kioukia *et al.*, 1995). Some experiments have attempted to determine whether infected cells are arrested in a particular phase by quantifying DNA levels using flow cytometry (Braunagel *et al* 1998; Ikeda and Kobayashi 1999). However, because these measurements cannot distinguish between cellular and viral DNA, this approach is limited to very early time points (before viral DNA synthesis).

The most widely used cell lines for baculovirus expression are Sf-21 and the preferred clonal cell line derived from it, Sf-9. These cell lines have excellent characteristics in terms of growth and virus production, and generally give good product levels, though not necessarily the highest for all products (for example, see Davies *et al.,* 1993). Sf-9 and Sf-21 grow well in suspension culture in a range of readily available media, including serum-free media. Growth in suspension is an important factor in choice of cell line, desirable for monitoring growth and viral infection, and essential for ease of scale-up. For specific applications other cell lines may be useful. For example, the High-five cell line (Invitrogen) may give higher titres of secreted products (Davies *et al.,* 1993); however, this cell line is more difficult to grow reliably in suspension, with a tendency to clump. In addition, virus titres are often lower from High-five than from Sf-9 (Sf-21), requiring scale-up of virus stocks in a cell line different to the system for protein production. The majority of work on expression of recombinant proteins using the baculovirus system has been carried out in Sf-9 or Sf-21 cells, and there is an extensive body of literature describing the use of the baculovirus system in these cell lines. The Sf-9/Sf-21 cell lines are recommended for all initial work with the baculovirus system, and it is on these cell lines that the remainder of the discussion focuses.

5. Culture of Cells

Cells sourced from a reliable Culture Collection such as American Type Cell Culture (ATCC) or European Collection Animal Cell Cultures (ECACC) will ensure that the authentic cell line is obtained, and will avoid problems with contaminants including mycoplasma. Cell banks should be prepared at the earliest opportunity, and it is well worth the additional effort of preparing a Master Cell Bank (MCB), from which all working cell banks will be made. Using this method will minimise variations in the properties of the culture stocks. For many years, insect cell lines were grown routinely in media containing serum (e.g. TNM-FH or IPL-41 supplemented with 10% foetal bovine serum). However, serum has significant disadvantages as a medium supplement, including batch-to-batch variation, potential viral and mycoplasmal contaminants, undefined enzyme activities (e.g. proteases), and a high protein content. Fortunately, a range of serum-free media is now readily available for the culture of insect cells. While there are many commercial serum-free media, for most purposes the supplemented IPL-41 formulation of Maiorella *et al.,* (1988) is satisfactory. The basal IPL-41 is supplemented with Yeastolate, Pluronic F-68, and lipids. The Pluronic supplement is of particular importance, because it gives a significant measure of protection against shear forces: Sf-9 cells are very susceptible to shear damage, especially when cultured in serum-free medium (Murhammer and Goochee 1988; Kioukia *et al.*, 1996,). Pluronic also helps to maintains cellular integrity in insect cell cultures that are exposed to bubble disengagement (before and during viral infection) that occurs as a consequence of sparging or surface entrainment (Kioukia *et al.*, 1992; Kioukia *et al.*, 1996). It appears that Pluronic alters the properties of the cell membrane, though the precise mechanism of action is unknown.

Sf-9 and Sf-21 cells are tolerant to a broad range of temperature, though they grow most rapidly in the range 26-28°C and the most commonly used temperature is 27°C. In standard media at this temperature, the culture doubling time (T_d) is approximately 18-22 hours. The cell lines can be maintained at low temperature (18°C) in order to reduce culture maintenance: at 18°C the Td is increased to approximately 70-80h. Media for insect cell culture is normally in the pH range 6.0-6.4, and utilises a phosphate buffer system. There is therefore no requirement to maintain cultures in an atmosphere containing carbon dioxide, as is the case for most mammalian cell media. Under standard conditions, the pH of the medium does not change by more than about 0.2 units during growth or infection, due mainly to the efficient metabolism of the insect cells, and the low level of lactate accumulation (Bedard *et al.,* 1993, Neerman and Wagner, 1996). Standard media for insect cell culture contain amino acids, carbohydrates, lipids, vitamins, and minerals: all are provided in excess to prevent nutrient limitation of laboratory scale cultures. If growing cultures to higher cell concentrations (above 3 x 10^6 cells/mL), glucose and glutamine may become depleted, and additional amounts of these substrates may be required (Bedard *et al.,* 1994). Insect cell metabolism appears to be very efficient compared to that of mammalian cells, and under most conditions lactate does not accumulate to high levels. In addition, glutamine is not essential for growth of Sf-9 cells provided ammonia (as NH_4Cl) is provided (Ohman *et al*, 1996). Oxygen is the nutrient most likely to become limiting during routine culture because of its limited solubility, and care must be taken to ensure an adequate supply (Wang *et al.,* 1993). At laboratory scale, oxygen supply can be improved by increased agitation (high stirrer rate, or use of shake flasks), loose caps on culture vessels, and an increased surface area:volume ratio. At large scale, increased oxygen concentration in the gas supply, agitation and sparging are the main ways to increase oxygen supply.

Sf-9 (Sf-21) cells are not anchorage dependent and may be transferred between monolayer and suspension cultures repeatedly without noticeable losses in viability or growth rate (Kioukia *et al.*, 1995). Sf-9 cells attach very strongly to standard tissue culture plastics, particularly in serum-free media: once a monolayer of cells has formed, daughter cells are budded off into the supernatant and can be collected for subculture. However, it may be impossible to detach more than a fraction of the adherent population without scraping. Suspension culture has significant advantages over adherent culture, especially the ease of estimating cell concentrations and viabilities, and is the method of choice wherever possible. In addition, suspension culture is the simplest and most practical way to scale up the baculovirus expression system, and the following discussion is limited to this culture method.

Cells may be cultured in stirred vessels (e.g. Techne), though stirrers which generate significant levels of shear (such as those having bearings within the culture fluid) should be avoided because of the sensitivity of Sf-9 cells to shear damage. The preferred alternative to stirred vessels is the shake flask, which are easier to use and give improved oxygen transfer rates. In order to produce both high titre virus stock and reproducibly high levels of recombinant product, it is essential to use cells which have been maintained in logarithmic growth. It is therefore important to estimate cell concentrations regularly (at least every two days), and to ensure that stock cultures are maintained between 2×10^5 and 3×10^6 cells/mL at all times. Growth of cultures in standard media above 3×10^6 cells/mL, or dilution to below 2×10^5 cells/mL, is likely to induce a lag phase before subsequent regrowth.

6. Production and Assay of Baculovirus Stocks

The production of high quality, accurately quantified, virus stocks is critical to the successful use of the baculovirus system. Expansion of baculovirus stocks is readily achieved by infection of Sf-9 cells with the virus at a low MOI (0.01-0.05 infectious particles per cell). It is important to infect at low multiplicity of infection to minimise the reproduction of defective interfering particles (DIPs), viruses that have an incomplete genome and require assistance from a normal baculovirus in order to replicate. Co-infection of a single cell with a DIP and a standard baculovirus will allow over-expression of the defective virus. Initial infection of 1-5% of the cell population will minimise the probability of co-infection, and the propagation of DIPs. Conversely, infection at high MOI will lead to a rapid increase in the proportion of defective particles, and a concomitant deterioration in per1formance (Kool *et al.,* 1991; Wickham *et al.,* 1991).

Baculovirus stocks are routinely produced in medium containing serum, which improves their stability and allows them to be kept for several years without significant deterioration. In serum-free medium, the virus titre is likely to decrease significantly (approx 1 log over a 1-2 month period). Where necessary, virus stocks can be prepared in serum-free medium for immediate use (within 2-3 weeks). Preparation of baculovirus stocks for large-scale expression may be carried out at laboratory scale, or if larger amounts of virus are required can be done in bioreactors.

The standard method for estimation of baculovirus titre is the plaque assay (Kost *et al.*, 2000), which determines the number of foci of infection on a plate of cells infected with a known dilution of the virus. This assay can give good and reproducible estimates of virus titre, but requires considerable practice and attention to detail in order to do so. In addition, the plaque assay adds a significant delay to the whole process, requiring approximately six days to obtain an estimate. A modified plaque assay, which detects the virus coat protein gp64 on the surface of infected cells by use of an antibody, gives some improvement by reducing the timescale to two days (Kitts and Green 1999). Because of the variation inherent in such assays, not least the serial dilution to 10^{-7} or 10^{-8}, the virus titre should be regarded as being within two-fold of the result of the assay, which is sufficiently accurate for all routine purposes.

Within SmithKline Beecham, we have developed a simple, rapid (18 hour), and reproducible assay, based on the inhibition of cell growth by baculovirus infection (Venkat *et al.*, 2000). Because infection prevents subsequent cell division, the inhibition of growth in test cultures infected at a range of virus dilutions can be used to quantify the number of infectious particles. Simplistically, infection of a culture containing 2 x 10^8 cells with 1 x 10^8 infectious particles would be expected to give 50% growth inhibition compared to a non-infected control. In practice, the assay requires a slight modification to account for the increasing probability of co-infection as the ratio of virus:cells increases. The growth inhibition assay gives results that are within two-fold of the plaque assay (table 1).

Table 1: Comparison of Growth Inhibition Assay with Plaque Assay

Baculovirus	Titre by Growth Inhibition	Titre by Plaque Assay
Virus-A	17 x 10^8	10 x 10^8
Virus-B	11 x 10^8	8 x 10^8
Virus-C	19 x 10^8	11 x 10^8
Virus-D	3.7 x 10^8	1.8 x 10^8
Virus-E	5 x 10^8	6 x 10^8

7. Production of Recombinant Product

For production of recombinant proteins, exponentially growing cultures should be infected at a high MOI, to ensure synchronous infection of the majority of cells. At any combination of cells and virus, there is a probability that a fraction of the population will not be infected. The proportion likely to remain non-infected may be determined by the probability function known as the Poisson distribution. Thus at an average of one infectious particle per cell (MOI of 1) only 63% of cells are likely to be infected, whereas 95% of cells will be infected at an MOI of 3. In practice, it is preferable to infect at a nominal MOI of 5, to take account of any variation in the assay of virus. Increasing the MOI above that required to infect essentially all the population has no significant effect on the progress of infection or the level of recombinant protein production (Bedard *et al.,* 1994). During the first day post-infection, there is an increased demand for nutrients, and in particular for oxygen. It is important to anticipate this demand and to be able to meet it, if necessary by infecting at lower cell concentration, because nutrient limitation may cause a drastic reduction in product yield (Tom *et al.,* 1995). To optimise the production of the required product, it is necessary to carry out a time course evaluation. For the majority of proteins produced using the baculovirus system, the optimum time for harvest is between two and three days pi. However, some products have an optimum harvest time as early as 40h pi, usually because of degradation of the specific product observed at later times. Other products may continue to accumulate after day three.

For downstream processing, cells and supernatant are separated by centrifugation (or filtration). To recover cell-associated products, the cells are disrupted by homogenisation, detergent, or other methods, and the relevant cell fraction progressed for purification or assay. The product is immediately available in concentrated form in a relatively small volume of buffer. In contrast, secreted products are initially available only in dilute form in a large volume of medium, which may be a significant disadvantage in terms of analysis or purification.

7. 1 SCALE-UP OF PRODUCTION SYSTEMS

The baculovirus expression system may be scaled up readily to 100L or more provided appropriate equipment is available (Agathos 1991). The parameters that are important for successful scale-up of the baculovirus system are the same as those required at small scale: sterility, supply of adequate nutrients, maintenance of healthy log-phase cultures, minimisation of shear forces, and use of high quality baculovirus stocks at suitable MOI. The culture of insect cells and expression of recombinant protein are carried out in essentially the same way in reactors as at laboratory scale, though there are opportunities to increase the cell concentration, and therefore the productivity, of the bioreactor systems. The term "large-scale" refers primarily to the methodology used rather than to an actual volume, though in practice large-scale operations are likely to be above 5 L. Thus one key element of large-scale is the requirement for aseptic inoculation and of vessels that could not practicably be transferred to laminar flow cabinets for these operations. In addition, standard large-scale operation would include *in-situ* sterilisation; a mechanism such as sparging to increase the supply of oxygen; and a system for control of dissolved oxygen concentration. The simplest and most widely used bioreactors for baculovirus expression are animal cell stirred tank reactors fabricated from stainless steel, and it is this system which is described below. Complete systems for culture of animal cells are available from a large number of manufacturers.

Bioreactors must have a suitable system for sterilisation, usually from an external clean steam supply (sometimes from in situ generation of steam) capable of providing steam at 1.5-2 bar (up to 135°C), and therefore should be rated at 2 bar working pressure minimum. Preferably, gas inlet and outlet filters should be capable of steam sterilisation both at the same time as the complete sterilisation and independently of the main body of the reactor, to allow for replacement of filters during a culture run. The bioreactor should also include systems for aseptically taking samples from the reactor and for making sterile additions to it (for example, by steamable connections). In addition to the bioreactor, other sterilisable vessels will be required for preparation and storage of medium. Generally, medium for large-scale use will be prepared from powder basal mixture and supplements, and sterilised immediately by filtration through a 0.1 (0.2) μm filter into a sterile vessel, from where it can be transferred as required to the bioreactor. Liquid transfers are frequently achieved by pressurising with gas, though pumping is also an option. The bioreactor and other vessels will require temperature control, including the ability to cool vessels for media storage (may also be required to maintain cultures at 27°C, depending on the environment).

Mixing in the bioreactor is required to maintain cells in suspension and to give an even distribution of nutrients, particularly oxygen. In animal cell bioreactors, mixing is usually achieved by using a central impeller, located near the bottom of the vessel. Impellers, such as the four-bladed 45° angled impeller or marine impeller, are designed to give lift in the middle of the reactor (vertical mixing) as well as rotational mixing. Such systems have low shear over the usual range of agitator speeds (50-250 rpm), and are very satisfactory for the serum-free culture of insect cells provided the medium contains pluronic F-68 (0.1-0.2%). Supply of adequate amounts of oxygen, which may be a limiting factor even at laboratory scale, becomes more difficult with increasing volumes of culture, because of the decreasing surface area per unit volume. The most efficient method for supplying oxygen to stirred tank bioreactors is to sparge air, or air enriched with oxygen, into the culture on demand. Typically, the sterile gas mixture is supplied at a slow rate (0.001-0.005 volume/reactor volume/min) via a microporous material to the bottom of the reactor, below the impeller. A stream of very small gas bubbles is produced, giving efficient exchange of oxygen with the liquid phase. This process requires a good system for control of dissolved oxygen (dO_2), to prevent oscillation between high and low oxygen concentrations. It is also important to provide a flow of air at higher flow rate over the surface of the culture, to assist removal of carbon dioxide and excess oxygen (thereby improving dO_2control). Oxygenation can also be improved by increased agitation. Sparging generates additional shear forces, but as previously noted, inclusion of pluronic F-68 gives very good protection against the level of shear forces encountered in a normal sparged reactor (Kioukia *et al.*, 1996, Murhammer and Goochee, 1988). Antifoam is routinely added to the medium of sparged bioreactors, to prevent the build up of foam, which is detrimental to cells.

Dissolved O_2 levels between 30% and 60% of saturation are satisfactory for cell growth and protein production (Schmid, 1996). The rate of oxygen utilisation of Sf-9 cells is generally accepted to be in the range 4-8 mmol per 10^9 cells per day for non-infected cells: most authors report an increase of between 30% and 100% immediately after infection, decreasing at later stages of the process (Schmid, 1996). The sparged bioreactor readily provides the amounts of oxygen needed at large scale. Direct scale-up of the baculovirus system from laboratory scale, infecting at concentrations of up to 3 x 10^6 cells/mL, gives results comparable to laboratory scale on a volumetric basis. In addition, because of the improved transfer of oxygen that can be achieved in sparged culture, it is possible to grow insect cells to much higher concentrations in bioreactors. Provided nutrients such as glucose or glutamine are added to the culture (fed batch process), it is possible to increase the cell concentration at infection to 10^7 cells/mL, with a substantial increase in volumetric productivity (Bedard *et al.*, 1994). Although production cultures may be grown to higher cell concentrations, the expression protocol remains the same as the laboratory scale process, based on infection at high MOI with the time of harvest determined from small-scale optimisation studies.

The baculovirus expression system is therefore a rapid and flexible system for production of recombinant proteins. The process is effective at laboratory scale, and can readily be scaled to 100L, or greater, with substantial increases in productivity.

8. References

Agathos, S. N. (1996) Insect cell bioreactors. *Cytotechnology* **20,** 173-189.

Ailor, E. and Betenbaugh, M. J. (1999) Modifying Secretion and Post-translational Processing in Insect Cells. *Current Opinion in Biotechnology* **10,** 142-145.

Ailor, E. and Betenbaugh, M. J. (2000) Engineering Post-translational Processing of Recombinant Proteins Produced in Insect Cells. *This book.*

Ailor, E., Pathmanathan, J., Jongbloed, J.D.H. and Betenbaugh M.J. (1999a) A Bacterial Signal Peptidase Enhances Processing of Single Chain Antibody Fragment in Insect Cells. *Biochem. Biophys. Res. Comm.* **255,** 444-450.

Altmann, F., Staudacher, E., Wilson, I.B.H., Marz, L. (1999) Insect Cells as Hosts for the Expression of Recombinant Glycoproteins. *Glycoconjugate Journal* **16,** 109-122.

Bedard, C., Tom, R., and Kamen, A. (1993) Growth, nutrient consumption, and end-product accumulation in Sf-9 and BTI-EAA insect cell cultures: insights into growth limitation and metabolism. *Biotechnology Progress* **9,** 615-624.

Bedard C, Kamen A, Tom R, Massie, B. (1994) Maximisation of recombinant protein yield in the insect cell/baculovirus system by one time addition of nutrients to high density batch cultures. *Cytotechnology* **15,** 129-138.

Braunagel, S.C., Parr, R., Belyavskyi, M., Summers, M. D. (1998) Autographa californica Nucelopolyhedrovirus Infection Results in Sf9 Cell Cycle Arrest at G_2/M phase. *Virology* **244,** 195-211.

Brown, M. and Faulkner, P. (1975) Factors affecting the yield of virus in a cloned cell line of *T. ni* infected with NPV. *J. Invertebr. Pathol.* **26,** 251-257.

Cieplik, M., Klenk, H.D., Garten, W. (1998) Identification and characterization of *Spodoptera frugiperda* furin: a thermostable subtilisin-like endoprotease. *Biol. Chem.* **379,** 1433-1440.

Davis, T.R., Trotter, K.M., Granados, R.R., and Wood, H.A. (1992) Baculovirus expression of alkaline phosphatase as a reporter gene for evaluation of production, glycosylation, and secretion. *Bio/Technology* **10,** 1148-1150.

Davis, T.R., Wickham, T.J., McKenna, K.A., Granados, R.R., Shuler, M.L., and Wood, H.A. (1993). Comparative recombinant protein production of eight insect cell lines. *In Vitro Cell Dev. Biol.* **29A,** 388-390.

Griffiths, C.M. and Page, M.J. (1997) Production of Heterologous Proteins Using the Baculovirus/Insect Cell System Methods in Molecular Biology (eds J.W. Pollard and J.M. Walker). *Basic Cell Culture Protocals* **75**, 427-440.

Hefferon, K.L., Oomens, A.G., Monsma, S.A., Finnerty, C.M., Blissard, G.W. (1999) Host cell receptor binding by GP64 and kinetics of virion entry. *Virology* **258,** 455-468.

Hink, W.F. (1982) Production of AcNPV in cells from large-scale suspension cultures. *Microb. and Viral Pesticides* (ed. Kurstak), pp.492-506.

Hooker, A.D., Green, N.H., Baines, A.J., Bull, A.T., Jenkins, N., Strange, P.G., James, D.C. (1999) Constraints on the Transport and Glycosylation of Recombinant IFN-γ in CHO and Insect Cells. *Biotechnol. Bioeng.* **63** (5), 559-572.

Ikeda, M. and Kobayashi, M. (1999) Cell-Cycle Peturbation in Sf9 cells Infected with Autographa californica Nucleopolyhedrosis virus. *Virology* **258** (1), 176-188.

King, L.A. and Possee, R.D. (1992) "*The Baculovirus Expression System: A Laboratory Guide*" Oxford University Press, Oxford, UK

Kitts, P.A. and Green, G. (1999) An Immunological Assay for determination of Baculovirus Titres in 48 hours. *Analytical Biochemistry* **268,** 173-178.

Kitts, P.A. and Possee, R.D. (1993) A Method for Producing recombinant Baculovirus Vectors at High Frequency. *Biotechniques* **14,** 810-817.

Kioukia, N., Nienow, A. W., Emery, A. N. and Al-Rubeai, M. (1992). The impact of fluid dynamics on the biological performance of free suspension animal cell culture: Further Studies. *Trans IchemE*, Vol **70,** Part C, 143-148.

Kioukia, N., Al-Rubeai, M. Zhang, Z., Emery, A. N, Nienow, A. W., and Thomas, C. R. (1995). A study of uninfected and baculovirus-infected *Spodoptera frugiperda* cells in T-spinner flasks. *Biotechnology Letters*, Vol **17,** No. 1, 7-12.

Kioukia, N., Nienow, A. W., Emery, A. N. and Al-Rubeai, M. (1995). Physiological and environmental factors affecting the growth of insect cells and infection with baculovirus. *Journal of Biotechnology* **38,** 243-251.

Kioukia, N., Nienow, A. W., Al-Rubeai, M. and Emery, A. N. (1996). Influence of agitation and sparging on the growth rate and infection of insect cell. *Biotechnology Progress* **12**(6), 779-785.

Kool. M., Voncken, J.W., Van Lier , F.L.J., Tramper, J., and Vlak, J.M. (1991) Detection and analysis of Autographa californica NPV mutants with defective interfering properties. *Virology* **183,** 739-746.

Kost, T. (1997) Gene expression systems in the genomics era. Current Opinion in Biotechnology **8** 539-541.

Kost, T., Klein, J. L. and Condreay, J. P. (2000) Application of Recombinant Baculoviruses in Biopharmaceutical Research. *This book.*

Maiorella, B., Inlow, D., Shauger, A., and Harano, D. (1988) Large scale insect culture for recombinant protein production. *Biotechnology* **6,** 1406-1410.

Murhammer, D.W., and Goochee, C.F. (1988) Scale-up of insect cell cultures protective effects of pluronic F-68. *Biotechnology* **16,** 1411-1418.

Neerman, J. and Wagner, R. (1996) Comparative analysis of glucose and glutamine metabolism in transformed mammalian cell lines, insect and primary liver cells. *Journal of cellular physiology* **166**, 152-169.

Oomens, A.G. and Blissard, G.W. (1999) Requirement for GP64 to drive efficient budding of *Autographa californica* multicapsid polyhedrosis virus. *Virology* **254,** 297-314.

O'Reilly, D.R., Miller, L.K., and Luckow, V.A. (1994) "*Baculovirus Expression Vectors: A Laboratory Manual*" (Freeman, New York).

Schmid, G. (1996) Insect cell cultivation: growth and kinetics. *Cytotechnology* **20,** 43-56.

Tessier, D.C., Thomas, D.Y., Khouri, H.E., Laliberte, F., and Vernet, T. (1991) Enhanced secretion from insect cells of a foreign protein fused to the honeybee melittin signal peptide. *Gene* **98,** 177-183.

Tom, R.L., Debanne, M.T., Bedard, C., Caron, A.W., Massie, B, and Kamen, A.A. (1995). Improved yields of the extracellular domain of the epidermal growth factor receptor produced using the baculovirus system by medium replacement following infection.. *Applied Microbiology and Biotechnology* **44,** 53-58.

Tramper, J., Joustra., D and Vlak., J. M. (1987) Bioreactor design for growth of shear sensitive insect cells in "*Plant and animal cells process possibilities*". (Editors C Webb and F Mavituna) Chapter **8**. 125-136.

Venkat, R.V., Mannix, C.J., and Kane, J.F. (2000). Alternative Methods for Baculovirus Titration. *Cell Culture Engineering meeting* VIII.

Wang, M.Y., Kwong, S., and Bentley, W.E. (1993). Effects of oxygen/glucose/glutamine feeding on insect cell baculovirus expression: a study on epoxide hydrolase production. *Biotechnology Progress* **7,** 355-361.

Wickham, T. J., Davis, T., Granados, R. R., Hammer, D. A., Schuler, M. L., and Wood, H. A. (1991). Baculovirus defective interfering particles are responsible for variations in recombinant protein production as a function of multiplicity of infection. *Biotechnology Letters* **13,** 483-488.

TRANSIENT GENE EXPRESSION IN MAMMALIAN CELLS BASED ON THE CALCIUM PHOSPHATE TRANSFECTION METHOD

M. JORDAN
Laboratory of Cellular Biotechnology
Swiss Federal Institute of Technology
CH – 1015 Lausanne
Fax +41-21-693-6142
e-mail Martin.Jordan@epfl.ch

1. Introduction

Supercoiled plasmid DNA purified from *E. coli* can be transferred into cultured cells with a variety of transfection methods. Whenever the DNA makes its way into the cell nucleus, transcription of the genes is initiated and recombinant proteins can be detected as quickly as a few hours after transfection. In successful transfections, a high percentage of the cells will express the recombinant protein. High transfection efficiencies lead to significant expression levels for 3 to 6 days followed by a rapid drop in production. Such short-term expression, referred to as transient expression, does not require an active selection for production. It is therefore the fastest method to produce recombinant proteins using mammalian cells (Wurm and Bernard, 1999).

Transient expression cannot easily be prolonged, because extrachromosomal plasmids can only be replicated when certain additional elements are provided (see section 6.4). Usually, transfected plasmids are lost after a few cell divisions. In order to produce more protein, more cells must be transfected. This involves scale-up from 1 ml to 500 ml for attached cells and from several L up to beyond 100 L for suspended cells. It is obvious that for the industrial scale, simple protocols are needed as well as inexpensive, non-toxic reagents that result in high expression levels.

The classical calcium phosphate (CaPi) technique is established for rapid protein expression e.g. using the adenovirus-transformed human embryonic kidney cell line, 293 (Graham *et al.*, 1977; Paborsky *et al.*, 1990). Recently, the technique was also performed successfully at the bioreactor scale (Jordan *et al.*, 1998). The attractiveness of the method is its general efficiency for a wide variety of cell types and its simplicity: calcium and phosphate are mixed in the presence of DNA and form a spontaneous precipitate. DNA is immediately adsorbed to the forming precipitate. The solution containing the CaPi-DNA complexes is then added to the cell culture medium. CaPi boosts the uptake of exogenous DNA into the cells. CaPi-DNA complexes are taken up by phagocytosis and DNA is released when the precipitate dissolves within the cells (Loyter *et al.*, 1982). This leads to high transfection efficiencies with minimal toxic side effects.

M. Al-Rubeai (ed.), Cell Engineering, 56–79.

If transient expression is used to purify proteins on a regular basis, it is very desirable to maximize expression levels for every run. While in general the CaPi technique is a simple and robust method, several critical parameters need to be carefully optimized in order to obtain reproducible transfection efficiencies. This review emphasises basic physico-chemical aspects of formation and dissolution of the precipitate. It also describes strategies to adapt the transfection protocol for adherent and suspension cells.

2. Transfection of cells

2.1 HISTORICAL DEVELOPMENT OF CaPi AND OTHER TECHNIQUES

The CaPi technique was first described as a method to quantify the infectivity of isolated adenovirus DNA (Graham *et al.*, 1973), as an alternative to the DEAE-dextran method (Cutchan *et al.*, 1968). Except for small modifications that improved the efficiency of the CaPi precipitation method, the protocol used today is very similar to the originally published method, using the same concentrations of calcium and phosphate. The popularity of the CaPi technique increased dramatically when transfection efficiencies for many different cell lines could be improved by optimizing the pH of the precipitation buffer and the CO_2 level of the incubator (Okajama *et al.*, 1987). Recently, kinetic studies of precipitate formation showed an additional improvement that is based on a reduction of the precipitation time from 20 - 30 minutes to 1 minute resulting in a small and effective precipitate (Jordan *et. al* 1996). Other groups suggested to form the precipitate at 65°C instead of at room temperature (Kjer and Fallon 1991) or to replace the calcium with the more soluble strontium (Brash *et al.*, 1987). The pH of the solutions and the DNA concentration are the most widely suggested parameters to optimize the formation of CaPi-DNA precipitates (Chen *et al* 1987; Chu and Sharp, 1981; Graham *et al.*, 1973; O'Mahoney and Adams 1994; Steven *et al.*, 1995). In some cases, replacing it with carrier DNA such as salmon sperm DNA reduces the quantity of the plasmid DNA (Graham and Bacchetti, 1983).

Absolute transfection efficiencies depend on the vector and cell line that is used. For some cell lines, satisfying expression levels are only obtained when the cells are treated with an osmotic shock using either glycerol or DMSO (Graham and Bacchetti, 1983; Wilson and Smith, 1997).

Recently, numerous commercially available transfection reagents have dramatically increased the choice of transfection methods. Nevertheless, no single method fulfils the needs for all kinds of cell types and/or applications. When tested in our laboratory with 293 cells, the CaPi technique was among the most efficient techniques. In Figure 1, it is shown that expression of green fluorescent protein (GFP) from CaPi-DNA precipitate treated cells is equivalent to results obtained with Fugene™, Lipofectamine™ or other reagents. Since commercially available transfection reagents are very expensive if used at the liter scale, calcium phosphate and PEI are the most attractive reagents for routine large-scale applications.

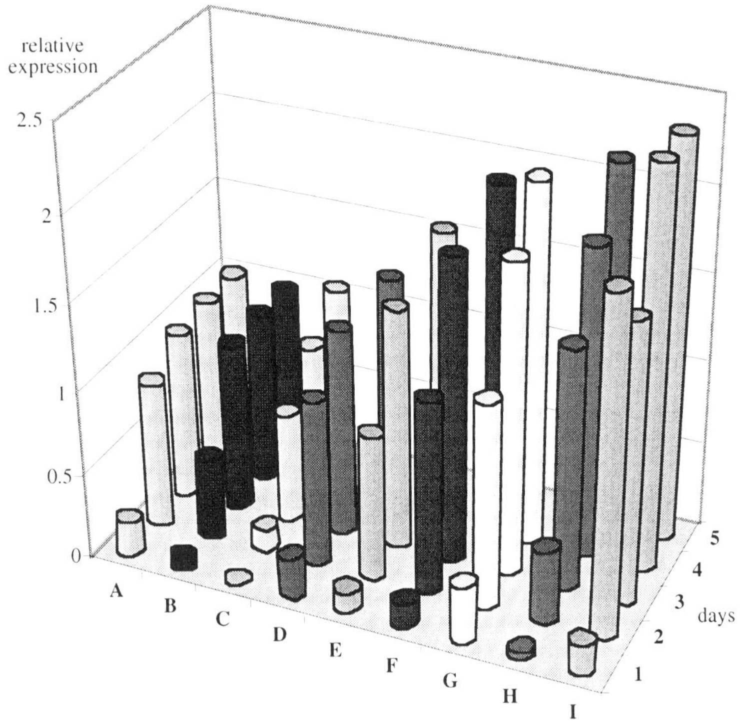

Figure 1. Comparison of various transfection methods. 9 different methods have been optimized for 293 cells at the 1 ml scale (A: Superfect™, B: Clonfectin™, C: Transfectam™, D: DOSPER, E: Fugene™, F: DOTAP, G: Lipofectamine™, H: PEI, I: CaPi). The expression levels were compared relative to method A that was set to 1. For each method, the protocol was first optimized and then used in this comparative study. For most reagents, optimizing the supplier's protocol significantly improved expression levels.

2.2 GENERAL TRANSFECTION PROCEDURE USING CALCIUM PHOSPHATE

Preparative work for a transfection includes purifying a sufficient quantity of DNA and subcultivation and scale up of the necessary quantity of cells. The transfer of DNA into the cells, done at production scale, can be divided into 3 major steps (Fig. 2).

i) Per milliliter of cell culture, 100 µl of a suspension containing 2.5 µg of plasmid DNA bound to CaPi precipitates is prepared as follows: DNA is diluted into 50 µl of calcium containing solution (solution A). Since DNA is stable in solution A, this can be done in advance. When the cells are ready, 50 µl of the HEPES buffered phosphate solution (solution B) are added and mixed quickly. After one minute of incubation at room temperature (RT), the precipitate is added to the cells, either adherent cells in plates/flasks or suspended cells in spinners/reactors.

ii) Cells are exposed to the precipitate for several hours, typically in DMEM/F12 medium containing 1-10% serum. At this stage, the calcium concentration in the culture medium is elevated to 12 mM. The concentration of phosphate remains more or less constant. Under these conditions, the precipitate undergoes further maturation (e.g. growth in particle size) and interacts with cells and other medium components such as serum proteins.

For most cell types, an exposure time of about 4 hours results in good transfection efficiencies with only minimal loss of cells. If cells are exposed to the precipitate for longer periods, the cell number has been observed to decrease due to cytotoxicity. It can be speculated that the toxicity is somehow related to a continuous uptake of CaPi-particles, which interferes with the intracellular calcium regulation.

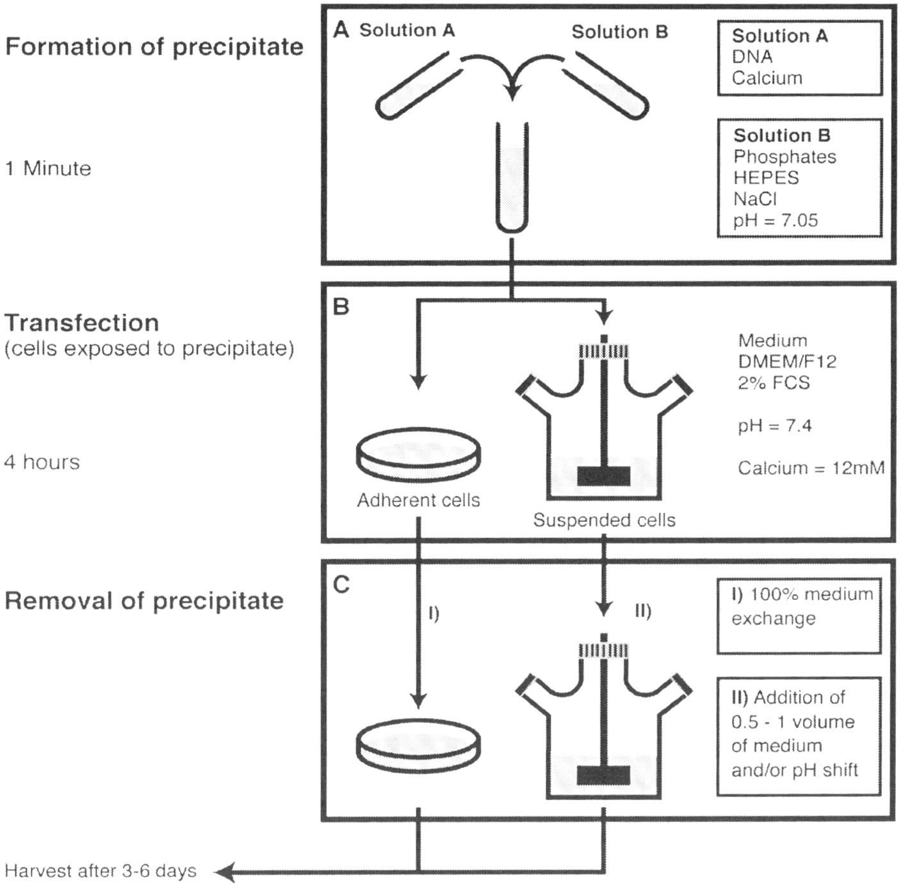

Figure 2. Outline of transient expression for adherent or suspended cells describing the three different steps: formation of precipitate, transfection and removal of precipitate. The most critical points are the formation of the precipitate and the exact conditions during the transfection. Solution A: 250 mM $CaCl_2$, plasmid DNA; solution B: 140 mM NaCl, 1.5 mM Na_2HPO_4, 50 mM HEPES, pH = 7.05.

iii) Removal/elimination of the precipitate is an important step. With adherent cells this is easily achieved by a medium exchange: most of the precipitate is removed and the concentration of calcium is brought back to 1 mM. In suspension, especially at larger scale, medium exchange is less convenient. Here, increasing the solubility can eliminate CaPi precipitates. Complete dissolution is achieved either by a reduction of the calcium concentration (addition of fresh medium), a pH shift, or a combination of both. In contrast to the adherent system, an elevated concentration of calcium will remain in the medium during the production phase. We did not observe negative effects on expression levels or cell viability due to this.

In both systems, typically more than 50% of the cells express the protein of interest for a few days.

2.3 ADHERENT VERSUS SUSPENSION CELLS

Although the transfection procedure briefly described above appears to be almost identical for adherent and suspended cells, the following major differences have to be noted. Adherent cells generally have a larger surface area than round cells in suspension. This means that attached cells principally can be covered with a high dose of precipitate. Under optimal conditions, small precipitate particles settle down slowly onto cells during the transfection. Therefore, adherent cells are continuously exposed to an increasing number of particles. However certain criteria in terms of mass and density of CaPi-precipitates have to be met. If the particles were too small, they would remain in the supernatant for several hours without interacting with cells, which are distributed on the surface of the growth substrate. On the other hand, particles that are too large rapidly cover the cells, while not being taken up at a high efficiency (Loyter *et al.*, 1982).

Cells in suspension are constantly mixed. Therefore contacts between cells and precipitate, even with very small particles, occur fast. These interactions are sufficient to perform transfections at high efficiencies. In some suspension cultures, especially under serum-free conditions, the "stickiness" of CaPi-precipitates creates a new problem: within a few minutes after the addition of CaPi-DNA complex to serum-free medium, the cells form aggregates. In the presence of serum, cell aggregates can be observed as well, but they remain smaller and are formed less rapidly. It is both the presence of CaPi precipitates and the elevated concentration of calcium that stimulate the cells to form aggregates (Peshwa et al., 1993). Thus, in suspension the formation of aggregates during or after transfection cannot be completely avoided. It could be shown, however, that most cells within aggregates express GFP at high levels, indicating that these cells contacted CaPi-DNA complexes before becoming part of an aggregate (personal communication P. Meissner).

2.4 SCALING UP THE TRANSFECTION TO 100 L

Tansfections up to a scale of 100 L are technically feasible. Based on data obtained at scales from 1 – 5 L (Meissner *et al.*, 1999), yields per batch at the 100 L scale should be in the range of gram amounts, particularly when considering recombinant antibodies that appear to be easily expressed with the method described here. Some modifications of the protocol established for adherent cells are necessary for the larger volumes. A linear scale up of the protocol would mean that per 100 L suspension culture in a bioreactor, 10 L of precipitate containing 250 mg DNA are required. 5 L of solution A would be mixed with 5 L of solution B and added within one minute to the cells (see Fig. 2). For such volumes, manual mixing and addition is no longer feasible. Another bottleneck is the preparation and purification of the required quantities of DNA. Reducing the volume of precipitate would address both issues. However, a smaller volume of the precipitation mixture also results in a lower concentration of calcium during the transfection. Since about 12 mM calcium seems to be desirable during transfections, lower calcium concentrations, especially at neutral pH values, bear the risk to dissolve the precipitate too fast.

Another issue is the time needed to transfer the precipitate containing solution into the reactor. Continuous mixing and addition by a peristaltic pump could control the formation time of the precipitate. But the remaining problem is that the first part of precipitate added to the reactor would rapidly dissolve due to the low initial concentration of calcium in the medium. In order to address this problem an additional step is suggested. First, the solubility is changed by either increasing the calcium concentration in the reactor and/or by shifting the pH up. The CaPi-DNA complexes are then added in a second step. This provides more flexibility on timing of the addition of the precipitate. It also eliminates the need to add large volumes of precipitation mixture for saturation of the medium with calcium thus less DNA is required.

2.5 REPORTER GENES

Optimization of transfections is no longer a time-consuming and difficult task since a number of sensitive and quantitative reporter systems are available (Table 1). Some are based on non-secreted proteins such as chloramphenicol acetyltransferase (CAT), luciferase, ß-galactosidase, ß-lactamase and the green fluorescent protein (GFP); others on secreted ones such as secreted alkaline phosphatase (SEAP).

TABLE 1. A comparison of frequently used reporter proteins

Reporter	Secretion	Detection of positive cells by microscopy	Detection in living cells by flow cytometry	Quantification
CAT	-	-	-	++
ß-galactosidase	-	+++	++++	+++
GFP	-	++	+++	+
Luciferase	-	-	-	++++
ß-lactamase	-	+++	++++	++++
SEAP	+++	+	-	+++

The most sensitive reporters are luciferase and ß-lactamase (Bronstein *et al.* 1994, Zlokarnik *et al.* 1998). Luciferase can be quantified in cell lysates by a chemiluminescent assay. ß-galactosidase can be used to detect individual transfected cells after fixation and a colorimetric staining. Alternatively, it can be also quantified in lysates based on a substrate-specific enzymatic reaction. Quantification of GFP does not involve any enzymatic reaction. However, GFP is the least sensitive reporter. The advantage for GFP is that, when expressed at high levels, it can be detected noninvasively within living cells either under a microscope, by flow cytometry or even with a fluorescence plate reader. The flexibility and the possibility for repeated measurements of the same cell population make it very attractive for experiments that do not require high sensitivity. Another advantage of the GFP-reporter over the luciferase-reporter is the long half-life of GFP that allows the amount of accumulated GFP to be measured even several days after transfection. In our laboratory, GFP is regularly used as a reporter. We had good experiences using it both for transient as well as for stable expression (Hunt *et al.*, 1999).

2.6 OPTIMIZING TRANSFECTIONS

Transfections can be compared using either transfection efficiency or expression levels of proteins. High transfection efficiency, defined as the percentage of positive cells, is a desirable feature for transient expression, yet satisfying levels of expression can only be obtained when positive cells show a high specific productivity as well. The transfection efficiency, especially when detected by a sensitive reporter system, appears not to be related to specific productivity. Thus, measuring the absolute amount of protein per volume of culture or per µg of input DNA is more appropriate for optimization experiments geared towards high levels of expression.

Establishing and optimizing a protocol for a cell line means adjusting parameters in a transfection experiment: a dose response curve to the DNA concentration, the effect of seeding density and eventually a variation of the exposure time are the most obvious parameters to be optimized. Adherent cells should be transfected 10 – 20 hours after plating at a cell density that allows at least one more cell doubling. Depending on the cell type, the optimum seeding density for transfection ranges from 1×10^5 to 4×10^5 cells per ml. When the cell density is too low, the risk that the precipitate becomes toxic to the cells is elevated, in particular after a prolonged residence time. For densities that are too high, the rapid acidification of the culture medium could lead to dissolution of the precipitate too early.

Batches of freshly prepared transfection solutions should be tested before use. After adjusting the phosphate of the solutions by checking the OD values (see section 4.2 for turbidity assay) the fine tuning of parameters is done in a transfection experiment. For large-scale transfections, it is also wise to test each new batch of purified plasmid DNA in a small-scale transfection, because we have seen important, as-yet unexplained variations.

3. Formation and dissolution of CaPi

3.1 PHOSPHATE IN AQUEOUS SOLUTIONS

Knowing some basic chemical aspects of phosphate molecules in aqueous solutions is important to understand the strong pH-dependence of the transfection method. In a physiological pH-range, the inorganic phosphate is present mainly in two forms: $H_2PO_4^-$ (monobasic) and HPO_4^{-2} (dibasic). Only a negligible fraction is present as PO_4^{-3}

(tribasic), the most insoluble version of phosphate in the presence of calcium. The equilibrium between these ions is defined by the pK-values (equations 1-3). The individual concentrations are a function of the pH of the solution.

$$H_3PO_4 \leftrightarrow H_2PO_4^- + H_3O^+ \quad pK_1 = 2.15 \quad (1)$$
$$H_2PO_4^- \leftrightarrow HPO_4^{-2} + H_3O^+ \quad pK_2 = 7.2 \quad (2)$$
$$HPO_4^{-2} \leftrightarrow PO_4^{-3} + H_3O^+ \quad pK_3 = 11.9 \quad (3)$$

Tribasic calcium phosphate is practically insoluble at neutral pH and only slightly soluble under acidic conditions. Nevertheless, small amounts of CaPi precipitates can be dissolved in pH-buffered cell culture media where PO_4^{-3} will become immediately protonated to HPO_4^{-2}. Since the pH value determines the actual fraction of phosphate that is present as PO_4^{-3}, it is a critical parameter for the solubility of the CaPi-DNA precipitate. The effect of pH on the concentrations of the different phosphate species is represented in figure 3 (more pH effects: see section 5.1 and 6.5)

The standard protocol for transfections prescribes a precipitation mixture containing 125 mM calcium and 0.75 mM phosphate at a pH of 7.05. The phosphates in this mixture are present at concentrations of 5.5 x 10^{-6} mM H_3PO_4, 0.44 mM $H_2PO_4^-$, 0.31 mM HPO_4^{-2} and 4.4 x 10^{-6} mM PO_4^{-3}.

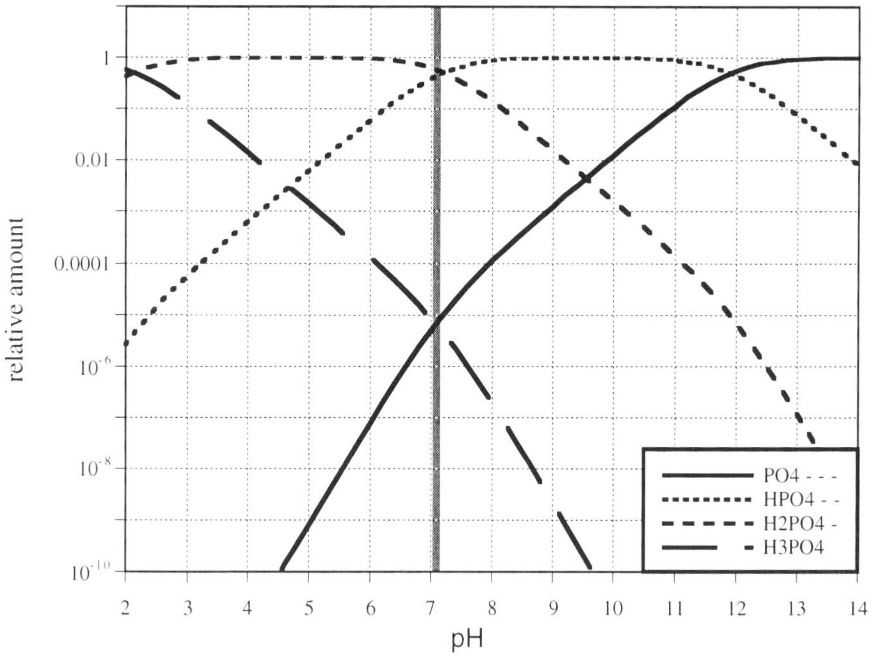

Figure 3. Effect of pH on the phosphate ion equilibrium in water (relative fractions). The sum of all the species corresponds to the total amount of phosphate. The actual concentration can be calculated by multiplying the value of this graph with the molar concentration of phosphate. The vertical line indicates the pH of the precipitation buffer used for transfections.

3.2 PRECIPITATION: NUCLEATION AND CRYSTAL GROWTH

Precipitate formation is initiated by mixing the solution containing calcium and DNA with the phosphate containing solution. DNA is readily adsorbed by the forming CaPi precipitate. The spontaneous precipitation occurs in two steps:

i) nucleation
ii) crystal growth

i) If no crystals are present, the mixture can exist as a supersaturated solution and no precipitate will form (region B, Fig. 4). Nucleation does not occur unless conditions at or above the metastable limit are achieved (region C). The metastable limit is very sensitive to impurities. Therefore, it is extremely difficult to quantify or to control nucleation.

ii) The second step, crystal growth, can be manipulated more easily by control of parameters affecting the solubility: Undersaturated conditions result in crystal shrinkage or in complete dissolution (region A), while supersaturated conditions induce further growth (region B). In both cases, an equilibrium between growth and dissolution will be achieved eventually. These conditions are described by the solubility curve that separates regions A and B in figure 4. As long as the precipitation mixture fulfils the criteria for the solubility curve, crystals are stable.

Temperature is one of the many parameters that affect solubility. Unlike most other salts, CaPi precipitates show a decreased solubility at higher temperatures.

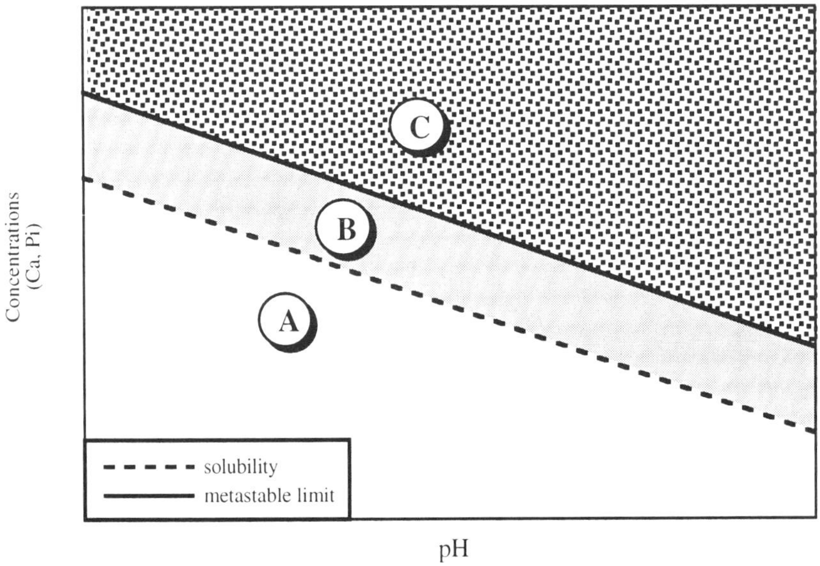

Figure 4. Scheme explaining the metastable limit and the solubility. Nucleation (region C), crystal growth (region B and C = supersaturated) and dissolution of crystals (region A = undersaturated) occur under certain physico-chemical conditions.

Precipitation reactions are typically fast, poorly controlled reactions. The following arguments should be considered:

i) Crystal-growth slows when approaching the solubility curve. In most cases, nucleation is expected just during mixing. Seconds after mixing, the presence of small CaPi precipitates and the decrease of the phosphate concentration will prevent further nucleation events.

ii) More crystal nuclei are expected for conditions that are above the metastable limit in comparison to points that are exactly on the metastable limit. Therefore, a starting point clearly above the curve for a metastable limit creates a great number of small crystal nuclei that can react with DNA and/or can grow further.

iii) Because phosphate is the limiting factor, its depletion slows the crystal growth until the equilibrium at the solubility curve is achieved.

iv) Even if no net growth of crystals appears, the precipitate can undergo further changes, e.g. aggregation with other crystals or adsorption of more DNA.

v) DNA is compacted during the co-precipitation and becomes resistant to DNase activity (Loyter *et al.*, 1982).

DNA - a flexible, thin, elongated linear or circular molecule - has a geometry that differs from compact crystals. While the mass of the final crystals exceeds the mass of the DNA, early crystals and, in particular, the crystal nuclei are much smaller than the DNA molecules. How the DNA interacts with nuclei, early or late crystals and at what stage the DNA is incorporated into the forming CaPi precipitate, is not known.

3.3 DIFFERENT COMPOSITIONS OF CaPi PRECIPITATES ARE POSSIBLE

Although hydroxyapatite (HAP) is frequently mentioned in the context of CaPi precipitates, it is not really clear what constitutes the precipitate. Mechanisms of precipitation in solutions containing calcium and phosphate are complex. The actual conditions that are used for transfections allow several different types of CaPi to be formed if "supersaturation" is considered as the basic criterion.

TABLE 2. Calcium phosphates with low solubility

Amorphous calcium phosphate	ACP	$Ca_x(PO_4)_y$
Dicalcium phosphate dihydrate	DCPD	$CaHPO_4 \times 2H_2O$
Dicalciumphosphate anhydrous	DCPA	$CaHPO_4$
Octacalciumphosphate	OCP	$Ca_8H_2(PO_4)_6 \times 5H_2O$
Tricalciumphosphate	TCP	$Ca_3(PO_4)_2$
Hydroxyapatite	HAP	$Ca_{10}(PO_4)_6(OH)_2$

Previously described CaPi precipitates with low solubility are listed in table 2 (Brown and Constanz, 1994). At neutral pH, solubility decreases in the order DCPD >DCPA >> TCP >> HAP. Having the lowest solubility, HAP is the most stable version of CaPi and the most difficult form to dissolve. Nevertheless, it is unlikely that DNA-CaPi-complexes contain pure HAP. In multi-component precipitations of sparingly soluble salts, the component with the lowest supersaturation is expected, mainly as a result of favourable kinetics. A direct precipitation of HAP is described only for temperatures above 200°C (Brown and Constanz, 1994).

If precipitation is done in impure solutions, calcium ions can potentially be substituted by Mg^{2+}, Sr^{2+}, Pb^{2+}, Zn^{2+}, Na^{+}, K^{+} etc., whereas phosphate ions can be replaced by CO_3^{2-}, $P_3O_7^{4-}$ etc. In HAP, Cl^- or F^- may take the place of the hydroxyl group. In cell culture medium, a partial substitution is expected when the precipitate is added, since most of these ionic candidates are present.

It will be a challenge for the future to i) understand the structure ii) to control the creation of precipitates with a defined crystals size and DNA binding capacity and also iii) to control the net charge of the crystals. Each of these parameters could be relevant for transfections. New experimental approaches are needed to investigate such questions. Extending our knowledge beyond "solubility" and crystal "size" could lead to further improvements of the method.

4. Kinetics of the CaPi formation in the presence of DNA

4.1 CENTRIFUGATION ASSAY TO MEASURE DNA CO-PRECIPITATION

Precipitate formation is the most critical step for efficient transfection. Thus, it is not surprising that many publications have reported on parameters affecting this first step (Graham and Van der Eb, 1973; Chen and Okayama, 1987; O'Mahoney; Adams, 1994; Jordan *et al.*, 1996). Since naked plasmid DNA is not capable of transfecting cells at an acceptable efficiency, it is evident that DNA needs to be co-precipitated (or just bound to the precipitate) in order to enter the cells by an energy-dependent endocytic pathway (Chang, 1994; Loyter *et al.*, 1982). A simple centrifugation step gives insights into quantities of DNA being integrated or absorbed to CaPi-crystals. The precipitate is separated from the liquid by a brief centrifugation step (e.g. 30 s at 16 000 g in a 1.5 ml eppendorf™ tube). Unbound DNA stays in the supernatant and can be quantified either by UV measurements or sensitive DNA-specific dyes such as Picogreen™ (Molecular Probes). It was found that, under conditions leading to a visual precipitate, DNA was adsorbed within 30 seconds, the minimal time required for the centrifugation step. Thus it appears that DNA is bound within seconds during the formation of CaPi precipitates. The actual amount of bound DNA can be impressive: assuming that 0.75 mM phosphate yields maximally about 100 μg of CaPi precipitate per ml, a complete binding of up to 50 μg of DNA was found. This means that DNA contributes a significant part to the total mass of the CaPi-DNA complexes. It is not surprising that under such conditions DNA also affects the kinetics of the precipitation step.

25 μg DNA per ml of precipitate is proposed by most protocols as the optimal DNA amount to achieve high levels of transient expression or a high number of stable clones. This quantity of DNA can be precipitated instantly (Fig. 5a). Increasing the concentration of DNA however, not only slows down the binding of DNA, but also lowers the amounts of precipitated DNA. In the presence of 50 μg/ml DNA, less than 5 μg DNA was bound after 20 minutes, representing only 10%. Such "negative" effects of DNA on precipitate formation can be overcome by changing the solubility: increasing the calcium concentration decreases solubility, thus 50 μg DNA/ml can be precipitated at 250 mM calcium at RT (Fig. 5b). Lowering the temperature increases the solubility.

Therefore, for a given calcium concentration less DNA is precipitated at 0°C than at RT.

It can be speculated that DNA is also a parameter that changes the metastable limit and/or the solubility. Since CaPi-DNA precipitates are formed under conditions only slightly above the metastable limit, transfections are sensitive to any factor that influences nucleation or the kinetics of precipitation.

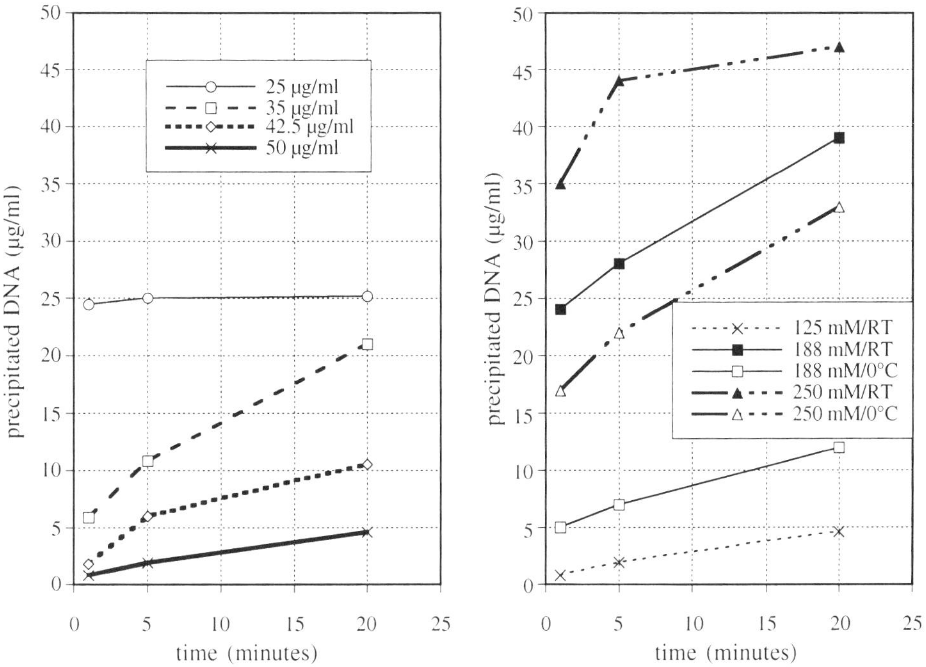

Figure 5. Kinetics of the incorporation of DNA into a forming precipitate. At the indicated time points, the precipitate was pelleted by centrifugation and the DNA remaining in the supernatant was quantified. (a) Increasing amounts of DNA were precipitated with optimized solutions A and B. (b) Increasing the calcium concentration (increasing the supersaturation) or decreasing the temperature to 0°C (decreasing the supersaturation by increased solubility).

4.2 CaPi PRECIPITATION MONITORED BY TURBIDITY

While the centrifugation assay can help in understanding precipitation, alternative tools are needed to monitor the formation of CaPi-DNA precipitates. Therefore a second assay, based on turbidity measurements, has been developed.

Small precipitates render the solution visibly opaque. This turbidity can be quantified by measuring the absorbance. A signal is obtained at any wavelength, yet shorter wavelengths give significantly higher OD values. 320 nm is the shortest wavelength at which DNA and other soluble components of the precipitation mixture have a negligible absorbance. Therefore, turbidity measurements were performed at 320 nm.

As a centrifugation step is not needed, turbidity is well suited for measuring multiple samples and to study the kinetics of precipitate formation. In addition, turbidity assays can be done with or without DNA.

Turbidity studies confirmed the effect of DNA on the precipitation reaction itself (Fig. 6). Increasing the amount of DNA from 0 to 8.25 and 25 μg/ml leads to an increase in absorbance. For 50 μg DNA/ml however, the absorbance at any specific time is lower than the control. Turbidity data correlate with results obtained by DNA incorporation experiments (Fig 5). These also correspond to previous data (Graham and van der Eb, 1973) showing that addition of carrier DNA up to a total of 30 μg DNA/ml increases the infectivity of 2 μg/ml of virus DNA, while higher doses reduce infectivity.

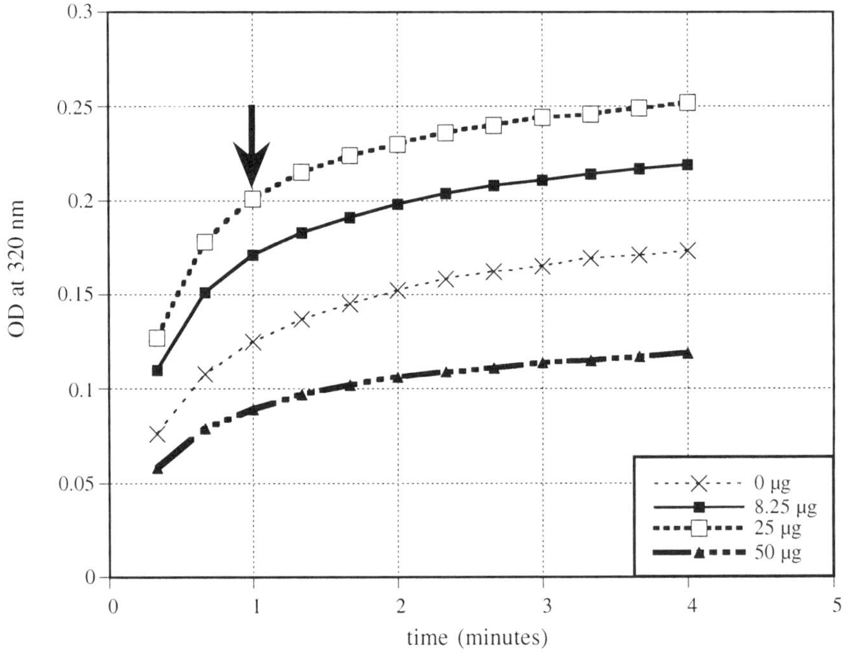

Figure 6. In an experiment completely independent from 5, the turbidity (OD value at 320 nm) was measured during the formation of precipitate. DNA concentration up to 25 μg per ml had a positive effect and increased the turbidity, while 50 μg had a clear negative effect on turbidity resulting in an absorbance below the DNA-free control. Arrow indicates standard conditions for transfections.

DNA affects the formation of the precipitate not only at slightly supersaturated conditions, but also when the calcium or phosphate concentration is increased. While the DNA binding assay (described above) no longer differentiates once DNA is completely bound to CaPi, the turbidity assay yields more detailed information (Fig. 7). Different conditions are compared by measuring the absorbance exactly 1 minute after mixing. At the lowest phosphate concentration, 0.6 mM, all DNA concentrations had a minor negative effect on absorbance. At the standard concentration of 0.75 mM phosphate, turbidity was significantly increased for 6 and 12 µg DNA/ml, but not for 25 µg DNA/ml. At higher levels of phosphate, DNA had a comparable dose dependent positive effect on the absorbance. Negative effects or a decrease of absorbance can still be provoked, but significantly more DNA is needed to do so.

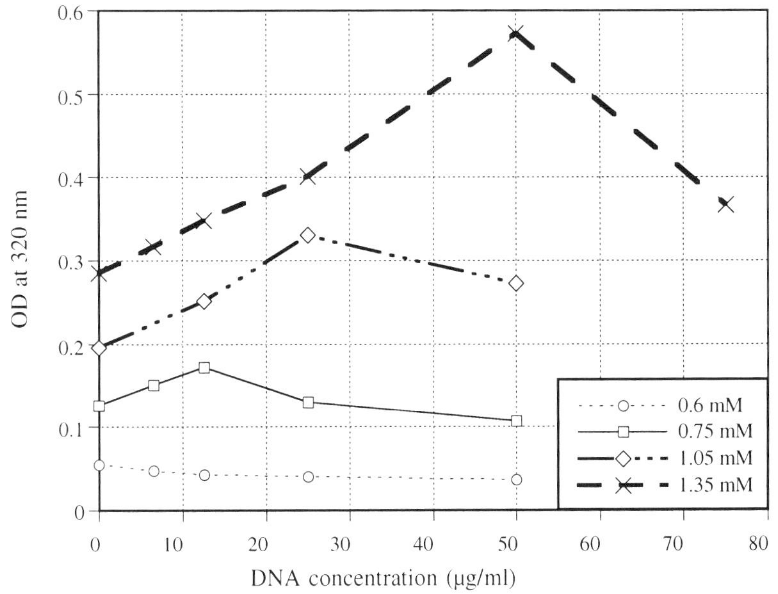

Figure 7. Effects of DNA and phosphate concentration on the formation of CaPi-DNA precipitates (calcium fixed at 125 mM). Turbidity was measured after a reaction time of 1 minute. At low phosphate concentrations, addition of DNA slightly reduces turbidity. At higher phosphate concentrations, the turbidity increases until reaching a critical point, beyond which DNA has a negative effect.

Optimized solutions A and B are sensitive to changes in a single parameter. Even a minor modification can lead to a reduction in expression levels. Therefore it can be necessary to optimize several parameters at once, as demonstrated by a whole set of combinations of calcium and phosphate concentrations leading to the formation of an efficient CaPi-DNA precipitate (Fig. 8). Different concentrations of calcium and phosphate have been screened by DNA binding and turbidity assays. Combinations of calcium and phosphate concentrations that precipitate 50% of the DNA (50 µg DNA/ml) within a reaction period of one minute are plotted in figure 8. The fitted curve reflects the conditions relatively close to the metastable limit that can be found for a pH of 7.05 at RT. The conditions defined by the curve are expected to lead to physically comparable precipitates.

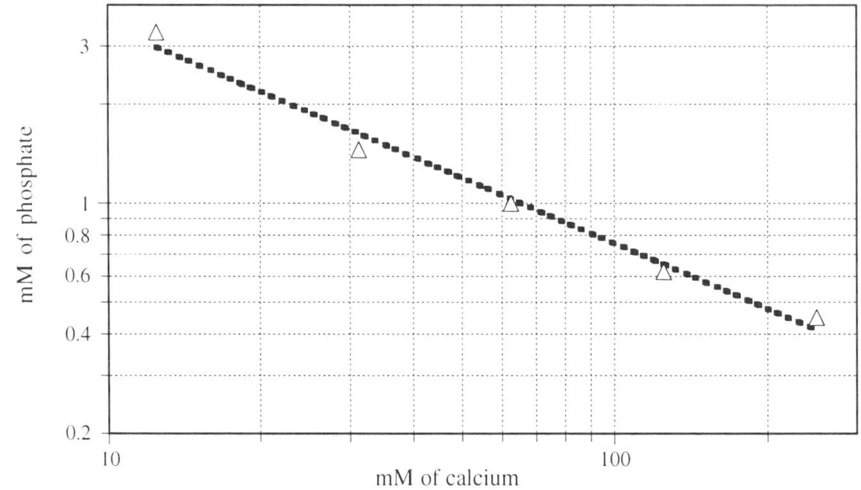

Figure 8. Different ratios of calcium and phosphate that precipitate 50 μg DNA/ml at comparable kinetics. At a temperature of 23°C and a pH of 7.05, a set of calcium concentrations in combination with different phosphate concentrations was tested. Points on the graph reflect the amount of phosphate needed to precipitate 50 % of the DNA within exactly 1 minute. Optimized precipitate formation conditions should be close to these different combinations.

5. Transfection: exposure of cells to CaPi-DNA

Immediately after the precipitate is formed, it is transferred to the cell culture medium. At this stage, the cell culture medium is supersaturated, but most likely still below the metastable limit. Thus, new nucleation events are not expected. Yet the added CaPi-DNA precipitate undergoes further maturation processes that depend on the solubility. The CaPi-DNA complexes slowly settle onto the cells and can be visualized under a microscope as dark particles during the first few hours of transfection (Fig. 9).

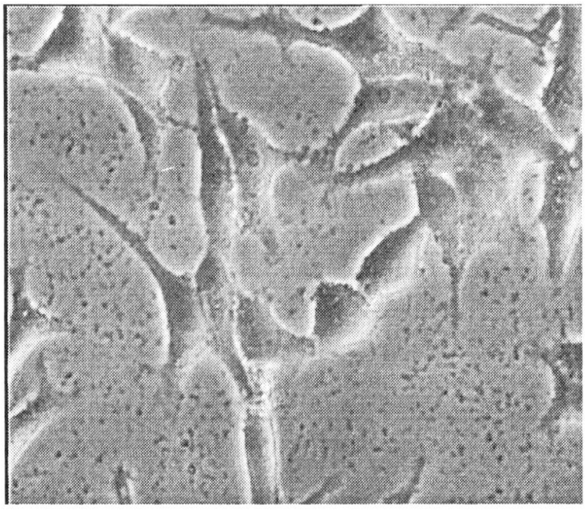

Figure 9. Attached 293 cells covered with precipitates. CaPi-DNA complexes can be visualized by phase contrast microscopy as dark dots, recognizable on the plastic surface that is not covered by the cells.

5.1 SOLUBILITY OF DNA CaPi CO-PRECIPITATE IN CELL CULTURE MEDIUM

CaPi precipitates are soluble in all the standard media used for cell culture, mainly because the concentrations of calcium and phosphate are low, about 1 mM for most media. In certain media, the calcium concentration can be reduced to 0.5 mM or less (Peshwa *et al.*, 1993). A calcium concentration significantly higher than 1 mM is avoided in commercially available media. Increasing the concentration of calcium could lead to a spontaneous formation of precipitates, especially at basic pH values. Precipitates in the medium would not allow sterilization by filtration and they are not desired during storage or when the medium is used. However, transfections based on the CaPi technique depend on higher concentrations of calcium in the medium in order to prevent crystal instability.

The solubility of CaPi precipitates in culture medium is affected by several parameters such as serum concentration, media composition and ionic strength and, of course, by the pH of the medium (see Fig. 3). The pH itself cannot be strictly controlled in many popular cultivation systems such as T-flasks, roller bottles, shaker flasks or spinner flasks. Although the $NaHCO_3$ buffer system combined with CO_2-incubators is routinely used, liquid surface area, diffusion, mixing, CO_2 production by the cells, etc. affect the equilibrium of the buffer system and cause pH shifts over time. In such systems, the transfection efficiency can vary when cell density, buffer system and other pH-affecting parameters are not optimized for each cell line. On the other hand, modifying the pH provides an excellent tool to change the size of the CaPi-DNA complexes. It can be speculated that pH might also affect the particle charge and the interaction between cells and precipitate.

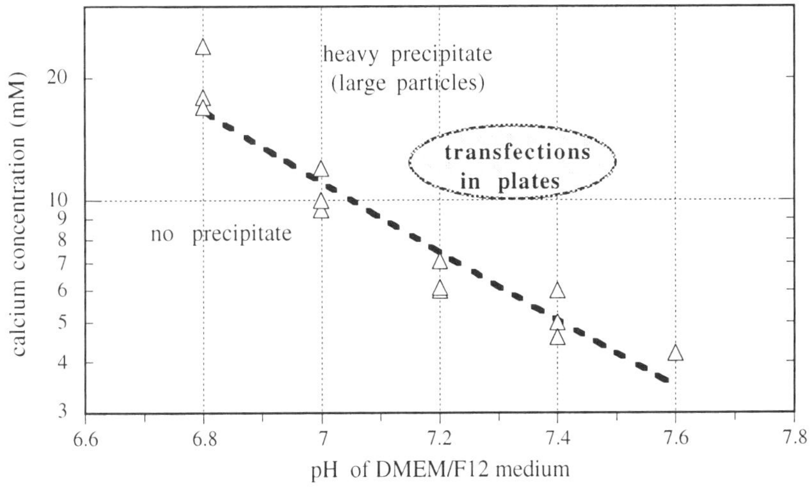

Figure 10. Solubility of the precipitate in serum-free, HEPES buffered DMEM/F12 medium as a function of pH. Cell free medium was supplemented with increasing concentrations of calcium and incubated overnight at 37°C. Then plates were visually checked for the presence of CaPi precipitates. Points reflect the lowest concentration of calcium yielding visible particles. Different points were obtained in independent experiments.

A solubility curve in DMEM/F12 medium at the standard phosphate concentration was established using the following approach: HEPES buffered serum-free DMEM/F12 medium was adjusted to a range of pH values. Subsequently, various volumes of a concentrated calcium stock solution were added. After incubation overnight at 37°C, the presence of a precipitate was checked under a microscope at 400 x magnification. The minimal concentration of calcium resulting in a visible precipitate is plotted in figure 10. Both pH and calcium can be efficiently used to change solubility.

Adherent cells in serum containing DMEM/F12 medium are transfected under clearly "supersaturated" conditions. It seems that in the static system, a critical mass is needed for the settling of precipitate complexes onto the cells. Serum also has a certain influence. In the presence of serum, particle size tends to be smaller. Without serum extensive supersaturation leads to a heavy precipitate that is not optimal for transfection.

Since many serum proteins directly interfere with calcium, serum is expected to have additional effects on transfection. Serum proteins do also bind to forming precipitates. This can be checked with a centrifugation assay by replacing DNA with serum. One can assume that proteins in cell culture medium interact with CaPi-DNA precipitates, especially if the crystal surface is not fully covered with DNA. This interaction is expected to change the composition as well as the charge of the CaPi precipitate and therefore influence its reactivity towards cells.

6. Starting transfections

6.1 TESTING OF NEW SOLUTIONS AND DNA PREPARATIONS

It is recommended to compare newly prepared batches of transfection solutions against an "old" reference in a transfection. In addition, "bad batches" can be avoided by using the turbidity test even before transfections are started. The formation of crystals can be observed as an increase in turbidity. Strong OD signals result from both an increasing number of particles and an increasing particle size. The same batch of solutions mixed at a fixed ratio results in a reproducible absorbance (Fig. 11). Therefore the rapid measurement of turbidity can be used as a quality control tool to compare batches of solutions. If the new batch does not result in the desired absorbance, the phosphate concentration can be adjusted. It is not necessary to include DNA in these quality control experiments. The OD at 320 nm after 1 minute with pure water as reference (blank) should be between 0.1 and 0.14. Visually this can be recognised as a slight cloudiness of the solution. Higher OD values are typical for mixtures with elevated concentrations of phosphate. If DNA is present during the precipitation, the OD will be affected (see Fig. 7).

A series of precipitations with increasing amounts of DNA should result in a similar curve for different batches of purified DNA. When the turbidity is different from the expected value, the expression levels could be affected as well. Problems with transfection can be expected if the DNA concentration is not correct or if DNA is not pure. The concentration of DNA can be easily confirmed as well as the absence of some impurities such as RNA. Other impurities might be "visible" only in a turbidity assay, since precipitation is sensitive to certain contaminants. Thus, the turbidity assay can also help to check the purity of the plasmid DNA.

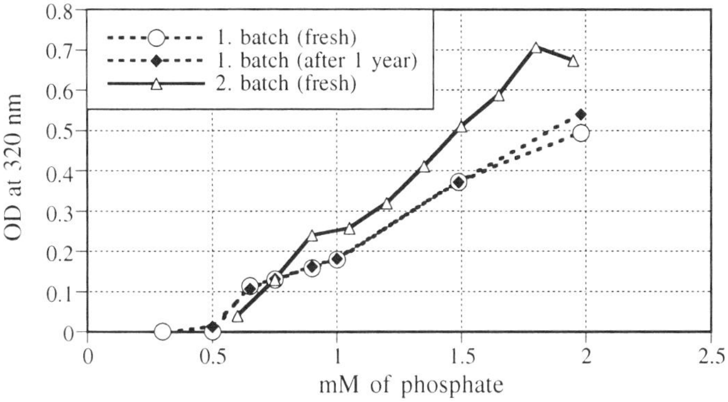

Figure 11. Turbidity test with variable concentrations of phosphate applied to two batches of transfection solutions. The response of batch 2 differs slightly from the response of batch 1, prepared earlier.

6.2 STABILITY OF SOLUTIONS AT DIFFERENT TEMPERATURE

In most laboratories solutions A and B are kept frozen as suggested by some authors or by instructions of commercially available transfection kits (Wilson *et al.*, 1995, ProFection™ from Promega). However, we keep our solutions at room temperature for a year or longer. We performed a stability study over a period of six months. For this study, sterile aliquots of solutions were filled into well-closed 15 ml polypropylene tubes and stored at different temperatures (Fig. 12). After 6 months, the solutions were investigated in a turbidity assay as well as in a transfection using 293 cells and GFP as a reporter. No significant effects resulting from the different storage temperature could be observed after this period of time.

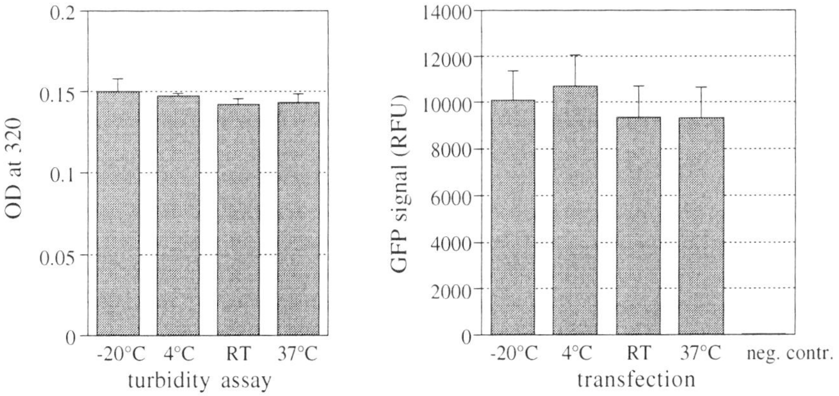

Figure 12. Stability of solutions A and B stored at different temperatures for half a year: turbidity data and expression of green fluorescent protein show that stability and effectiveness of solutions is not affected by storage temperature.

6.3 CELL LINES

The CaPi technique is compatible with most cell lines. With CHO cells and different 293 cell lines we observed that the transfection efficiency and the transient expression levels, as well as the optimal transfection conditions depend on the cell type.

The CaPi precipitation method is not limited to immortalized cell lines. A transfection efficiency of 35% was achieved after optimizing the protocol for primary chromaffin cells (Wilson *et al.*, 1995). Other cell types successfully transfected with CaPi-DNA complexes include insect cells (Kjer and Fallon, 1991), primary cardiac myocytes (Xu *et al.*, 1991), primary adult rat hepatocytes (Rippe et al., 1990) and many other current cell lines (Gorman *et al.*, 1983; Chen and Okayama, 1988). For some cell types, especially primary ones, transfection causes changes in morphology or strongly affects cell survival. The optimized calcium phosphate method also compares favourably to other methods in this respect (Hilliard *et al.* 1996).

The calcium phosphate method is not limited to transient expression. The method works well for the creation of stable cell lines expressing recombinant proteins at high levels (Wurm *et al.*, 1996). Typically, multiple DNA molecules are inserted at one single random site of the genome (Wurm and Petropoulos, 1994). Thus several genes can be co-transfected as separate plasmid vectors and subsequently they are co-expressed by most clones.

6.4 EPISOMAL PLASMIDS

Transient transfection of 293 cells can rapidly generate analytical amounts of proteins required for biochemical or physiological studies (Gorman *et al.*, 1990; Paborsky *et al.*, 1990). High levels of expression can be achieved when an episomal replication system is used (Shen *et al.*, 1995; Mazda *et al.*, 1997). For example, this can be based on the large T antigen from SV 40 virus or the EBNA from the Epstein Barr virus. The plasmids transfected carry the origin of replication corresponding to, for example, the large T antigen etc. One can use cell lines stably expressing the large SV 40 T antigen or EBNA. Alternatively, for cells expressing none of these virus-derived antigens constitutively, its sequence can be provided together with the expression vector.

6.5 CULTURE MEDIUM

The CaPi method works well in DMEM/F12 based media. Other media should work as well, as long as the calcium and phosphate concentrations are similar to that of DMEM/F12. RPMI medium, which has 5 times more phosphate, is not recommended for transfections because a very heavy precipitate that is difficult to remove will appear. Attention also has to be paid to the composition of complex serum-free medium: polyanions such as heparin or dextran sulfate potentially interfere with transfections and should be avoided (Mounkes *et al.* 1998).

Although high transfection efficiencies have been reported in serum-free medium (Wilson *et al.*, 1995), 1% - 10% serum is usually present in the medium during the transfection.

For DMEM/F12 medium, the most important parameters are calcium concentration and medium pH during transfection (Fig. 13). Transfection efficiency, measured as ß-galactosidase positive CHO cells per μg of DNA, showed a strong pH dependency. Based on the solubility (see Fig. 10), such a pH dependency is expected. It has to be mentioned that in this experiment the concentration of calcium was varied by increasing the total amount of added precipitate. Thereby increasing the calcium also increased the amount of DNA added per well. Although additional dose effects cannot be completely excluded, the pH dependency is fully supported by the data.

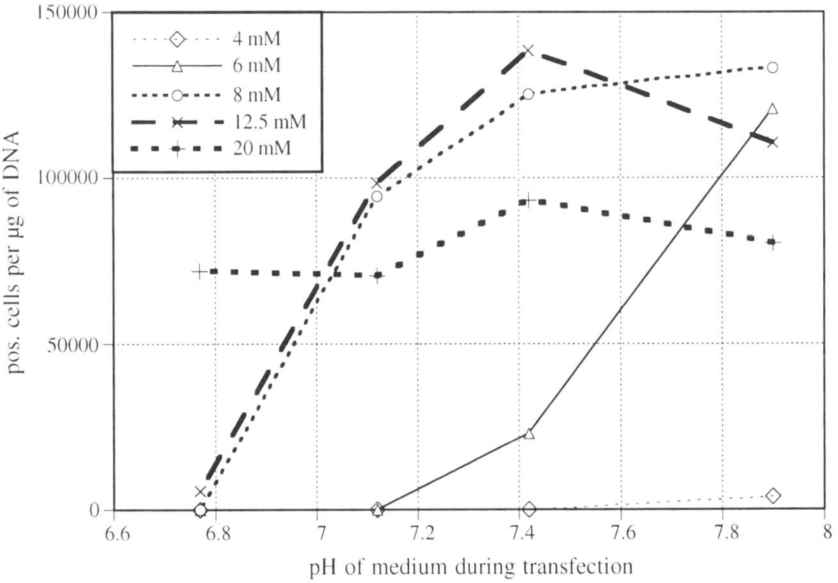

Figure 13. Optimization of pH in 12-well plates. A "standard" precipitate was used at different concentrations of calcium combined with various pH conditions during the transfection period. The number of positive cells was measured one day later after staining of intracellular ß-galactosidase with X-gal (Sanes *et al.* 1986).

6.6 TRANSFECTION OF CELLS CORRELATES WITH CO-PRECIPITATION OF DNA

The essential role of DNA co-precipitation for a successful transfection was confirmed both for adherent and suspended 293 cells (Fig. 14). Phosphate concentrations that neither increased the turbidity nor caused binding of DNA to CaPi did not yield detectable levels of expression. Conditions that led to co-precipitation of DNA yielded comparable levels of expression in both systems. Having more than 1 mM of phosphate during the precipitate formation negatively affected the expression, mainly due to the presence of a much coarser precipitate for which a lower DNA content (relative to crystal mass) is expected. Transfection of adherent cells at the 1 ml scale gives an estimation of yields at larger scales (200 ml suspension culture). This is important when novel proteins (on standard expression vectors or new vector constructs) are transiently expressed.

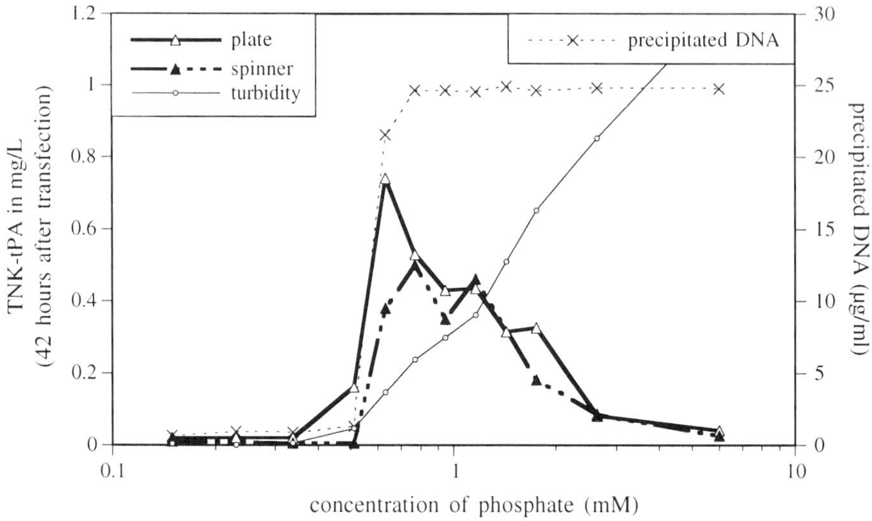

Figure 14. A phosphate optimization experiment for attached cells in 12-well plates or suspended cells in spinner flasks. The phosphate of solution B ranged from 0.2 mM to 6 mM. Good expression coincides in both systems with co-precipitation of DNA. Increasing the phosphate concentration beyond 1 mM phosphate had a negative effect on expression levels.

6.7 TRANSIENT PRODUCTION OF PROTEINS IN BIOREACTORS

Transfections in bioreactors can yield from a few mg up to 20 or more mg/ml of secreted proteins such as antibodies or a soluble receptor (Fig. 15). 293-EBNA cells were inoculated at 3 x 10^5 cells/ml and transfected 2 hours later. 1 mg of precipitated plasmid DNA was added to the reactor by syringe injection.

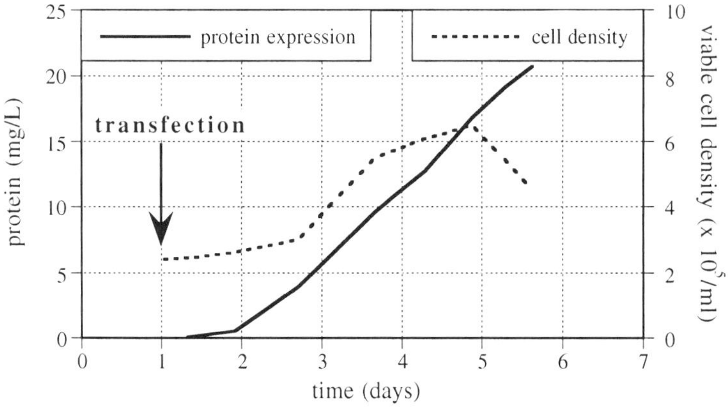

Figure 15. Cell number and expression levels of 293 EBNA cells transfected in 2 L bioreactors and cultivated for 5 days.

Cells started to secrete the protein at maximal rates one day after transfection. Viability of the cells was not affected by the high calcium concentrations during transfection. After the transfection, still at an elevated concentration of calcium, cells grew to a density of 6 x 10^5 cells per ml. This cell density is not considered a maximal density achievable for 293 cells. Improving the cell growth after transfection could contribute to a further increase of recombinant protein expression.

7. Conclusions

The need to rapidly express larger amounts of proteins can be fulfilled by transient expression. While for transfections at the ml-scale a number of powerful methods exist in the form of protocols or transfection kits, only a few techniques seem suitable for large scale transfections (Blasey *et al.* 1996; Jordan *et al.*, 1998; Schlaeger *et al.* 1999). CaPi is a promising technique for bioreactor applications. Understanding critical points mentioned in this article will help in establishing the technique. Even if at the moment CaPi-DNA precipitation is not fully understood, parameters to optimize the precipitation are well known. Monitoring the turbidity allows consistent and reproducible formation of CaPi-DNA complexes.

The next challenge will be to control the transfectability of cells. Various factors such as the history of pre-culture, effects of cell passage number and variations of the growth behaviour can cause a variability that is not necessarily related to the transfection technique. We have preliminary evidence that, for example, cell cycle is an important factor for transfectability of cells: effects of cell synchronization on transfection efficiency are comparable to those reported for electroporation (Yorifuji *et al.* 1989, Schwachtgen *et al.* 1994).

In addition, the type of protein to be expressed and the plasmid vectors used have a major impact on expression levels. Certainly, some proteins can be already ruled out as candidates for large-scale expression based on productivity at small scale. Only the routine application of transfections at industrial scale can and will lead to additional improvements. Further adaptation of transfection protocols in combination with established animal cell culture techniques in bioreactors will provide a competitive alternative to stable expression when only gram amounts of protein are needed.

8. Acknowledgements

The author would like to thank Prof. F.M. Wurm for discussions and input to this article, P. Girard for the contribution of Fig. 1/9 and P. Meissner for the contribution of Fig. 15.

9. References

Blasey H.D., Aubry J.P., Mazzei G.J. and Bernard A.R. (1996) Large Scale Transient Expression with COS Cells. *Cytotechnology* **18,** 183-192.

Brash D.E., Reddel R.R., Quanrud M., Yang D., Farrell M.P. and Harris C.C. (1987) Strontium Phosphate Transfection of Human Cells in Primary Culture: Stable Expression of the Simian Virus 40 Large-T-Antigen Gene in Primary Human Bronchial Epithelial Cells. *Mol. Cell. Biol.* **7,** 2031-2034.

Bronstein I., Fortin J., Stanley P.E., Steward G. and Kricka J. (1994) Chemiluminescent and Bioluminescent Reporter Gene Assays. *Anal. Biochem.* **219,** 169-181.

Brown P.W. and Constanz B. eds. *Hydroxyapatite and Related Materials.* CRC press, Boca Raton 1994.

Chang P.L. (1994) Calcium Phosphate-Mediated DNA Transfection. In: *Gene Therapeutics: Methods and Applications of Direct Gene Transfer.* Wolff J.A. ed. Birkhäuser Boston.

Chen C. and Okayama H. (1987) High-Efficiency Transformation of Mammalian Cells by Plasmid DNA. *Mol. Cell. Biol.* **7,** 2745-2752.

Chu G. and Sharp P.A. (1981) SV40 DNA Transfection of cells in suspension: analysis of the efficiency of transcription and translation of T-antigen. *Gene* **13,** 197-202.

Gorman C., Padmanabhan R. and Howard B.H. (1983) High Efficiency DNA-Mediated Transformation of Primate Cells. *Science* **221,** 551-553.

Gorman C.M., Gies D.R. and McCray G. (1990) Transient Production of Proteins Using an Adenovirus Transformed Cell Line. *DNA and Protein Engin. Techn.* **2**, 3-10.

Graham F.L. and Bacchetti S. (1983) DNA mediated gene transfer using the calcium technique. *Nucleic Acid Biochem.* **B506**, 1-14.

Graham F.L., Smiley J., Russell W.C. and Nairn R. (1977) Characteristics of a Human Cell Line Transformed by DNA from Human Adenovirus Type 5. *J. gen. Virol.* **36**, 59-72.

Graham F.L. and Van der Eb, A.J. (1973) A new Technique for the Assay of Infectivity of Human Adenovirus 5 DNA. *Virology* **52,** 456-467.

Hilliard C.M. Fletcher S. and Yeoh G.C. (1996) Calcium phosphate transfection and cell-specific expression of heterologous genes in primary fetal rat hepatocytes. *Int. J. Biochem. Cell Biol.* **28**, 639-650.

Hunt L. Jordan M. De Jesus M. and Wurm F.M. (1999) GFP-Expressing Mammalian Cells for Fast, Sensitive, Noninvasive Cell Growth Assessment in a Kinetic Mode. *Biotech. Bioeng.* **65**, 201-205.

Jordan M. Koehne C. and Wurm F.M. (1998) Calcium-phosphate mediated DNA transfer into HEK-293 cells in suspension: control of physicochemical parameters allows transfection in stirred media. *Cytotechnology* **26,** 39-47.

Jordan M., Schallhorn A. and Wurm F.M. (1996) Transfecting mammalian cells: optimization of critical parameters affecting calcium-phosphate precipitate formation. *Nucleic Acids Res.* **24**, 596-601.

Kjer K.M. and Fallon A.M. (1991) Efficient Transfection of Mosquito Cells is Influenced by the Temperature at Which DNA-Calcium Phosphate Coprecipitates are Prepared. *Insect Biochem.* **16,** 189-200.

Loyter A., Scangos G.A., Juricek D., Keene D and Ruddle F.H. (1982) Mechanism of DNA entry into mammalian cells. *Exp. Cell. Res.* **139,** 223-234.

Loyter A., Scangos G.A. and Ruddle F.H. (1982) Mechanisms of DNA uptake by mammalian cells: Fate of exogenously added DNA monitored by the use of fluorescent dyes. *Proc. Natl. Acad. Sci.* **79**, 422-426.

Mazda O., Satoh E., Yasutomi K. and Imanishi J. (1997) Extremely efficient gene transfection into lympho-hematopoietic cell lines by Epstein-Barr virus based vectors. *J Immunol. Meth.* **204,** 143-151.

McCutchan J.H. and Pagano J.S. (1968) Enhancement of the infectivity of simian virus 40 deoxyribonucleic acid with diethylamino-dextran. *JNCI* **41,** 351-357.

Meissner P., Girard P., Kulangara A., Tsao M.C. Jordan M and Wurm F.M. (1999) Process development for transient gene expression in mammalian cells at the 3 L scale: 10-50 mg of r-protein in days. In: *Animal Cell Technology: Products from Cells, Cells as Products.* Bernard A. Griffiths B. Noé W. and Wurm F., eds. Kluwer. 351-357.

Mounkes L.C., Zhong W., Cipres P.G., Heath T.D. and Debs R.J. (1998) Proteoglycans mediate cationic liposome-DNA complex-based gene delivery in vitro and in vivo. *J Biol. Chem.* **273,** 26164-70.

O'Mahoney J.V. and Adams T.E. (1994) Optimization of Experimental Variables Influencing Reporter Gene Expression in Hepatoma Cells Following Calcium Phosphate Transfection. *DNA and Cell Biol.* **13**, 1227-1232.

Paborski L.R., Fendly B.M. Fisher K.L., Lawn R.M., Marks B.J., McCray G., Tate K.M., Vehar G.A. and Gorman C.M. (1990) Mammalian cell transient expression of tissue factor for the production of antigen. *Protein Engineering* **3**, 547-553.

Peshwa M.V., Kyung Y.S., McClure D.B. and Hu W.S. (1993) Cultivation of Mammalian Cells as Aggregates in Bioreactors: Effect of Calcium Concentration on Spatial Distribution of Viability. *Biotech. Bioeng.* **41**, 179-187.

Rippe R.A., Brenner D.A. and Leffert H.L. (1990) DNA-Mediated Gene transfer into Adult Rat Hepatocytes in Primary Culture. *Mol. Cell. Bio.* **10**, 689-695.

Sanes J.R., Rubenstein J.L.R. and Nicolas J.F. (1986) Use of a recombinant retrovirus to study post-implantation cell lineage in mouse embryos. *EMBO* **5**, 3133-3142.

Schlaeger E.J. and Christensen K. (1999) Transient gene expression in mammalian cells grown in serum-free suspension culture. *Cytotechnology* **30**, 71-83.

Shen E.S., Cooke G.M. and Horlick R.A. (1995) Improved expression cloning using reporter genes and Epstein-Barr virus ori-containing vectors. *Gene* **156**, 235-239.

Schwachtgen J.L., Ferreira V., Meher D. and Kerbiriou N.D. (1994) Optimization of the transfection of human endothelial cells by electroporation. *Biotechniques* **17**, 882-887.

Wilson S.P. and Smith L.A. (1997) Addition of Glycerol during DNA Exposure Enhances Calcium Phosphate Transfection. *Anal. Biochem.* **246**, 148-150.

Wilson S.P., Liu F., Wilson R.E. and Housley P.R. (1995) Optimization of Calcium Phosphate Transfection for Bovine Chromaffin Cells: Relationship to Calcium Phosphate Precipitate Formation. *Anal. Bioch.* **226**, 212-220.

Wurm F.M. and Petropoulos C.J. (1994) Plasmid Integration, Amplification and Cytogenetics in CHO Cells: Questions and Comments. *Biologicals* **22**, 95-102.

Wurm F.M., Johnson A., Ryll T., Köhne C., Scherthan H., Glaab F., Lie Y.S., Petropoulos C.J. and Arathoon W.R. (1996) Gene Transfer and Amplification in CHO cells. *Annals of the New York Academy of Sciences.* **782**, 70-78.

Wurm F. and Bernard A. (1999) Large-scale transient expression in mammalian cells for recombinant protein production. *Curr. Opin. Biotechnol.* **10**, 156-159.

Xu H., Miller J. and Liang B.T. (1992) High-efficiency gene transfer into cardiac myocytes. *Nucleic Acids Res.* **20**, 6425-6426.

Yorifuji T., Tsuruta S. and Mikawa H. (1989) The effect of cell synchronization on the efficiency of stable gene transfer by electroporation. *FEBS Lett* **245**, 201-203.

Zlokarnik G., Negulescu P., Knapp T.M., Mere L., Burres N., Feng L., Whithney M., Roemer K. and Tsien R.Y. (1989) Quantification of Transcription and Clonal Selection of Single Living Cells with ß-Lactamase as Reporter. *Science* **279**, 84-88.

ADENOVIRUS VECTORS IN FUNCTIONAL GENOMICS

WAHIBA OUALIKENE[1] AND BERNARD MASSIE[1,2,3]
1: Institut de Recherche en Biotechnologie, Conseil National de Recherche Canada., 6100 Avenue Royalmount, Montréal, QC, Canada H4P 2R2.
Tel: (514) 496-6131- Fax: (514) 496-5143
2: INRS-IAF, Université du Québec, Laval, QC, Canada H7N 4Z3
3: Département de Microbiologie et Immunologie, Faculté de Médecine, Université de Montréal, Montréal, QC, Canada H3C 3J7.
email:Wahiba.Oualikene @nrc.ca, Bernard.Massie@nrc.ca.

1. Introduction

Among more than one hundred different serotypes of adenoviruses (Ads) known, 47 can cause usually mild respiratory or gastrointestinal infections in humans (reviewed in Horwitz, 1996). Although they show tropism only for epithelial cells during a regular course of infection, they are able to infect a wide variety of cell types. One of the unique features that has motivated the study of Ads is their ability to replicate at extremely high titers since they can saturate the infected cells with more than $3X10^4$ infectious units (IU)/cell, yielding abundant material for biochemical analysis. After their initial discovery, substantial information has accumulated on their biology and molecular characteristics over the following three decades. This wealth of information has promoted their use as expression vectors initially for recombinant protein production, then as recombinant vaccines, and more recently as gene transfer vectors in animal experiments as well as in human trials (reviewed in Acsadi et al, 1995; Berkner, 1988; Graham and Prevec, 1992; Imler, 1995; Kovesdi et al, 1997; Randrianarison-Jewtoukoff & Perricaudet, 1995; Wiley, Gene Therapy web site).

Ad-based vectors (AdVs) are presently considered to be among the most efficient viral vectors, due to their following properties, some of which are unique to this group of vectors: (i) AdVs do not require cell proliferation for transgene delivery and expression (i.e., are effective in non-dividing or terminally differentiated cells), so they are applicable over a wide range of host cells; (ii) AdVs do not integrate their DNA into the chromosomes of the cell, so their effect is transient and is unlikely to interfere with the cell's normal functions; (iii) AdVs show a transducing efficiency of almost 100% in a variety of animal cultured cells and in several organs of various species *in vivo*; (iv) AdVs usually possess an ability to replicate to high titers of 10^{10} IU/ml (up to 10^{12} UI/ml after concentration), a feature important for the preparation of vector stocks suitable for the achievement of efficient transduction *in vivo*; (v) AdVs can accommodate large inserts of exogenous DNA

M. Al-Rubeai (ed.), Cell Engineering, 80–154.

and their genome is stable through successive replication cycles; (vi) there are no known associations of human malignancies or other serious health problems with Ad infections and live replicative Ads have been safely used as a human vaccine. Given the numerous reviews on the applications of AdVs for vaccination and gene therapy, this chapter will focus more specifically on the various uses of recombinant AdVs as a tool to study gene function (functional genomics).

2. Adenovirus Vectors (AdVs)

2.1. MOLECULAR BIOLOGY OF AdS

Ads are non-enveloped linear double-stranded DNA viruses first isolated in 1953 (Rowe et al, 1953). The 47 different human Ad serotypes are classified into 6 subgroups (A-F) (Horwitz, 1996), of which subgroup C Ad2 and Ad5 are by far the most extensively characterized. Ads have been studied for a long time because, as mammalian DNA viruses, they have provided an attractive model for studying the molecular biology of the eukaryotic cell (DNA replication, transcriptional regulation and translation), and also because they express oncogenic proteins. Two main groups exist, the mastadenoviruses and aviadenoviruses. Most of them have a natural tropism for epithelial cells of various origins, such as respiratory and digestive tract.

Structurally, the Ad virion is composed of a capsid, itself composed of pentons (fiber and penton base) and hexons, which contains a 34-36 Kb genome. The capsid is an icosahedral structure (20 facets and 12 vertices) with a diameter of 60-90 nm. It is composed of 252 capsomers with 240 hexons (12 per facet) and 12 pentons. The penton bases are located at each vertex from which protrude the antenna-like fibers (Horwitz, 1990). Ad cell entry (Fig.1) is mediated by at least two cell surface molecules: integrins alpha v beta 3 and alpha v beta 5 (Mathias et al, 1994; Wickham et al, 1993; Li et al, 1998a) and the coxsackie and Ad receptor (CAR) (Bergelson et al, 1997). Initial binding of Ad to cells takes place via the high-affinity attachment of the globular head domain of the protruding fiber to the CAR. The subsequent entry into the cell is mediated by the Arg-Gly-Asp (RGD) sequence in penton base, which binds to the secondary integrin cell receptors. After entry in the host cell, Ads migrate in clathrin-coated vesicles (Dales and Chardonnet, 1973) and then in endosomes. The acidity of that compartment promotes the activation of the protease which disrupt the capsid. Disrupted virions then escape to the cytosol and localize to nuclear pores (Fig.1). The viral genome with only one polypeptide (terminal protein) attached to the terminal regions of the DNA (ITR) enters into the nucleus by an unknown mechanism (Greber et al, 1993; 1996).

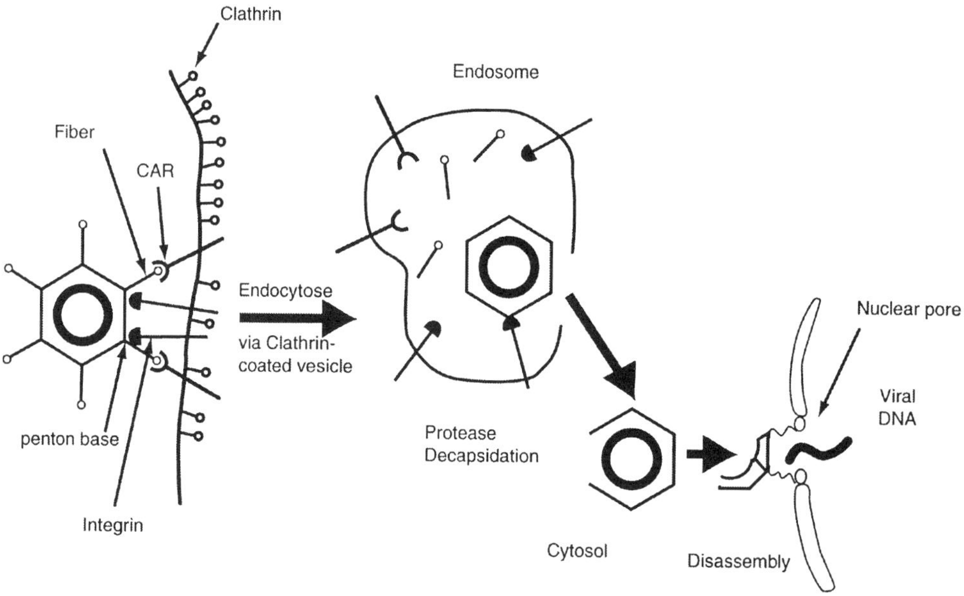

Figure 1: Adenovirus cell entry. After attachment on the cell surface to the primary (CAR) and secondary (integrins) receptors, the virus enters the cell by endocytose. The decapsidation begins inside the endosome and several structural peptides are cleaved by the Ad protease. The virion then migrates to the cytosol and to the nuclear pores through which viral DNA is translocated inside the nucleus.

Following the pattern of other DNA viruses, the replication and packaging of the viral DNA takes place in the nucleus of the infected cells (reviewed in Bridge and Petterson, 1995). The replication cycle has two distinct phases, early (E) and late (L), and is completed within 24 to 48 hours. The early transcription units, designated E1A&B, E2A&B, E3 and E4, are mainly active before the onset of DNA replication (Fig.2). Genes from E1, E2 and E4 encode essential regulatory proteins involved in the transactivation of viral and cellular promoters, DNA replication, cell cycle and apoptosis. E3 encodes genes involved in the interaction with the host such as escaping cell defenses by preventing T cell cytotoxicity and tumor necrosis factor (TNF) action. It is therefore dispensable for replication in cell culture. In the late phase of infection, the late genes are transcribed at very high levels providing the proteins required for packaging the viral genome. The late region spans more than 28 Kb starting with the major late promoter (MLP) at 16 map units. Although slightly active during the early phase, the MLP becomes very active after the onset of DNA replication, and drives the synthesis of a primary transcript which is processed to yield five sets (L1-5) of progenitor mRNAs with identical 5' terminus made of a 200 bp long sequence known as the tripartite leader (Shaw and Ziff, 1980). The tripartite leader sequence renders the mRNAs more efficiently translated than cellular transcripts and increases their export from the nucleus (Huang et al, 1998). The late mRNAs have variable 3' ends due to differential cleavage and polyadenylation. As for the early transcripts, the progenitor mRNA sets give rise to a wide variety of mRNAs by splicing. The late genes encode mainly structural and non-structural proteins involved in capsid assembly (reviewed in Shenk, 1996).

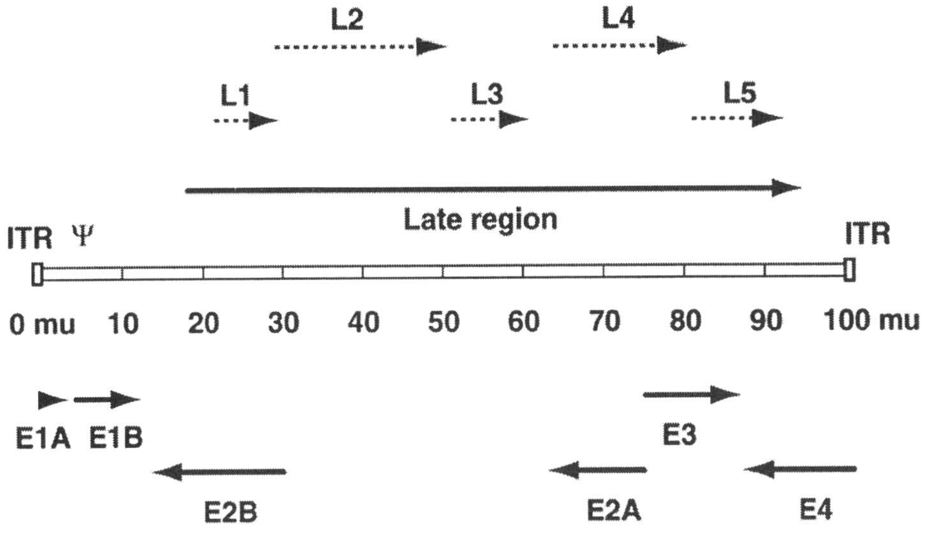

Figure 2: Ad transcription map: Simplified transcription map of the Ad genome with the major transcription units. The Ad transcription takes place in two successive periods: Early and Late. The six primary early transcripts are E1A, E1B, E2A, E2B, E3 and E4. One major late transcript driven by the major late promoter is subsequently spliced into L1-L5 (-----). At both ends of the genome are the inverted repeats (ITR) involved in DNA replication and at the 5' end the encapsidation signals (ψ).

In fully permissive cells, up to 10^5 virions/cell can be produced. However, only a small fraction of these particles are infectious; from 40% to as low as 0.1% depending on the serotype, the host cell and the culture conditions (Mittereder et al, 1996; Shenk, 1996; our own unpublished data). At the end of the lytic cycle, the infected cells die by a process of attrition and an E3-encoded protein, called adenoviral death protein (ADP), allows for efficient release of viral particles by cell lysis (Tollefson et al, 1996a, b).

2.2. DIFFERENT GENERATIONS OF AdVS: ADVANTAGES AND LIMITATIONS

The knowledge of the genomic organization and replication cycle of Ads has allowed their manipulation to create vectors which have been derived mainly from the Ad2 and Ad5 serotypes, because they are not associated with severe illness nor tumorigenic in rodents, by contrast to some other serotypes (Fig. 3). There are several efficient procedures for the construction of recombinant Ads (reviewed in Hitt et al, 1999). They involve either the *in vivo* recombination, in mammalian cells or *E. coli*, of a plasmid bearing a transgene expression cassette (transfer vector) with an Ad viral genome, or *in vitro* ligation and cloning in *E. coli* of a transgene expression cassette with a large plasmid containing the entire AdV genome, followed by transfection in permissive mammalian cells (Mizuguchi and Kay, 1999). Two types of AdVs have been developed, non-defective and defective. Replication-competent AdVs are ideally suited for applications in which high level transient expression of transgenes is both sufficient and desirable like in the case of vaccination, whereas defective AdVs are mainly used for protein production and functional studies. The encapsidation capacity of Ad being up to 105% of the genome, 2 Kb can be accommodated without deletion (Bett et al, 1993). Thus, combining the deletion of the dispensable E3 region (2.7 Kb) with the Ad packaging capacity allowed for the insertion of up to 4.7 Kb of foreign DNA into a replication-competent AdV. Insertion of larger transgenes requires the deletion of essential regions which renders AdVs replication-defective.

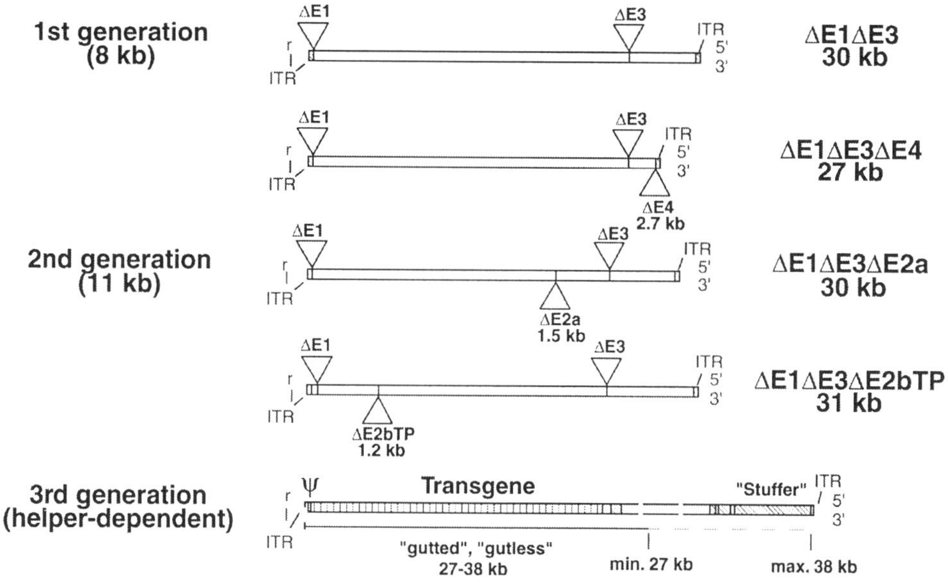

Figure 3: Different generations of AdVs. First-generation AdVs are E1- and E3-deleted, they are thus replication defective and can accommodate up to 8 Kb of foreign DNA. In second-generation vectors deleted for E4, 2.7 more Kb can be accommodated, while E2A deletion allows for 1.5 Kb cloning space, and E2B for 1.2 Kb. The third-generation vectors are gutless and helper-dependent. With these vectors, 27-38 Kb can be cloned because the whole genome but the ITRs and encapsidation signals is deleted.

The vast majority of gene transfer experiments have been done using E1-deleted replication-defective AdVs. These vectors are typically propagated without helper in the 293 cells which constitutively express the E1A and E1B genes (Graham et al, 1977). First generation AdVs, also deleted in the E3 region, can accommodate up to 8 Kb of foreign DNA. Any additional deletions in essential genes which can be complemented in stable cell lines have produced a variety of second generation AdVs with a maximum cloning capacity of 11 Kb. To generate AdVs with stable, unrearranged genomes, the minimal packaging size is about 75% of the genome's size or 27 Kb (Parks and Graham, 1997). Third generation AdV involves the replacement of almost the entire adenoviral DNA with the desired recombinant gene(s), which could then be as long as 38 Kb. Only the inverted terminal repeats with a short packaging sequences remain intact in this vector as minimal sequences for DNA replication and encapsidation (Fig.3). Since these 'gutless' vectors are unable to replicate by themselves, the presence of a helper virus (first generation AdV) is required. Recent advances using the Cre-lox helper-dependent system, has allowed to produce helper-dependent (HD) AdV reasonably free of helper virus (for an extensive review, see Hitt et al, 1999; Wivel et al, 1999).

The main advantage of the first generation AdVs is that they are easier to construct and produce than second and third generation vectors, yet they have a fairly high cloning capacity (8 Kb) which can accommodate most of the transgenes. One problem however for their production, is that replication competent Ad (RCA) will appear during multiple passages in 293 cells. RCA are found at very low level ($<1X10^{-8}$) in the early passage of AdV stocks (<5), presumably because of the very low frequency of the recombinational events which generated them. However, the population of RCA tends to increase faster during the following cycles of amplification required to produce the very large stocks of AdV needed for *in vivo* applications (Hehir et al, 1996; Lochmüller et al, 1994; Zhu et al, 1999). This situation is worsen when AdVs express toxic transgenes (Bruder et al, 2000). Even low level replication can jeopardize the efficacy of the gene transfer and the safety of the patients in clinical trials (Yang et al, 1994). The tolerable level of the RCA has been set at 1 per dose by the FDA, a fairly strict requirement. To solve that problem, new E1-complementing cell lines have been developed by several groups, including us (Fallaux et al, 1996; Gorziglia et al, 1996; Imler et al, 1996a; Massie, 1996). They are currently being evaluated in large-scale production.

A more serious problem is the toxicity of first generation AdVs that is mainly attributable to the leaky expression of the E2 and E4 regions which can ultimately result in low level production of infectious particles in some permissive cells (Brand et al, 1999; Dedieu et al, 1997). Furthermore, at high multiplicities of infection (MOIs), the leaky expression of some of the E4 ORFs can induce cell cycle dysregulation and apoptosis (Brand et al, 1999; Marcellus et al, 1998; Wersto et al, 1998), as well as turning on the expression of some key signaling genes (Ramalingam et al, 1999). The E2 region encodes three major proteins involved directly in DNA replication: a 72 kDa DNA binding protein (DBP), a 140 kDa DNA polymerase and the 80 kDa terminal protein (TP). Basal expression of the E2 promoter is very low in absence of E1A transactivation but can be increased by both the E4 ORF6/7 and the DBP (reviewed in Swaminathan and Thimmapaya, 1995). The E4 region is involved in DNA replication, late gene expression and splicing, as well as host cell shutoff (Halbert et al, 1985). Although the precise role of the E4 ORFs remains to be fully elucidated, the ORF4, 3 and 6 are involved indirectly in DNA replication by modulating the expression of the E2 genes (Medghalchi et al, 1997). Also, E4 ORF4 autoregulate E4 transcription by participating in the dephosphorylation of multiple cellular transcription factors involved in E1A transactivation of the E4 promoter (Bondesson et al, 1996) and is somehow involved in p53-independent induction of apoptosis in transformed cells (Lavoie et al, 1998; Marcellus et al, 1998; Shtrichman and Kleinberger, 1998). Curiously, a novel role for E4 ORF4 has recently been documented in the protection of Ad-infected cells from lysis by cytotoxic T lymphocytes (CTLs) by an unknown mechanism

(Kaplan et al, 1999). E4 ORF3 is a transcriptional activator of heterologous promoters such as the CMV immediate-early and the RSV LTR promoters, and is required for persistent transgene expression at high levels *in vivo* (Armentano et al, 1999; Gorziglia et al, 1999; Lusky et al., 1999; Yew et al, 1999). Interestingly, it has been demonstrated that the deletion of E4 but ORF6 allows the resulting AdVs to fully replicate in 293 cells (Armentano et al, 1995). Although the E4 ORF6 is potentially oncogenic (Moore et al, 1996), an E1-deleted AdV expressing only E4 ORF6 is much less toxic at high MOIs and is no longer interfering with normal cell cycle control (Wersto et al, 1998).

The main objective for the development of the second generation AdVs was to try to minimize basal viral gene expression in order to tentatively decrease immune response and thus to obtain a more sustained transgene expression. These vectors would also provide increased biosafety features and more cloning space. Some AdVs that do not need special complementing cell lines for propagation were designed, such as the Ad2 deleted for E4 except ORF6 mentioned above. This was not sufficient to prevent immune response after repeated administrations (Kaplan et al, 1996). Another strategy has been to construct a temperature-sensitive AdV mutated in E2A 72 kDa DBP and that express lower levels of late genes at 37 °C (Engelhardt et al, 1994a). With this vector, CTL infiltration was reduced and transgene expression increased, but remained however only transient (Engelhardt 1994a; 1994b). Other second generation vectors require specific cell lines to provide the missing proteins in *trans*. There are several examples such as: E4 (Armentano et al, 1995; Krougliak and Graham, 1995), E2A (Gorziglia et al, 1996; Zhou et al, 1996), E2B pTP (Schaack et al, 1996; Amalfitano et al, 1997) or DNA polymerase (Amalfitano et al, 1997; 1998). Although second generation AdV are in general beneficial in reducing toxicity and immunogenicity (Amalfitano et al, 1998; Christ et al, 2000; Gorziglia et al, 1999; Moorhead et al, 1999), controversial results from *in vivo* studies have been reported, and optimization in the design of these vectors is on going to ensure transgene expression as efficient as with first generation vectors.

One of the main advantages of the HD AdV is their large cargo space. Furthermore, removal of all of the viral genes has resulted in reduced toxicity and improved expression of transgenes *in vivo* (Chen et al, 1997b; Morsy et al, 1998; Schiedner et al., 1998), although the long term *in vivo* stability of these gutless AdVs is still controversial (Lieber et al., 1996). Moreover, due to packaging constraints, if the inserted transgene is smaller than 27 Kb, noncoding stuffer DNA is required, and this raises the question of what is the best sequence to be used in HD vectors (Parks et al, 1999). Also, although this system is very flexible in terms of the insert size, the presence of the helper virus could compromise the *in vivo* applications. Up to now, their production, which relies on a helper virus to provide all of the lacking gene products, has been much less efficient than for the helper-independent vectors while the purity of the HD vectors from helper contamination has been difficult to ensure. Another problem is that rearrangements of HD vectors can occur during their propagation due to homologous recombinations between helper and HD vectors. However, recent optimization in the design of both helper and vector backbone has lead to significant improvements in their production as well as in their *in vivo* potency (Sandig et al, 2000). Nevertheless, the long-term efficient transgene expression *in vivo* with these optimized HD vectors remains to be further assessed.

Recently, we have developed a new type AdV ideally suited for short-term high level expression of transgenes (Oualikene et al, 2000). This vector is completely replication-disabled in the absence of Ad E1 proteins but is capable of a single-round of replication when Ad E1 is present. This was made possible by deletion of the Ad protease gene (PS), which is essential for many steps of Ad life cycle. The PS-deleted virus can be propagated in 293-derived cell lines engineered to express PS (Fig.4).

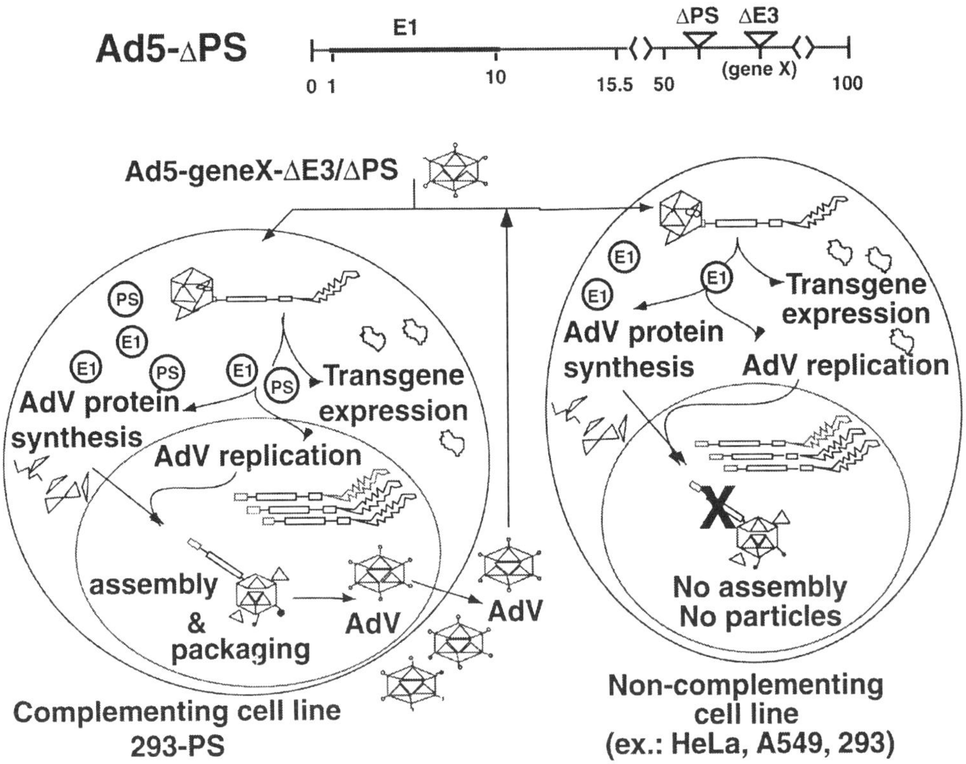

Figure 4: Single-round replication AdVs.Vectors deleted for a late gene not directly involved in DNA replication (like the protease) can fully replicate in complementing cell lines. In non-complementing cells, the DNA replicates, and the proteins encoded by the genome are expressed at levels similar to those of the wild-type. However, as some viral proteins are not completely processed, assembly and release of infectious virions can not occur.

With this vector, the efficiency versus safety dichotomy is defined using a different paradigm. By deleting an essential gene that is not involved in the early steps of viral replication, it is possible to obtain a localized replication with a total block thereafter, and thus obtain a greatly enhanced expression of the transgene without risk of viral shedding. Such a vector is suitable for applications in which enhancement of the transgene expression is required like in the case of vaccination, *in situ* therapy for tumors, protein production, or the large-scale production of other viral vectors such as adeno-associated virus (AAV). In another application, we have shown that using a PS-deleted genome the ectopic expression of the PS gene in the transfer vector allow for positive selection of recombinant AdV with a 100% efficiency (Fig. 5). This feature is an asset for the efficient construction of AdV libraries, which will have multiple applications in functional genomics.

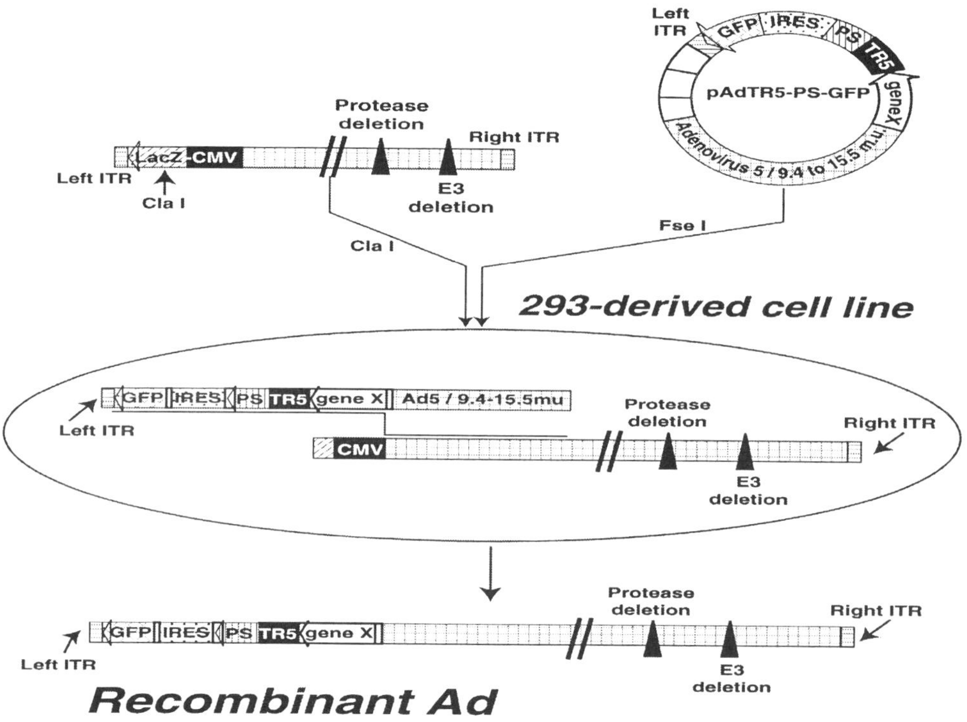

Figure 5: Schematic representation of the molecular cloning performed to generate recombinant adenoviral vectors by positive selection with Ad protease. The recombinant represented here featured a first generation AdV. A shuttle vector, containing Ad5 9.4 to 15.5 mu part of the genome to allow recombination, also harbored a double expression cassette containing, among others, the protease gene and a foreign gene of interest (X) in place of the E1 region. After linearization, the shuttle vector was cotransfected in a 293-derived cell line with a protease-deleted Ad genome cleaved in E1. Due to protease deletion, only genomes for which recombination, and thus the rescue of the protease gene, has occurred, produced viral plaques. The resulting recombinant viruses harbored no protease gene in L3 region, but the E1-cloned gene and the protease are ectopically expressed from the E1 region.

2.3. AdVS FOR *IN VIVO* GENE TRANSFER: CIRCUMVENTING THE IMMUNE RESPONSE

Efficient *in vivo* gene delivery is one of the hallmarks of AdVs. However, AdV-mediated gene delivery into various animals has been shown to induce immune reactions of varying magnitude depending on the dose, the route of administration, the target tissue, the transgene, and the animal (reviewed in Kovesdi et al, 1997; Trapnell and Gorzilia, 1994; Wivel et al, 1999). Although these immune reactions do not generally cause adverse clinical symptoms, they are partially responsible for the transient expression of the transgene. Three types of reactions have been observed. 1) A transient dose-dependent local inflammatory response, 2) a humoral antibody response compromising the repeated administration of AdV, and 3) a CTL response that is thought to be responsible for the elimination of transduced cells. Furthermore, it was shown that the innate immune mechanism (early) dominates the adaptive immune-related clearance of AdV (Worgall et al. 1997). In addition to the loss of transgene expression, administration of high doses of AdV has been associated with inflammatory responses which, depending on the sites, may alter organ function (Petrof et al, 1996; McClane et al, 1997). These inflammatory processes are in part due to a delayed-type hypersensitivity response (Russi et al, 1997).

Besides vector modification (see above), the short-term use of immunosuppressants is currently being explored as a practical solution to prevent the overall immune response. Thus, glucocorticoids and cyclosporin have been used to control acute inflammation (Fang et al, 1995; Adesanya, 1996; Russi et al, 1997). Also, very promising results were obtained with the immunosuppressive drug FK506 (Vilquin et al, 1995) to prolong AdV-mediated expression into mouse and dog muscle (Lochmuller et al, 1995; 1996; Mc Howell et al, 1998), rat liver (Ilan et al, 1997; Kuriyama et al, 2000), rat brain (Durham et al, 1997) as well as the mouse sciatic nerve (Jani et al, 1999). Alternatively, transient subversion of CD40 ligand function with antibodies to CD40L (Yang, et al. 1996a), anti-$CD4^+$ antibodies (Byrnes, 1996; DeMatteo et al, 1996; Kolls et al, 1996; Lei et al, 1996; Petrof et al, 1996; Poller et al, 1996), anti-IL-8 antibodies (Amin et al, 1995), or anti-T cell receptor antibodies (Sawchuk et al, 1996; Zsengellér et al, 1997) at the time of vector administration as well as macrophage depletion in the target organ prior to AdV administration (Kuzmin et al, 1997; Wolff et al, 1997), can substantially enhance both the efficiency of gene transfer and persistence of transgene expression. Finally, combination of transient immunomodulation with anti-CD40 antibodies, or FK506, with CTLA4 Ig was shown to facilitates primary and secondary gene transfer in mouse liver (Kay et al, 1997) as well as *mdx* mouse muscles by substantially reducing the inflammation (Guibinga et al, 1998).

To circumvent the humoral immunity precluding AdV readministration, coadministration of IL-12 which reduces the formation of neutralizing antibodies (Yang et al, 1995), or the use of AdV derived from different serotypes were recently shown to be viable solutions (Kass-Eisler et al, 1996; Mastrangeli et al, 1996; Mack et al, 1997). The efficacy of 15-deoxyspergualin was also demonstrated to decrease humoral response (Cichon et al, 1998). In another approach, Chen et al. (2000) have demonstrated that changing the route of administration from intranasal and intravenous to intramuscular, allowed repeated administration in spite of high titer neutralizing antibodies. Finally, O'Riordan et al (1999) showed that covalent attachment of PEG to the AdV surface protected the virus from neutralizing antibodies while retaining its infectivity.

3. Factors Influencing Transgene Expression in AdV

3.1. EFFICIENCY AND VERSATILITY OF AdVS

High-level transgene expression is another hallmark of AdVs. It requires, in addition to an efficient expression cassette, efficient cell entry, which is modulated by the presence of the appropriate receptors on the host cells, and high copy number genomes resulting from viral DNA replication or infection at high MOIs (Massie et al, 1998b). In a permissive infection, Ad hijacks the cell machinery to produce its capsid proteins at a level exceeding 30-40% of the total cell proteins (TCP). Therefore, it was soon realized that appropriate manipulations of the Ad genome could potentially result in the construction of Ad recombinants expressing foreign genes at very high levels. Thus, a large number of expression cassettes have been constructed with a wide combination of viral and/or cellular promoter-enhancer elements, splice sites, and polyadenylation signals.

One of the most interesting features of AdVs is their great versatility. On the one hand, expression cassettes in Ad transfer vectors can be used in three different scenarios: i) for high-level transient expression, preferably in 293 cells; ii) to establish stable cell lines when co-transfected with a selectable marker; and iii) to generate recombinant Ad. On the other hand, when an expression cassette is recombined into the Ad genome, the same AdV can be used either to produce large amount of protein or to express the gene product in a wide variety of mammalian cells both *in vitro* and *in vivo*. Finally, the green or blue fluorescent protein (GFP or BFP) have been cloned in Ad transfer vectors for co-expression in a bicistronic configuration or with separate expression cassettes. Co-expression of GFP or BFP in the vectors accelerates the sorting of Ad recombinants and facilitates the functional characterization of genes both *in vitro* and *in vivo* by allowing easy quantification of gene transfer and expression (Massie et al, 1998b).

3.1.1. HIGH-LEVEL CONSTITUTIVE EXPRESSION CASSETTES

With an optimized transcription unit, very high levels of recombinant protein expression can be reached in AdV-infected cells (Garnier et al, 1994; Massie et al, 1995; Massie et al, 1998a; b). It was demonstrated in our laboratory that this system could rival or surpass the baculovirus system for the production of functional proteins (Massie et al, 1995). Indeed, amounts reaching 90 mg of protein per liter of culture volume have been obtained, corresponding to 10-20% TCP. Moreover, the over-expression of recombinant proteins can take place in the absence of the production of the abundant late viral proteins and constitute as much as 35% TCP (Massie et al., 1998b). This latter characteristic is clearly an asset to facilitate the subsequent purification of recombinant proteins (Lamarche et al, 1996; Langelier et al, 1998).

High-level expression cassettes are also valuable tools to reach desired protein expression thresholds at lower MOIs, thereby minimizing the deleterious effects AdVs can have on cells at higher MOIs. For example, the optimized CAG promoter (a modified chicken beta-actin promoter with CMV-IE enhancer) has been successfully used to correct OCT deficiency *in vivo* with minimal toxicity in the liver due to its improved expression at lower MOIs (Kiwaki et al, 1996).

3.1.2. INDUCIBLE EXPRESSION CASSETTES.

Inducible gene expression is required to express proteins that have deleterious effects on the cell or on Ad production. Interestingly, inducible promoters can be as efficient as constitutive promoters can. Thus, with an optimized tetracycline-regulatable expression cassette, it was possible to construct AdV able to express proteins that are either cytotoxic or which interfere with Ad replication at levels of 10-15% TCP (Massie et al, 1998a; b). Inducible expression cassettes can be desirable as well for some particular applications in which regulating transgene expression *in vivo* improve its safety and/or efficacy. For example, using a chimerical promoter that contains glucocorticoid response elements, AdV-mediated delivery of thrombopoietin cDNA to the liver has allowed intermittent, repetitive up-regulation of platelet levels (Narumi et al, 1998). These results showed that after administration of a transgene by an AdV, it is possible to tightly regulate the pharmacological levels of a protein by injecting adequate doses of a drug such as dexamethasone.

Since the demonstration of the suitability of the bacterial tetracycline-regulated system for inducible expression in mammalian cells (reviewed in Gossen et al, 1994), several systems based on bacterial regulatory elements have been successfully used in AdVs. Thus, the *in vivo* efficacy of the tet system for tight regulation of human tyrosine hydroxylase following AdV gene transfer in the brain of a rat model of Parkinson's disease was demonstrated (Corti et al, 1999). Although satisfactory for many applications, initial studies have shown that the original tet system is not as tightly regulated as it could be (Massie et al, 1998a; our own unpublished results), and optimization work is in progress to further improve its performance. One such improvement was done by Chen et al. (1998a) who demonstrated that coupling the tandem repeats form the enhancer region of the glial fibrillary acidic protein gene to a minimal CMV promoter to drive the tetracycline transactivator allowed glial-specific tightly repressible expression of transgenes. Specific expression can also be obtained by other strategies. Tomanin et al. (1997) developed a system for expression of toxic genes using the bacteriophage T7 binary system. Experiments performed to evaluate the leakiness of the Ad-T7 system detected very little expression from the T7 gene 10 promoter in the absence of T7 polymerase suggesting that this system may be useful for the cloning and expression of genes encoding cytotoxic proteins. Alternatively, with the cre-loxP system from bacteriophage P1, it is possible to disrupt or activate genes *in vivo* in a controlled manner using AdVs (Anton et al, 1995; Sakai et al, 1995; Wang et al, 1996) and thus to control gene expression. This system has been useful for numerous functional studies in transgenic mice. For example, a mutant allele flanked by loxP sites in a transgenic mouse can be disrupted by infusing a Cre-expressing AdV (Kobayashi et al, 1997; Lee et al, 1997; Liu et al, 1998a; Shibata et al, 1997; Stec et al, 1999). Another strategy with this system has been the use of dual AdVs of which one expresses the Cre recombinase, and the other carries an inactivated expression cassette that needs Cre recombination to activate gene expression. Because both AdVs are required in the same cell, this technique has allowed tumor-specific expression by co-injecting them *in situ* (Sato et al, 1998; Kijima et al, 1999).

The growing needs for either more specific, more efficient or less immunogenic inducible cassettes has led to the characterization of several new systems. Burcin et al. (1999) recently presented a new system allowing inducible tissue-specific expression. A chimerical transactivator, GLp65 was constructed, consisting of a mutated progesterone receptor-ligand binding domain fused to the yeast GAL4 DNA binding domain with part of the activation domain of the human p65 protein (component of the NF-kappa B complex). By using a liver specific promoter, the authors were able to demonstrate the specificity and the inducibility of transgene expression in the liver upon exposure to the drug mifepristone. Another similar system, containing only human proteins in the regulatory components, and thus potentially less immunogenic, has recently shown promising results *in vivo*. The rapamycin-controlled expression cassette proved efficient for long-term tightly regulated expression of a transgene in the AAV context, although its use within an AdV led only to short-term expression due to immune response to AdV (Rivera et al, 1999). The ecdysone system (No et al, 1996) was also successful in the AdV context (Johns et al, 1999; Holt et al, 1999). Although the usefulness of these various inducible systems has been established, their relative performance for specific applications has yet to be assessed in comparative studies.

3.2. AdV CELL ENTRY TARGETING

While AdVs have extensively been used in a wide variety of cells and tissues, the need for targeted expression has grown over the years for several reasons. First, not all cells are equally susceptible to AdVs (Fechner et al, 1999). Second, the wide expression of a gene which is normally expressed in only one cell type can lead to unpredictable side effects in therapeutic applications. Third, the programmed deleterious effect of a protein in a particular cell type (tumor, tumor blood vessels) may have to be avoided in others (normal tissue). Finally, in most tissues, the efficiency of gene delivery increases with increasing doses of AdV particles and/or exposure times, however, those optimal conditions are sometimes difficult to reproduce *in vivo*.

To overcome these drawbacks, specific targeting is an efficient and sometimes mandatory way of expression. The targeting of AdVs can be modulated at the level of cell attachment and entry, or by transcriptional targeting with specific and/or inducible promoters. As a virus, AdVs use cell receptors for entry and modulating this interaction opens the possibility to extend or restrict its cell tropism. Two main approaches have been used to modulate AdV tropism. The first one consists of direct modification of the capsid proteins while the second one involves chemical coupling of the purified viral particles with antibodies or polymers such as biotin or polyethylene glycol (PEG) linked to ligands with binding specificities towards chosen receptors (Table 1).

3.2.1. EXTENDED TROPISM

The relative abundance of Ad receptors on cells is a major determinant of the transduction efficiency. In that respect, CAR appears to be the most important receptor since its absence on some tissues such as mature muscle fibers (Nalbantoglu et al, 1999) or alveolar macrophages (Kaner et al, 1999) is responsible for the very poor AdV-mediated gene transfer in those tissues. Moreover, it was shown that CAR is associated with efficient AdV transduction in several melanoma cell lines (Hemmi et al, 1998) and human bladder cancer cells (Li et al, 1999a). Similar results were obtained as well with integrin (alpha-v beta3 and alpha-v beta5) expression levels (Hashimoto et al, 1997; Takayama et al, 1998). However, recent findings showed both receptors to be necessary but not sufficient *in vivo* for AdV-mediated transduction, and that other factors play important roles (Fechner et al, 1999). Nevertheless, the knowledge of the primary and secondary cell receptors of Ads paved the way for new targeting strategies.

Shortly after the discovery of the integrins (alphav beta3 and alphav beta5) as secondary receptors (Mathias et al, 1994; Wickham et al, 1993), the first retargeting experiment involving modifications of the penton was reported (Wickham et al, 1995). Then, several preliminary studies subsequently showed that the modification of Ad fiber could be successfully done and lead to improved gene delivery, via alternative receptors, of cells otherwise poorly susceptible to Ad infection (Krasnykh et al, 1996; Michael et al., 1995; Wickham et al, 1997b). Thus, it was demonstrated that the inclusion of an RGD-containing peptide in the HI loop of the fiber or a polylysine moiety at its carboxy-teminal end, significantly improves the transduction of numerous tumor cells such as ovarian cancer cells (Dmitriev et al, 1998; Vanderkwaak et al, 1999), gliomas (Yoshida et al, 1998), and head and neck cancer cells (Kasono et al, 1999). Similar results have been obtained in a variety of normal cells and tissues like macrophages, endothelial smooth muscles, fibroblasts, T cells (Wickham et al, 1997a), the colon epithelium (Wirtz et al, 1999), and striated muscles (Bouri et al, 1999). Consistent with those observations, Reynolds et al (1999) showed that systemic administration of such vectors resulted in an altered profile of transgene expression *in vivo*. Retargeting of Ad37 to human conjunctiva cells was also achieved by a single amino acid mutation in the fiber (Huang et al, 1999a).

After the discovery of CAR as the primary receptor, subsequent studies showed that it is involved in the entry of A, C, D, E and F Ad serotypes (Roelvink et al, 1998). Although the receptor for group B Ads is still unknown, exchanging the group C Ad5 fiber for group B Ad3 turned out to be an interesting approach to retarget AdV to human monocytes, squamous carcinoma, embryonic lung fibroblast (Stevenson et al, 1997), and human B cells (Von Seggern et al, 2000). Curiously, even though group D Ad17 also uses CAR, its binding strategy must be somewhat different than the one of Ad2 or 5 since fusing its fiber knob to the Ad2 fiber tail yielded an AdV with an altered transduction profile displaying improved gene delivery to primary neurons (Chillon et al, 1999) and human airway epithelia (Zabner et al, 1999). Further works to determine the exact binding domain of fiber to CAR are under way (Kirby et al, 1999; 2000; Freimuth et al, 1999; Bewley et al, 1999; Roelvink et al, 1999; Tomko et al, 2000), and this will eventually allow for more accurate and specific targeting. It is worth mentioning that the fiber plays a role also in intracellular trafficking of virions (Miyazawa et al, 1999). This feature will have to be taken into account for further retargeting strategies involving fiber modifications. Recently, retargeting of AdV has been demonstrated by addition of a ligand to the hexon, opening a completely new avenue for altering Ad tropism (Vigne et al. 1999).

3.2.2. RESTRICTED TROPISM

Another strategy for targeted entry of AdV into some cell types makes use of bi-specific antibodies. By coupling, a monoclonal antibody with high affinity to the fiber (Douglas et al, 1996) or the penton (Wickham et al, 1996), to a Fab with affinity to a different receptor, not only is the AdV retargeted to a new receptor but also, its endogenous CAR or integrins binding can be ablated, and thereby its tropism restricted. This was successfully done for tumor cells using the folate receptor (Douglas et al, 1996) or the fibroblast growth factor (FGF2) receptor (Goldman et al, 1997; Gu et al, 1999; Printz et al, 2000; Rogers et al, 1997). The inclusion in the fiber HI loop of a specific motif to which an antibody can bind also allowed specific targeting without ablation of CAR binding (Krasnykh et al, 1998; Wickham et al, 1997b).

TABLE 1- AdV- tropism modifications

Strategy	Target	References
Natural tropism	CAR/Integrin expressing cells	Bergelson et al, 1997
Penton modification	Monocytes, lymphocyes, cancer cells.	Wickham et al, 1995
Fiber modification	Multiple cell types (macrophage, endothelial, smooth muscle, fibroblaste and T cells)	Wickham et al, 1997b
	Human monocytes, Squamous carcinoma, Embryonic lung fibroblast	Stevenson et al,1997
	Tumor cells	Dmitriev et al, 1998; Yoshida, et al, 1998; Vanderkwaak et al, 1999; Kasono et al, 1999
	Primary neurons	Chillon et al, 1999
	Human airway epithelia	Zabner et al, 1999
	Conjunctival cells	Huang et al, 1999a
	Muscle	Bouri et al, 1999
	Inflamed colon	Wirtz et al, 1999
	B cells	Von Seggern et al, 2000
Hexon modification	Human vascular smooth muscle	Vigne et al, 1999
Bi-specific antibodies	Cancer cells- folate receptor	Douglas et al, 1996
	Various, Cancer cells- FGF2 receptor	Rogers et al, 1997; Goldman et al, 1997; Gu et al, 1999, Printz et al, 2000
	T cells via CD3	Wickham et al, 1997a
	Gliomas- EGF	Miller et al, 1998
	Adenocarcinomas- Ep-CAM	Haisma et al, 1999
	Dendritic cells- CD40	Tillman et al, 1999
Dual-binding oligopeptides	CAR-lacking cells	Michael et al, 1995
Biotinylation	Hematopoeitic cells	Smith et al, 1999a
PEG-conjugated peptides	Several cell lines	Romanczuk et al, 1999
Pseudo-receptors	Generic	Douglas et al, 1999; Einfeld et al, 1999

As listed in Table 1, several other ligands were used in bi-specific antibodies to retarget AdV to a variety of new receptors (Wickham et al, 1997a; Miller et al, 1998; Haisma et al, 1999; Tillman et al, 1999). Another similar strategy making use of dual-binding oligopeptides linked to the carboxyl terminus of the fiber was also reported (Michael et al, 1995). Recently, new conjugation strategies have been experimented such as the biotinylation (Smith et al, 1999a) or the PEGylation (Romanczuk et al, 1999) of AdVs, and they both proved to be efficient to redirect AdVs specifically to new receptors. An indirect additional advantage of the PEGylation method was that PEG protect AdVs against neutralizing antibodies and helps to overcome some of the immune response (O'Riordan et al, 1999), thereby improving the overall efficiency of gene transfer (Romanczuk et al, 1999).

The ideal retargeted AdV would have a unique binding specificity without the endogenous binding to both CAR and integrins. Growing such a vector requires a complementing cell line that would provide a generic receptor independent of the ligand incorporated in AdV capsid for retargeting purposes. As a first step towards that goal, new pseudoreceptors were engineered in 293 cells as targets for new ligands introduced in the fiber HI loop (Douglas et al, 1999; Einfeld et al, 1999). They will allow the complete ablation of the endogenous binding by engineering the appropriate mutations both in the penton base (Wickham et al, 1995) and the fiber knob (Bewley et al, 1999; Kirby et al, 1999; Roelvink et al, 1999), and thereby the generation of fully retargeted AdVs by capsid protein modifications.

3.3. SPECIFIC TRANSGENE EXPRESSION BY TRANSCRIPTIONAL TARGETING

Transcription targeting can ensure specific and often more efficient protein expression in particular cell types. For cancer therapy especially, there are several strategies for which transcription specificity is highly desirable such as the expression of the so-called suicide genes.

3.3.1. TUMOR-SPECIFIC PROMOTERS

Different strategies can be used for cancer therapy: expression in tumor cells (toxic genes, tumor suppressor genes), inhibition of tumor vascularization, abrogation of cancer cell proliferation by modification of the susceptibility to hormones, or immune up-regulation. Cell-specific expression is a powerful approach and it has been tested in a wide variety of tissues for cancer gene therapy. For this purpose, it is necessary to either find a tissue-specific promoter which can be used to transduce tumors of that particular type or to utilize a promoter more active in tumor than normal cells.

There are several examples of tissue- or tumor-specific promoters that have been tested for cancer cell expression. For example, most patients with hepatocellular carcinoma have an elevated alpha-feto-protein (AFP) level. This high level of AFP expression is transcriptionally controlled by the 5'-flanking sequence of the AFP gene. AdVs carrying genes driven by the alpha-feto-protein promoter (Kaneko et al, 1995; Kanai et al, 1996;

Arbuthnot et al, 1996) have allowed significant tumor regression by expression of the herpes simplex virus thymidine kinase (HSV-TK) gene and subsequent gancyclovir administration. Similarly, an increase of carcinoembryonic antigen (CEA) expression is noted in about 40% of patients with gastric cancer and the use of a carcinoembryonic antigen promoter also proved efficient to target gastric cancer cells (Lan et al, 1996; Tanaka et al, 1996, 1997a) and even liver cells (Brand et al, 1998). The osteocalcin promoter also allowed specific targeting for toxic gene expression (HSV-TK/ acyclovir) in the murine osteosarcoma model (Ko et al, 1996; Cheon et al, 1997). A prostate-specific antigen promoter-based strategy was developed as well for both androgen-dependent and androgen-independent prostate specific antigen (PSA)-secreting prostate cancer cells using the HSV-TK gene (Gotoh et al, 1998). Finally, pituitary hormone promoters (human GH or glycoprotein hormone alpha-subunit) were efficient to specifically transduce pituitary tumors with an AdV expressing an HSV-TK gene (Lee et al, 1999a).

In the aim of testing the ability of transcriptional regulatory elements of the human and mouse tyrosinase genes to promote high levels of pigment cell-specific transcription and to target effector genes to cancer cells throughout the body, including distant metastatic sites, melanoma specific expression was achieved by use of the human tyrosinase promoter in a murine model (Siders et al, 1996). Gene therapy using Fas ligand (FasL) for treatment of tumors is hampered because of the systemic toxicity of FasL. Neural cells were targeted with specific promoters to express FasL under the control of either the neuronal-specific enolase promoter or the astrocyte-specific glial fibrillary acidic protein promoter. In that work, abolishment of liver toxicity by tissue-restricted expression was also demonstrated (Morelli et al, 1999).

Many tumors, such as malignant gliomas, have disrupted retinoblastoma protein (pRb) function, either because of RB-1 gene mutations or as a result of mutations affecting upstream regulators of pRb such as cyclin D1 or p16/INK4a. Tumor suppression by pRb has been linked to its ability to repress E2F-responsive promoters such as the E2F-1 promoter. Parr et al (1997) demonstrated that AdVs containing transgenes driven by the E2F-1 promoter can mediate tumor-selective gene expression *in vivo*, allowing for eradication of established gliomas with significantly less normal tissue toxicity than seen with standard constitutive promoters. A similar approach made use of the hexokinase type II promoter to drive preferential expression of a toxic gene in a variety of cancer cell lines (Katabi et al, 1999).

Another strategy is to engineer a tumor-specific localized replication of AdV in order to enhance transgene expression while taking advantage of the cythopatic effect induced by the virus. Since the alphafetoprotein promoter is highly expressed in hepatocarcinoma cells (see above), tumor-specific replication of AdV could be obtained by driving E1 expression from this promoter (Hallenbeck et al, 1999). Similarly, the PSA promoter was used for selective replication of an AdV in prostate tumor cells (Rodriguez et al, 1997) as was the human glandular kallikrein (hK2) promoter, which contains an androgen response element to target expression in presence of testosterone in an attenuated replication competent AdV (Calydon virus 764), which apparently replicates similarly to wild-type Ad in prostate tumor cells but is attenuated in nonprostate tumor cells (Yu et al, 1999). A variation on that theme was to exploit the fact that many tumor cells are more responsive to IL-6 than normal cells, and to obtain preferential replication of an E1A-deleted AdV with an IL-6-inducible E1A substituting activity (Rancourt et al, 1999).

3.3.2. TISSUE-SPECIFIC PROMOTERS

A wide variety of gene transfer studies have been done with strong ubiquitous promoters such as the CMV immediate early promoter. However, many of these studies showed that specificity would be an asset. For example, since the immune response against the vector is due in part to a response against the transgene product, restricting its expression to the target tissues should reduce it (Pastore et al, 1999). Therefore several authors have explored the use of tissue-specific promoters to obtain a targeted transcription of genes with AdVs (Table 2).

Given the complexity of the Ad genome transcription regulation, insertion of expression cassettes with tissue-specific promoters in AdVs did not always result in tissue-specific expression of the transgene (Shi et al, 1997; Ring et al, 1996). However, the loss of tissue-specific expression observed in some AdV constructs could be avoided by the use of " insulator " sequences flanking the expression cassette (Vassaux et al, 1999) leading to successful results in many cases.

Tissue-specific targeted gene transcription was mainly attempted for transduction in the lung epithelia, liver, pancreas, muscles, neural cells, mammary glands for applications such as cystic fibrosis or hemophilia A gene therapy. Partial specific transduction was achieved with the cystic fibrosis transmembrane regulator (CFTR) promoter since only part of the natural CFTR gene regulation was restored in the AdV context (Imler et al, 1996b). Better results were obtained for hemophilia A with a liver-specific promoter to reach levels of Factor VIII expression close to therapeutic requirements (Connelly et al, 1996). Specific inducible and restricted expression in pancreas was obtained with the amylase promoter in neonatal and adult mice (Dematteo et al, 1997). Specific expression in muscle cells has been tested for applications in the treatment of Duchenne muscular dystrophy, in cardiomyocytes and in striated muscles. Two promoters have shown their ability for muscle-specific expression, namely the truncated muscle creatine kinase and the fast troponin I promoters (Larochelle et al, 1997). Several successful attempts have also been made to target cardiac cells. The ventricle-specific myosin light chain 2 and ventricle-specific alpha-myosin heavy chain promoters proved both efficient and specific (Rothmann et al, 1996; Franz et al, 1997). Griscelli et al (1998) studied two cardiac promoters and demonstrated by AdV-mediated gene transfer that the ventricle-specific cardiac myosin light chain-2v promoter was specific and efficient for expression in the heart. A similar strategy used the cardiac troponin T promoter (Inesi et al, 1998).

As mentioned above for cancer cells, in many disorders some proteins are highly overexpressed, and their specific promoters can be used to drive restricted expression in the corresponding organ. For example, the pancreatitis-associated protein I (PAP I) is a pancreatic secretory protein expressed in pancreas during acute pancreatitis but not in the healthy pancreas. The PAP I promoter therefore allowed specific expression in the diseased organ (Dusetti et al, 1997). Other authors have studied endothelium-specific promoters that could be useful to modulate vascularization. E-selectin expression is very low in normal adult blood vessels, but is significantly elevated in newly formed tumor capillaries. An AdV which express its transgene only in cells that transcribe the E-selectin gene was designed and it was demonstrated that the E-selectin promoter could be used to confer endothelium specificity to AdV-mediated transgene expression (Walton et al, 1998). The human transmembrane fms-like receptor tyrosine kinase (flt-1) promoter was also used for specific endothelial cell expression with a reporter gene in AdV (Morishita et al, 1995).

TABLE 2- Tissue-specific and tumor-specific promoters used with AdVs.

Promoter	Target cells/organ	References
Alpha feto protein	Hepatocellular carcinoma	Arbuthnot et al, 1996
Prostate-specific antigen	Prostate tumor	Gotoh et al, 1998; Steiner et al, 1999; Latham et al 2000
Human glandular kallikrein	Prostate	Latham et al 2000
hALA, oBLG	Breast tumor	Anderson et al, 1999
Hexokinase II	Tumor cells	Katabi et al, 1999
IL-6 modulated replication	Tumor cells	Rancourt et al, 1999
Astrocyte-specific glial fibrillary acidic protein	Glial cancer cells	Chen et al 1998a; Morelli et al, 1999; Ambar et al, 1999
Human tyrosinase	Melanoma	Siders et al, 1996
E2F1	Tumor cells	Parr et al, 1997
CFTR	Lung	Imler et al, 1996b
Human surfactant protein	Lung	Harrod et al, 1998
HBV core	Liver	Sandig et al, 1996
Albumin	Liver	Connelly et al, 1996; Pastore et al, 1999
Hepatocyte-specific	Liver	Guo et al, 1996
Pancreatitis associated I	Pancreas	Dusetti et al, 1997
Amylase	Pancreas	Dematteo et al, 1997
Creatine kinase, fast troponin	Muscle	Larochelle et al, 1997
Ventricle-specific myosin light chain 2	Cardiac muscle	Rothman et al, 1996; Franz et al, 1997; Griscelli et al, 1997; 1998
Ventricle-specific alpha-myosin heavy chain	Cardiac muscle	Franz et al, 1997
Cardiac troponin T	Cardiac muscle	Inesi et al, 1998
HSP70B	Chondrocytes	Arai et al, 1999
L7/PCP2 gene	Purkinje cells	Hashimoto et al, 1996
Myelin basic protein	Oligodendrocytes	Hashimoto et al, 1996
Neuron-specific enolase	Neuronal cells	Morelli et al, 1999; Navarro et al, 1999
Neurofilament heavy gene	Brain	Schwartz et al, 1994
Neuron-restrictive silencer elements	Neurons	Miyaguchi et al, 1999; Millecamps et al, 1999
Probasin	Prostate	Steiner et al, 1999
FMS-like receptor	Endothelium	Morishita et al, 1995
E selectin	Endothelium	Walton et al, 1998

Several functional studies have been performed with reporter's genes targeting the nervous system. Neural cells of various types were transduced with specific promoters such as those of the L7/PCP2 gene (highly restricted expression in cerebella Purkinje cells) and of the myelin basic protein gene (in oligodendrocytes) (Hashimoto et al, 1996). As well, targeted AdV-expression to neurons by a vector that expressed a reporter gene under the control of the rat neuron-specific enolase promoter was obtained (Navarro et al, 1999). As an alternative approach, transcription targeting was recently achieved by cloning neuron-restrictive silencer elements, which selectively represses the transcription of genes in non-neuronal cells. These elements cloned upstream the ubiquitous phosphoglycerate kinase promoter allowed the targeting of AdV expression specifically in neurons (Miyaguchi et al, 1999; Millecamps et al, 1999).

4. AdV Applications in Functional Genomics

Accounting for about 17% of the on-going human clinical trials, AdVs are among the most popular gene delivery systems in the field of gene therapy. However, one emerging application is their use as a powerful tool to assess the functional significance of newly discovered genes. Functional genomics with AdVs is done by either up- or down-regulating gene expression both *in vitro* and *in vivo* and assessing the phenotype with an appropriate assay. Gene knock-down by AdVs has been done using a number of strategies such as expression of antisense, ribozymes, dominant negative mutants or single chain intrabodies (targeted to intracellular proteins). The ability to readily generate AdVs and assess gene function *in vivo* in a manner analogous to transient transfection *in vitro*, is one of the main assets that makes them such a powerful tool in functional genomics. By analogy, gene knock-out in transgenic mice would be equivalent to the establishment of stable cell lines, a process much more time consuming. Also, gene knock-out cannot be done on genes essential in embryo development. Both limitations can be overcome by the used of AdVs.

The ability of AdVs to transfer genes in many cell types, including primary cells, allows for a more accurate assessment of gene function since the activity of the genes can be monitored in a variety of tissues. Hence, due to their great versatility, AdVs are excellent tools for target validation. We will briefly review here the wide variety of functional studies that have been performed with AdVs involving cytokines (Tables 3, 4), transcription regulation (Table 5), signaling pathways (Table 6), glucose metabolic pathways (Table 7), as well as cell cycle and apoptosis regulation (Table 8). It is worth emphasizing that the vast majority of these functional studies have been done with standard first generation AdVs in spite of their intrinsic limitations described above (sections 2.2 and 2.3).

4.1. CYTOKINES

Cytokines play major roles in cell maintenance, proliferation or differentiation. Table 3 presents a list of functional studies done with AdVs expressing interleukins and interferons while examples of studies with other cytokines are shown in Table 4. From these tables it is evident that a wide variety of major cytokines and interferons such as IL2, IL10, IL-12, and GM-CSF have been characterized in many applications involving the modulation of the immune response. The role of TGF-beta in tissue repair has also been studied extensively. Other specific applications are of special interest.

Neuronal death and survival has recently become a field of intensive research with AdVs (for a review, see Smith and Romero, 1999b). The expression of neurotrophic factors is a strategy to preclude or at least to delay neuronal death. *In vivo* suitability of the AdV could be established to express several neurotrophic factor genes and to show their efficacy in protecting neurons from cell death. For instance, amyotrophic lateral sclerosis (ALS) is a fatal neurodegenerative disease characterized by the progressive loss of motoneurons. The AdV-mediated expression of the brain-derived neurotrophic factor (BDNF) and the glial cell line-derived neurotrophic factor (GDNF) was shown to be beneficial *in vivo* in newborn rats (Gimenezyribotta et al, 1997). Similar results were obtained with the same neurotrophic factors and with the ciliary neurotrophic factor (Baumgartner and Shine, 1997; 1998a; 1998b).

To increase neuronal survival in the retina, injection of a BDNF-expressing AdV into the vitreous chamber of adult rats was attempted and resulted in selective transgene expression in Muller cells to temporarily rescue injured retinal ganglion cells (Di Polo et al, 1998). The successful AdV-mediated exogenous GDNF gene transfer also ameliorated ischemic brain injury after transient middle cerebral artery occlusion in association with the reduction of apoptotic signals (Kitagawa et al, 1999). As well, as a new strategy for the treatment of Huntington's disease, which is a genetic disorder leading to the degeneration of striatal GABA-ergic output neurons, BDNF was AdV-transferred. This resulted in a significant enhancement of neuronal survival (Bemelmans et al, 1999).

Substantial research is also underway addressing vascular diseases, mainly by expression of factors able either to promote or to inhibit vascularization. Two studies demonstrated the *in vivo* efficacy of AdV-mediated transduction of the vascular endothelial growth factor (VEGF) for induction of neovascularization (Magovern et al, 1997). These findings paved the way for ischemic vascular disease treatments (Rosengart et al, 1999a; 1999b). Similarly, a modified soluble secreted form of FGF-2 proved able to stimulate angiogenesis (Ueno et al, 1997). This was attempted because induction of angiogenic growth factor within either the narrowing arteries or jeopardized myocardium may promote angiogenesis *in vivo*, leading to the salvage of ischemic myocardium in coronary atherosclerosis. At the opposite, vascular proliferative disorders are characterized by the proliferation of vascular smooth muscle cells. The study of such disorders led Nakaoka et al (1997) to show by AdV-mediated expression that the bone morphogenic protein 2 was efficient to inhibit smooth muscle cell proliferation, a feature that could be beneficial for the prevention of vascular proliferative diseases. Along those lines, an AdV expressing a secreted version of the amino-terminal fragment (ATF) of murine urokinase (uPA) was used in murine models to assess the antitumoral effects associated with local or systemic delivery of ATF, a broad cell invasion inhibitor that antagonizes uPA binding to its cell surface receptor. Neovascularization within and near the injection site of AdV was suppressed, suggesting that AdV-expressed ATF inhibited primary tumor growth by suppressing angiogenesis (Li et al, 1998b).

TABLE 3- Cytokines 1: Interleukins and interferons studies with AdVs.

Application	Gene expressed	Target cells/organ	References
Cancer	IL-1	Non small cell lung	Esandi, 1998a
Cancer	IL-2	Various	Haddada et al, 1994; Cordier et al, 1995; Addison et al, 1995; Toloza et al, 1996; Huang et al, 1996; Heike et al, 1997; O'Malley et al, 1997; Stewart et al, 1997; 1999; Bui et al, 1998; Felzmann et al, 1997; Kircheis et al, 1998; Mulders et al, 1998; Bowman et al, 1998; Shiratori et al, 1998; Emtage et al, 1998; Ju et al, 1998; Addison et al, 1998; Gilly et al, 1999; Hirschowitz et al, 1999a; Motoi et al, 2000
Cancer, Studies of IL	IL-3	Non small cell lung	Esandi et al, 1998a; 1998b
Inflammation	IL-4	Various	Lukacs et al, 1997; Hogaboam et al, 1997; Woods et al, 1999; Gabaglia et al, 1999, Lubberts et al, 1999
Cancer	IL-4	Brain	Wei et al, 1998
Eosinophilia	IL-5	Airways	Wang et al, 1998d
Studies of IL	IL-6	Lung	Braciak et al, 1993; Xing et al, 1994; Xing et al, 1999
Cancer	IL-6	Anti-tumor immune response	Felzmann et al, 1997; Rancourt et al, 1999
Cancer	IL-7	Anti-tumor cytotoxicity	Finke et al, 1998; Miller et al, 2000; Willimsky et al, 2000
AdV studies	IL-8	In vitro	Hersch et al, 1995
Anti-inflammation	IL-10	Various	Boehler et al, 1998; Apparailly et al, 1998; Ma et al, 1998; Whalen et al, 1999; Jorgensen et al, 1999; Barbara et al, 2000; Henke et al, 2000; Kim et al, 2000
Atherosclerosis	IL-10	Endothelium	Pinderski et al, 1999
Transplantation	IL-10	Various	Drazan et al, 1995; Qin et al, 1997; Smith et al, 1997a; Brauner et al, 1997; Shinozaki et al, 1999

TABLE 3 (Continued)- Cytokines 1: Interleukins and interferons studies with AdVs.

Application	Gene expressed	Target cells/organ	References
TNF-alpha exp.	IL-10	Macrophages/ Endotoxemia	Xing et al, 1997
Allergic sensitization	IL-10	Airway	Stampfli et al, 1999
Cancer	IL-12	Various	Caruso et al, 1996; Bramson et al, 1996a,b ; Chen et al, 1997c; Siders et al, 1998; Gambotto et al, 1999; Addison et al, 1998; Fernandez et al, 1999; Emtage et al, 1999; Qiao et al, 1999; Nasu et al, 1999; Hirschowitz et al, 1999a, b; Mazzolini et al, 2000; Drozdzik et al, 2000; Narvaiza et al, 2000
Anti-inflammation	IL-12	Arthritis	Parks et al, 1998
Immune response (IR) to infection	IL-12	Cell-mediated IR, NO role	Greenberger et al, 1996; Gabaglia et al, 1999; Lasarte et al, 1999
Transplantation	IL-12p40	Pancreatic islets	Yasuda et al, 1998
Cancer	IL-12/B7-1	Tumors	Putzer et al, 1997
Granulopoiesis	IL-17	Spleen	Schwarzenberger et al, 1998
Cancer	IL-18	IFN-induction	Osaki et al, 1999
Cancer	IFN-alpha	Leukemia, Tumor suppression	Feldman et al, 1997; Ahmed et al, 1999
Cancer	IFN-beta	Various	Lu et al, 1999, Qin et al, 1998
Proliferative vascular dis.	IFN-beta	Porcine balloon injury model	Stephan et al, 1997
Cancer	IFN-gamma	Anti-tumor immunity	Zhang et al, 1996; Abe et al, 1996; Xu et al, 1996; Xu et al, 1997; Kircheis et al, 1998; Fathallah et al, 2000
Anti-microbial	IFN-gamma	Macrophage induction	Stoeckle et al, 1996

TABLE 4- Cytokines 2: Growth and survival factors studies with AdVs.

Application	Gene expressed	Target cells/organ	References
Vaccination/ Immune response	GM-CSF	Immune system	Kielian et al, 1999; Ozawa, 1999
Cancer	GM-CSF	Anti-tumor immunity	Nagai et al, 1998
GM-CSF study	GM-CSF	Lung	Zsengeller et al, 1998
Inflammation	GM-CSF	Lung	Lei et al, 1998
Angiogenesis	VEGF	Heart, Diabetes	Magovern et al, 1997; Mack et al, 1998; Poliakova et al, 1999; Rivard et al, 1999; Patel et al, 1999; Rosengart et al, 1999a, 1999b; Lazarous et al, 1999; Lee et al, 2000
Cell maturation	VEGF	Dendritic cells	Oyama et al, 1998
Antiangiogenesis	ATF UPA	Cancer (neovascularization)	Li et al, 1998b
Cancer	FGF	Melanoma	Nesbit et al, 1999
Angiogenesis	FGF	Heart, Muscle	Pili et al, 1997; Ueno et al, 1997; 1998; Safi et al, 1999; Garcia-Martinez et al, 1999
FGF signalling	FGF truncated	Smooth muscle cells	Miyamoto et al, 1998
Degenerative disc disease	TGF-beta	Intervertebral discs	Nishida et al, 1999
Collagenase 3 expression	TGF-beta	Gingival fibroblasts	Ravanti et al, 1999
VEGF expression	TGF-beta	Osteoclasts	Saadeh et al, 1999
Fracture repairs	TGF-beta	Osteogenesis	Mehrara et al, 1999
Dopaminergic cells survival	TGF-beta	Nigral grafts	Sanchez-Capelo et al, 1999
Fibrosis	Dom-negative TGF beta	Liver	Qi et al, 1999
Hematopoiesis	EGF	Hematopoietic cells	Takahashi et al, 1998b
Angiogenesis	Platelet-derived growth factor	Cancer, chronic wounds, arterial injury	Tanaka et al, 1997b; Deguchi et al, 1999; Liechty et al, 1999
Anemia, erythropoiesis	Erythropoietin	Kidney, general	Descamps et al, 1994; Setoguchi et al, 1994; Tripathy et al, 1994; Svensson et al, 1997; Osada et al, 1999

TABLE 4 (Continued)- Cytokines 2: Growth and survival factors studies with AdVs.

Application	Gene expressed	Target cells/organ	References
Vascular cells, Bone formation	BMP-2	Smooth muscle cells, Spine	Nakaoka et al, 1997;Riew et al, 1998; Okubo et al, 2000a, 2000b
Dwarfism, systemic action	Growth hormone	Various	Hahn et al, 1996; He et al, 1998
Neuronal death/ survival	Brain-Derived Neurotrophic Factor	Neurons	Baumgartner et al, 1997; Gimenezyribotta et al, 1997; DiPolo et al, 1998; Bemelmans et al, 1999; Isenmann et al, 1999; Bemelmans et al, 1999
Neuronal survival	Glial maturation factor	Neurons	Zaheer et al, 1999; Lim et al, 2000
Neuronal death/ survival	Glial-Derived Neurotrophic Factor	Glial cells	Baumgartner et al, 1997; 1998a, 1998b; Gimenezyribotta et al, 1997; Kitagawa et al, 1999
Megakaryocyte maturation	HST-1/FGF-4	Megakaryocyte progenitors	Konishi et al, 1996
Neuronal death/ survival	Ciliary neurotrophic factor	Neurons	Baumgartner et al, 1997; Cayouette et al, 1998
Neuronal death/ survival	Neurotrophin-3	Motor neurones, dendrites	Djikhuizen et al, 1997; Haase et al, 1998; Baker et al, 1998
G1 growth inhibition	Natriuretic peptide	Vascular muscle cells	Doi et al, 1997
Cancer	Antisense-EGF	Gastric cells	Hirao et al, 1999
Antiangiogenesis	Antisense-VEGF	Glioblastoma	Im et al, 1999
Vascular develop. disorders	Antisense-FGF	Embryonic heart, Artery injury	Hanna et al, 1997; Leconte et al, 1998, Neschis et al, 1998

4.2. TRANSCRIPTION REGULATION AND SIGNALING PATHWAYS

AdVs are widely used to explore transcription and signaling pathways. Several examples of transcription factors or promoter studies are depicted in Table 5. As an example, to study the controversial role of the transcription factor NF-Kappa B, an AdV was used to overexpress a specific inhibitor, the subunit I kappa-B alpha (Wrighton et al, 1996: Bondeson et al, 1999). The results obtained (beneficial reduction of inflammation and tissue destruction) show the suitability of AdV to define therapeutic targets by expressing inhibitors of transcription factors. This was efficient because during the inflammatory response, endothelial cells transiently up-regulate a set of genes encoding, among others, cell adhesion molecules and chemotactic cytokines that together mediate the interaction of the endothelium with cells of the immune system. This gene up-regulation is mediated predominantly at the transcriptional level and in many cases involves the transcription factor NF- Kappa B. In the study of atherogenesis, similar results were obtained on monocyte arrest and transmigration in inflamed endothelium by I kappa-B alpha transfer with an AdV. This strategy succeeded in decreasing monocyte infiltration in atherosclerotic plaques (Weber et al, 1999).

TABLE 5- Transcription regulation studies with AdVs.

Study/ Analysis	Gene expressed	Target cells/organ	References
Proliferative function	CCAAT/enhancer-binding proteins	Hepatocytes	Diehl et al, 1996
Promoter activities	Various promoters expressing hAAT	Liver	Guo et al, 1996
Thyroid hormone response	TRE-reporter	Hepatocytes	Menjo et al, 1999
Inflammation	Ikappa-Balpha	Inflamed endothelium	Wrighton et al, 1996
Inflammation	Ikappa-Balpha	Inflamed joints	Bondeson et al, 1999
Monocyte arrest and migration	Ikappa-Balpha	Inflamed endothelium	Weber et al, 1999
Cell survival	Dominant negative Ikappa-Balpha	Endothelium	Zen et al, 1999
Vascular proliferative disorders	c-myb ribozyme	Smooth muscle cells	Macejak et al, 1999
IL1-β/ TNF-α signaling	Dominant negative TRAF-2	Intestinal epithelial cells	Jobin et al, 1999
Differenciation	GATA-4/ GATA-6 antisenses	Cardiomyocytes	Charron et al, 1999
Cardiac hypertrophy	STAT3	Cardiomyocytes	Kunisada et al, 1998
Endothelial pathways	c-Jun, c-Fos	Endothelium	Wang et al, 1999b
Cell cycle	BRCA1	Breast cancer	MacLachlan et al, 2000
MDM2 expression	E2F-1	Cancer	Yang et al, 1999b
p21, p27 regulations	E2F-1	Cardiomyocytes	von Harsdorf et al, 1999
Granulosa cell survival	cAMP response element-binding protein	Rat ovary	Somers et al, 1999
Adherens junctions	Dominant negative Ras	Endothelial cells	Hegland et al, 1999
Fibrogenesis	TNF alpha	Lung	Sime et al, 1997
Morphogenesis	Decorin	Lung	Zhao et al, 1999

The role of CCAAT/enhancer-binding protein isoforms in the determination of hepatocyte phenotypes were explored by AdV-mediated gene transfer (Diehl et al 1996). This study showed that C/EBP alpha has a dominant antiproliferative function, and must interact with other factors to regulate hepatocyte-specific gene expression. Using a different strategy, the importance of cardiomyocyte transcription factors was studied by Charron et al (1999). Two members of the GATA family of transcription factors, GATA-4 and GATA-6, are expressed in the developing myocardium and are potent transactivators of several cardiac promoters, namely the atrial natriuretic factor and B-type natriuretic peptide promoters. The role of GATA-4 and GATA-6 in the maintenance of the expression the cardiac genetic program was demonstrated using antisense-expressing AdVs which significantly down-regulated the expression of the corresponding GATA transcription factor.

Finally, regulation of expression in the liver was examined by Guo et al (1996) who demonstrated that AdVs are suitable for promoter/enhancer studies. Transcription targeting was also recently used to study the responsive element itself. The triiodothyronine response element (TRE) sensitivity to 3,3,'5-triiodothyronine was studied by transducing hepatocytes with an AdV expressing a reporter gene driven by TRE (Menjo et al, 1999).

AdVs also proved efficient to explore several signal transduction pathways. In this aim, phosphatases, kinases and receptors have been expressed by AdVs (Table 6). In the field of cardiac diseases, Choukroun et al (1998; 1999) overexpressed an inhibitory kinase, the SAPK/ERK kinase 1 (a dominant inhibitory mutant of SEK-1), to block signal transmission. Their results allowed a detailed characterization of the agonist-induced type cardiac hypertrophy. Another signal transduction pathway has been studied with AdVs in the same kind of disease. Gp130, a signal transducer of the IL-6-related cytokines, is expressed ubiquitously, including in the heart. The activation of gp130 in cardiac myocytes was reported to induce myocardial hypertrophy. The downstream side of gp130 consists of two distinct pathways in cardiac myocytes, one is a Janus kinase/signal transducer and activator of transcription (JAK/STAT) pathway, and the other is a mitogen-activated protein kinase (MAPK) pathway. The JAK/STAT pathway and specifically the STAT3-dependent signaling pathway were studied using AdVs. The results obtained by expression of wild-type or mutated STAT3 cDNA suggest that the STAT3-dependent signaling pathway downstream of gp 130 promotes cardiac myocyte hypertrophy under stimulation with leukemia inhibitory factor (Kunisada et al, 1998).

In the field of immunology, Jobin et al (1999) have explored IL-1 beta and TNF-alpha signaling cascades leading to NF-Kappa B activation and IL-8 expression in human intestinal epithelial cells. The physiological role of TRAF proteins in cytokine signaling in intestinal epithelial cells (IEC) is unknown. AdV-mediated dominant negative TRAF-2 (Ad5dnTRAF-2) expression prevented TNF-alpha-induced, but not IL-1 beta-induced I kappa B alpha degradation and NF-kappa B activation. Induction of IL-8 gene expression by TNF-alpha was partially inhibited in Ad5dnTRAF-2-transduced HT-29 cells (Jobin et al, 1999). Recently, Matsukawa et al (2000) used AdV overexpression of monocyte chemoattractant protein-1 to study its role in the development of Th1 and Th2 type responses *in vivo*. The *in vivo* function of the CXC chemokines interferon-inducible protein-10 (IP-10) and monokine induced by gamma (MIG) was also examined using AdVs expressing human IP-10 or murine MIG (Palmer et al, 1999). Overexpression of RANTES using an AdV induced the tissue-directed recruitment of monocytes to the lung (Braciak et al, 1996).

In 1997, the role of TGF-beta-1 in lung fibrosis induction was examined using AdV-mediated gene transfer *in vivo*. The transfer of the cDNA of TNF-alpha to rat lung allowed the study of the effect of transient but prolonged (7 to 10 days) overexpression of TNF-alpha in normal adult pulmonary tissue. TGF-beta-1 was, as a consequence, transiently elevated from the period immediately preceding the onset of fibrogenesis (Sime et al, 1997). Involvement of the TGF-beta signaling in liver fibrosis was also established using AdVs. The blockade of TGF-beta signaling by AdV-mediated local expression of a dominant negative type II TGF-beta receptor in the liver of a rat model of persistent fibrosis allowed the prevention of liver fibrosis and dysfunction (Qi et al, 1999). This particular example shows how the direct exploration of the interaction of a transgene with a known pathway has led to a practical application through Ad-mediated gene transfer. Similar studies were carried out in vascular wall cells with AdV to abolish TGF-beta signaling (Yamamoto et al, 1996). As well, excessive TGF-beta signaling has been implicated in pulmonary hypoplasia associated with bronchopulmonary dysplasia, a chronic lung disease of human prematurely featuring pulmonary fibrosis. Decorin, an inhibitor of TGF-beta was tested as a therapeutic agent. Adverse effects of excessive expression of TGF-beta on lung morphogenesis could be precluded by AdV-expression of decorin (Zhao et al, 1999).

Rheumatoid arthritis (RA) is characterized by a chronic inflammation of the synovial joints resulting from hyperplasia of synovial fibroblasts and infiltration of lymphocytes, macrophages, and plasma cells. Recent studies have revealed the essential role of osteoclasts in joint destruction in RA. Src family tyrosine kinases are implicated in various intracellular signaling pathways, including osteoclastic bone resorption. Therefore, inhibiting Src activity could be a good therapeutic strategy to prevent joint inflammation and destruction in RA. The negative regulation of the Src family tyrosine kinases by C-terminal Src kinase AdV-mediated expression allowed the suppression of the multiplication of rheumatoid synoviocytes and the decrease in inflammation and bone destruction *in vivo* (Takayanagi et al, 1999; Miyazaki et al, 2000a; b).

Finally, using AdV-mediated expression of a dominant negative Ras in primary neurons, Vaillant et al (1999) demonstrated that the Ras-phosphatidyinositol 3-kinase-Akt pathway was implicated in neuronal survival by showing that this pathway was essential for approximately 50% of depolarization-mediated neuronal survival. In a similar study, AdV-mediated expression of wild-type and various mutants of the TrkB receptor in primary neurons, allowed to show that the TrkB-Shc site signals neuronal survival and local axon growth via MEK and PI 3-K pathways (Jasvinder et al, 2000).

TABLE 6- Signaling pathway studies (phosphatases, kinases and receptors) with AdVs.

Pathway/ disorder	Gene expressed	Target cells/organ	References
Tumor	MMAC1/PTEN phosphatase	Ovarian carcinoma	Minaguchi et al, 1999
Tumor	MMAC1/PTEN phosphatase	Glioblastoma cells	Cheney et al, 1999
Neurotrophic signalling	Sph-2	Neurons	Araki et al, 2000
VEGF-mediated signalling	HCPTPA	Endothelium	Huang et al, 1999b
Angiogenesis	Tie2	Endothelium	Lin et al, 1998
Cardiomyocyte hypertrophy	Calcineurin	Heart	Taigen et al, 2000
Fru-2,6-P2 metabolism	6-Phosphofructo-2-kinase/fructose-2,6-bisphosphatase	Gluconeogenic rat hepatoma cells	Argaud et al, 1995
Glucose transport	Human protein kinase C-zeta	Skeletal muscle	Etgen et al, 1999
Resistance to apoptosis	Protein kinase ciota	Cancer	Jamieson et al, 1999
csk distribution	Protein tyrosine kinase pp125FAK	Macrophages	De Nichilo et al, 1999
Rheumatoid arthritis	csk	Joints	Takayanagi et al, 1999
Cell migration	Calcium/calmodulin dependent protein kinase II	Vascular smooth cells	Lundberg et al, 1998
Cardiac hypertrophy	SAPK/ERK kinase 1	Cardiomyocytes	Choukroun et al, 1998; 1999
Cardiac hyperexcitability	Potassium channel	Cardiomyocytes	Nuss et al, 1999
IL1-β/ TNF-α signaling	Dominant negative TRAF-2	Intestinal epithelial cells	Jobin et al, 1999
Hepatoprotection	Hepatocyte nuclear factor gamma-3	Liver	Nakamura et al, 1999
Insulin secretion	Dominant negative-hepatocyte nuclear factor alpha-1	In vitro	Tanizawa et al, 1999
Phosphoenolpyruvate carboxykinase transcription	Dominant-negative Akt	Liver	Kotani et al, 1999
Cell cycle	Von-Hippel-Lindau-mediated factor	Cancer	Kim et al, 1998
Maturation	Cyclin D1	Hepatocytes	Albrecht et al, 1999
Cell cycle	SV40 small T antigen	In vitro	Howe et al, 1998
IL studies	IL6-soluble receptor	Various	Thibault et al, 1997
Cancer	Gastrin-releasing peptide receptor	Bombesin expression	Rosenfeld et al, 1997
Modulation of function	EGF receptor	Osteoclasts	Tanaka et al, 1998
JNK/thermotolerance	HSP70	Human lung, primary fibroblast	Volloch et al, 1998
TGF-beta antiproliferative role	TGF-beta type II receptor	Vascular wall cells	Yamamoto et al, 1996
Angiotensin II signaling	ATI receptor	Cardiomyocytes	Booz et al, 1999
Arthritis	IL1-TNF soluble receptors	Joints	Ghivizzani et al, 1998
Breast cancer	Dominant negative estrogen receptor	Breast cancer cells	Lazennec et al, 1999
Neuronal survival and local axon growth	TrkB receptor	Primary neurons	Jasvinder et al, 2000

4.3. DECOYS AND DOMINANT NEGATIVE RECEPTORS

Decoys competing with specific receptors for ligand binding can be used to block specific cascades. The direct AdV-mediated administration of IL-1 and TNF-alpha-1 soluble receptors proved efficient in a rabbit model of arthritis (Ghivizzani et al, 1998). AdVs were used to deliver genes encoding a soluble IL-1 type I receptor-IgG fusion protein and/or a soluble type I TNF-alpha receptor-IgG fusion protein directly to the knees of rabbits with antigen-induced arthritis. It was shown that knees receiving the soluble IL-1 receptor decoy had significantly reduced cartilage matrix degradation and white blood cell infiltration into the joint space while TNF-alpha receptor decoy was less effective. These preliminary results show that AdV expression of receptor decoys could be an interesting approach to study the role of these receptors in some diseases and provide insights to develop therapeutic interventions.

Dominant negative estrogen receptors are transcriptionally inactive, altered forms of the estrogen receptor (ER) that can dimerize with it to inactivate the biological functions of this receptor. Lazennec et al (1999) recently proposed a strategy to abolish sensitivity of breast cancer cells to estrogen. By AdV-mediated delivery of a dominant negative estrogen receptor gene, they succeeded in abrogating cell proliferation. In another application, AdV-mediated dominant negative TrkB receptor expression, revealed a requirement for presynaptic signaling in BDNF-induced synaptic potentiation in cultured hippocampal neurons (Li, 1998c). A final example is the study of Zhao et al (1998) who showed that the epithelium-specific AdV transfer of a dominant-negative mutant TGF-beta type II receptor stimulated embryonic lung branching morphogenesis in culture.

4.4. METABOLIC PATHWAYS

The use of AdVs in the field of metabolic diseases has opened new avenues to characterize more accurately some metabolic pathway disorders and to elaborate new therapies (for a review, see Antinozzi et al, 1999). Thus, important advances have recently been made in the fields of diabetes and obesity, as for instance the leptin gene therapy which could benefit both conditions (Muzzin et al, 1996; Liu et al, 1998b). Similar results were obtained in another non-related metabolic problem, Wilson's disease, which is an autosomal recessive disorder characterized by the excessive accumulation of copper in the liver. AdV-mediated expression of the WND (ATP7B) gene in LEC rat (an animal model for Wilson's disease) allowed restoration of holoceruloplasmin synthesis, thereby establishing the role of WND in copper transport (Terada et al, 1998).

4.4.1. GLUCOSE METABOLISM

Numerous enzymes and proteins involved in glucose homeostasis were studied through AdV-mediated expression (summarized in Table 7). This started in 1992 with one of the first successful AdV-mediated expression of a gene to study glycogen metabolism (Gomez-Foix et al), which stimulated several subsequent metabolic pathway studies. The study of the role of glucose-induced translocation of glucokinase in hepatic glycogen synthesis was performed by overexpression of this enzyme by an AdV (Agius et al, 1996). The role of this enzyme has recently been further characterized by Piston et al (1999). The efficiency of transduction of AdV has been an asset to characterize the overexpression of enzymes such as Glucose-6-phospatase which plays an important role in glucose homeostasis (Seoane et al, 1997; Trinh et al, 1997). The engineering of insulin secretion in pancreatic islets by AdV allowed the study of mechanisms of stimulus-secretion coupling (Noel et al, 1997) and this approach has provided new insights into the etiology of diabetes. In this matter, interesting results were also obtained by infusing a leptin receptor-expressing AdV in genetically deficient rats (Wang et al, 1998a; 1998b) and by expressing regulatory proteins involved in the PI 3-K activation (Frevert & Kahn, 1997). This pathway was also explored by Sharma et al (1998) by inhibition of the PI 3-K. The knowledge of glucose homeostasis was further improved by examining the consequences of overexpressing the protein targeting to glycogen (PTG) (Berman et al, 1998). Igarashi et al (1999) also recently used AdV to characterize p38 mitogen-activated protein kinase in glucose and diabetes pathways. Finally, some conditions such as type 1 diabetes are due in part to a deficiency in inducible nitric oxide synthetase. An AdV expressing this gene was efficiently used to reverse the phenotype (Wang et al, 1998b).

TABLE 7- AdV-mediated expression of proteins for glucose and lipids metabolic pathway studies.

Enzyme/ Protein	Target cells	References
Glycogen-phosphorylase	Muscle cells	Gomez-Foix et al, 1992; Baque et al, 1994
Leptin	Pancreatic islet cells	Chen et al, 1996; Koyama et al, 1997, 1998
Glucokinase	Hepatocytes, muscle cells	Agius et al, 1996; O'Doherty et al, 1996; Berman and Newgard, 1998; Baque et al, 1998; O'Doherty et al, 1999; Seoane et al, 1999; Jimenez-Chillaron et al, 1999
Hexokinase I	Hepatocytes	Becker et al, 1994; O'Doherty et al, 1996; Seoane et al, 1996
G-6-Phosphatase	Hepatocytes	Seoane et al, 1997; Trinh et al, 1997; Aiston et al, 1999
PDX-1	Pancreatic islet cells	Seijffers et al, 1999; Qian et al, 1999
Phospholipase C	Pancreatic islet cells	Gasa et al, 1999
Uncoupling protein-2	Pancreatic islet cells	Wang et al, 1999a
PTG	Hepatocytes	Berman et al, 1998
Leptin receptor	Pancreatic islet cells	Wang et al, 1998c
PEPCK	Hepatocytes	Scott et al, 1998
Glycerol kinase	Pancreatic islet cells	Noel et al, 1997
Melanoyl CoA decarboxylase	Pancreatic islet cells	Antinozzi et al, 1998
Sterol regulatory element binding protein-1c	Hepatocytes	Foretz et al, 1999a; 1999b
Negative mutant hepatocyte nuclear factor (HNF)-1alpha	Hepatoma, insulinoma cells	Tanizawa et al, 1999
Microsomal triglyceride transfer protein	HepG2	Liao et al, 1999
OB-Rb	Leptin-resistant islets	Wang et al, 1998b

4.4.2. LIPID METABOLISM

The study of lipid metabolic pathways could also benefit the therapy of obesity and related disorders. The AdV-mediated transfer of lipase in deficient mice allowed to perform an analysis of protein structure-function *in vivo* (Kobayashi et al, 1996a), and a series of papers dealt with the metabolism of low and high density lipoprotein plasma level regulation (Kobayashi et al, 1996a, 1996b; Föger et al, 1997; Oka et al, 1997; Heeren et al, 1999). Another strategy for the treatment of obesity has also recently been proposed by reversion of the adipocyte phenotype. This was achieved by AdV-mediated transfer of the leptin gene, which induced transient hyperleptinemia and depleted adipocyte fat while profoundly down-regulating lipogenic enzymes (Zhou et al, 1999).

4.5. CELL CYCLE AND APOPTOSIS

The processes of cell cycle regulation and apoptosis have been the subjects of intensive research over the recent years. From those studies emerge the notion that both processes are intimately interconnected through some of the key regulators like p53 and E2F1. Ad has provided insights on cell cycle regulation and apoptosis by itself, and, as a gene delivery tool, AdV-mediated expression of a wide variety of genes has allowed the characterization of numerous pathways and interactions (Table 8).

4.5.1. Ad GENES MODULATING CELL CYCLE AND APOPTOSIS

In order to efficiently complete their replicative cycle and to escape the host immune defense, Ads express many proteins modulating both cell cycle progression and apoptosis (reviewed in Chinnadurai, 1998; Shenk, 1996; Shisler and Gooding, 1998; Teodoro and Branton, 1997; White, 1998; Wold and Tollefson, 1998). For example, by relieving pRb inhibition of E2F1, E1A proteins induce cell entry in S phase to facilitate Ad replication but at the same time trigger apoptosis through both p53-dependent and independent pathways. E1A-induced apoptosis is inhibited by the E1B 19 kD, which is a functional homologue of Bcl-2 and Bcl-XL. As well, by physically interacting with p53, E1B 55 kD protein also plays an important role in the p53-induced apoptosis.

As a means of counteracting the host defense, TNF alpha- and Fas-induced apoptosis of Ad-infected cells are delayed by viral proteins such as E1B 19 kD, E3 14.7 kD and E3 10.5-14.5 heterodimer, albeit by different mechanisms. AdVs are also eliminated by mechanisms that involve the death of transduced cells. Modulation of apoptosis could thus allow increased persistence of transgene expression by this vector. For example, it was demonstrated that inhibition of NF-kappa B activation in combination with bcl-2 expression allowed for persistence of first generation AdVs in the mouse liver (Lieber et al, 1998). Although very important in a normal infection, E1- and E3-encoded proteins are not a practical concern for applications of first generation AdVs in the functional characterization of genes involved in cell cycle regulation and apoptosis. However, the remaining E4 region has recently been shown to interfere with different apoptosis pathways and cell cyclin expression leading to either cell death or G2 growth arrest (Brand et al, 1999; Wersto et al,

1998). For example, E4 ORF4 was shown to be a proapoptotic protein, which induces cell death through a p53-independent mechanism as mentioned in section 2.2.

As a practical solution to minimize vector-induced interference, infection at low MOIs, still sufficient to obtain 100% transgene expression, have been successfully used to alleviate AdV-mediated toxicity observed at higher MOIs (Brand et al, 1999; Langelier et al, 2000). This is presumably because the leaky expression of E4 and E2 regions at lower MOIs is not high enough to significantly affect the cell physiology. Although other E4 ORFs could also be involved in cell cycle dysregulation and apoptosis, an AdV expressing only E4 ORF6 is much less toxic, and this represents another practical alternative instead of the complete E4-deleted AdVs which are more difficult to grow at high titers in E1-E4 complementing cell lines (see section 2.2). Even though E4 ORF6 has been shown to interact with p53, in absence of the other Ad early proteins (E1A and E1B 55 kD), its influence on p53 activity appears to be minimal especially at a low level of expression (Steegenga et al, 1998; 1999).

4.5.2. CELL CYCLE AND APOPTOSIS STUDIES USING AdVS

Given their pivotal role in cell cycle control and apoptosis p53 and, to a lesser extent E2F1, have been studied in a wide variety of tumors using AdV-mediated gene transfer (reviewed in Gomez-Manzano et al, 1999). Interestingly, AdV-p53 transduction has resulted in tumor regression not only through direct apoptosis induction in tumor cells but also through some bystander effect involving the suppression of the neovascularization of the tumors. The role of p53 and E2F1 in primary neuron survival was also assessed using AdV-mediated gene transfer (see Table 8).

By comparison to apoptosis, fewer cell cycle regulation studies have been done with AdVs. One good example is that of the progression of hepatocytes through the G1 stage of mitosis which was obtained by AdV-mediated expression of cyclin D1 (Albrecht and Hansen, 1999). Other examples include the overriding of p53-mediated G2-arrest by constitutive activation of cyclin B (Park et al 2000) and the characterization of a p53-independent DNA damage checkpoint involving posttranscriptional regulation of cyclin A expression (Guo et al, 2000).

AdVs have been used to determine the effect of the expression of both apoptotic and antiapoptotic proteins (Table 8). For instance, expression of Bcl-Xs (a functional inhibitor of the Bcl-2 family of inhibitors of apoptosis which are overexpressed in many cancers) by AdV selectively induced cell death in cancer but not in normal cells (Clarke et al, 1995). In another study, *in vivo* AdV-mediated gene transfer of the antiapoptotic human Bcl-2 gene reduced ischemia-reperfusion injury in the mouse liver (ischemia-reperfusion injury has been demonstrated in a number of clinically relevant diseases such as myocardial infarction, cerebrovascular disease, sepsis, peripheral vascular disease, and organ transplantation) (Bilbao et al, 1999). Interestingly, using AdV-mediated gene transfer to obtain high expression level, Uhlmann et al (1998) demonstrated that Bcl-2 can also display a proapoptotic activity that is distinct from its antiapoptotic properties.

TABLE 8- Cell cycle and apoptosis studies with AdVs.

Application or pathway	Gene expressed	Phenotype	References
Cancer	p53	Growth arrest/proapoptotic	Katayose et al, 1995; Mujoo et al, 1996; Li et al, 1997a, b; Meeker et al, 1997; Overholt et al, 1997; Gotoh et al, 1997; Gomez-Manzano et al, 1997; Mukhopadhyay et al, 1998; Scheinman et al, 1999; Lang et al, 1999; Endo et al, 1999; Cirielli et al, 1999; Wolf et al, 1999a; Cascallo et al, 1999; Ameyar et al, 1999; Tatebe et al, 1999; Fukazawa et al, 1999; Li et al, 1999b; Stempien-Otero et al, 1999; Milas et al, 2000; Pearson et al, 2000; Huang et al, 2000; Mitry et al, 2000; Shao et al, 2000; Park et al, 2000; Ramondetta et al, 2000; Turturro et al, 2000; Sandy et al; 2000; Shatrov et al, 2000
Cancer	E2F1	Proapoptotic	Hunt et al, 1997; Fueyo et al, 1998; Dong et al, 1999; Liu et al, 1999; Yang et al, 1999b; Yang et al, 2000
Myocardium regeneration	E2F1	cell cycle reentry/Apoptosis	Agah et al, 1997
Primary neuron survival	E2F1 p53 Calcineurin	Proapoptotic Proapoptotic Proapoptotic	Hou et al, 2000 Cregan et al, 1999; Johnson et al, 1999 Asai et al, 1999
Cancer	p21	Cell cycle arrest	Prabhu et al, 1996; Joshi et al; 1998; Terasaki et al, 1999
p53 cell cycle regulation	Cyclin B	Overriding p53-mediated G2-M arrest	Park et al, 2000
Hepatocytes cycling	Cyclin D1	Cell cycle progression	Albrecht and Hansen 1999
p53-independent DNA damage checkpoint	Cyclin A	Overriding cell cycle arrest	Guo et al, 2000
Intestinal cell proliferation	Cdx1	Down-regulation of D-type cyclins	Lynch et al, 2000
Arterial restenosis/Cardiomyocytes proliferation	Gax	Growth arrest/ apoptosis	Maillard and Walsh, 1996; Smith et al, 1997b; Fisher et al, 1997; Perlman et al, 1999
Cancer	Promyelocytic leukemia gene	G1 arrest/apoptosis	Le et al, 1998
Cancer	p16	Cell cycle arrest	Craig et al, 1998
Cancer	Antisense to SV40 early gene region	Growth arrest and proapoptotic	Waheed et al, 1999

TABLE 8 (Continued)- Cell cycle and apoptosis studies with AdVs.

Application or pathway	Gene expressed	Phenotype	References
Hyperplasia of vascular smooth muscles cells	pRb	Growth arrest	Smith et al, 1997c
Cancer: esophageal carcinoma	p21dsi1	Differentition-dependent apoptosis	Kadowaki et al, 1999
Cancer	Bak	Caspase-dependent apoptosis	Pataer and al, 2000
Cancer	p21 sdi	senescence	Kagawa et al, 1999
Ischemia/ reperfusion injury	Bcl-2	Antiapoptotic (overexpressed: proapoptotic)	Bilbao et al, 1999; Uhlmann et al, 1998; Shinoura et al, 1999a
Cancer	Apoptotin	Proapoptotic	Pietersen et al, 1999
Cancer	Anti Bcl-2 ribozyme	Proapoptotic	Dorai et al, 1999; Gibson et al, 2000
Cancer	Intracellular anti-Bcl-2 antibodies	Proapoptotic	Piche et al, 1998
Apoptosis	Bcl-6	Proapoptotic	Yamochi et al, 1999
Cancer	Bcl-xs	Proapoptotic	Clarke et al, 1995; Dole et al, 1996; Ealovega et al, 1996
Apoptosis	Bax	Proapoptotic	Kagawa et al, 2000; Xiang et al, 2000
Apoptosis	Bcl-xl	Antiapoptotic	Matsuoka et al, 1999; Shinoura et al, 2000
Endothelial cells	c-Jun	Proapoptotic	Wang et al, 1999c
Cance: glioma	p33ING1+p53	Proapoptotic	Shinoura et al, 1999b
Cancer: glioma, prostate	Fas-ligand	Proapoptotic	Shinoura et al, 1998; Ambar et al, 1999; Bruder et al, 2000; Heldlund et al, 1999
Cancer	Fragile histidine triad gene	Proapoptotic	Ji et al, 1999
Endothelial cells	Inducible nitric-oxide synthetase	Antiapoptotic	Tzeng et al, 1997; Kullo et al, 1997
Nitric-oxide/cGMP effect	cGMP-dependent protein kinase	Proapoptotic	Chiche et al, 1998

TABLE 8 (Continued)- Cell cycle and apoptosis studies with AdVs.

Application or pathway	Gene expressed	Phenotype	References
Cancer	Anti sense cyclin D1	Proapoptotic	Sauter et al, 1999
NO release	A20	Antiapoptotic	Grey et al, 1999a, 1999b
Caspase inhibition	Inhibitors of apoptosis gene family (TAP)	Antiapoptotic	Simons et al, 1999
Caspase 3 activation	APP	Proapoptotic	Uetsuki et al, 1999
Caspase 8 inhibition	Herpes R1ribonucleotide reductase	Antiapoptotic	Langelier et al, 2000
Caspase 8 activation	Herpes R1(delta2-357)	Proapoptotic	Massie et al, 1998a; Langelier et al, 2000
Cancer	Inhibitors of cyclin-dependent kinases	Proapoptotic	Jin et al, 1995; Yang et al, 1996b; Chen et al, 1997a; Katayose et al, 1997; Tsao et al, 1999; Schreiber et al, 1999
p53-mediated Apoptosis	Fas/CD95/APO-1	Proapoptotic	Rakkar et al, 1999
Rb/cyclin D1/p16 pathway	p16INK4a	Proapoptotic	Todd et al, 2000
Rheumatoid arthritis	p16INK4a	Proapoptotic	Taniguchi et al, 1999
Cancer	p16INK4a	Proapoptotic	Jin et al, 1995; Schrump et al, 1996; Sandig et al, 1997; Gotoh et al, 1997; Chintala et al, 1997; Frizelle et al, 1998; Rocco et al, 1998; Lee et al, 1998; Naruse et al, 1998; Wolf et al, 1999b; Harada et al, 1999
Cardiomyocyte hypertrophy	p16INK4a	Proapoptotic	Tamamori et al, 1998

Several other apoptosis inducers were further explored by AdV-mediated overexpression. For example, the role of p16 in cell death was demonstrated in lung cells (Naruse et al, 1998), while gene transfer of Fas-ligand to glioma cells also induced this process (Shinoura et al, 1998). Overexpression of Bcl-6 was also used to study its role and to show that it can downregulate Bcl-2 and Bcl-XL (Yamochi et al, 1999). Gax role as a pro-apoptotic gene was also characterized with AdV (Fisher et al, 1997; Perlman et al, 1999). The expression of genes from the family of the inhibitors of apoptosis (IAP) was studied in neural cells with AdVs. These antiapoptotic genes, originally discovered in baculovirus, exist in animals ranging from insects to humans. With AdVs, it was demonstrated that forced expression of IAPs efficiently blocked potassium withdrawal-induced N-acetyl-Asp-Glu-Val-Asp-specific caspase activity and reduced DNA fragmentation. (Simons et al, 1999).

In addition to p16 mentioned above, other cell cycle regulatory proteins are important candidates as therapeutic tumor suppressors and, through AdV-mediated gene transfer, they were found to be apoptosis inducers in many tumor cells (Jin et al, 1995; Yang et al, 1996b; Chen et al, 1997a; Katayose et al, 1997; Tsao et al, 1999). In one study, a series of isogenic AdVs were constructed to overexpress cyclin-dependent kinase inhibitors p16, p18, p19, p21(WAF1/CIP1) and p27 under the control of a strong promoter. These vectors directed the efficient expression of each of the cyclin kinase inhibitors and induced growth arrest, inhibited DNA synthesis, and prevented phosphorylation of pRb in cell lines expressing a functional RB gene. It was also noted that three of the cyclin kinase inhibitors, p16, p18 and p27 were found to induce apoptotic death in transduced HeLa and A549 cells. Interestingly, expression of the cyclin kinase inhibitors in a polyomavirus middle T antigen model of murine breast carcinoma resulted in a delay in tumor formation that varied from several weeks for the p19 expressing vector, to greater than 25 weeks for the p27 expressing vector. However, when tumors were injected directly with the AdV-expressing the cyclin kinase inhibitors, only treatment with the vector expressing p16 resulted in a delay in tumor growth (Schreiber et al, 1999). Other studies have shown that the negative cell cycle regulator p21 (WAF1/Cip1) is often overexpressed in human gliomas and overexpression of this protein renders glioma cells resistant to chemotherapy agents. An AdV-mediated antisense gene transfer has therefore been tested to block this cyclin kinase inhibitor, and it was shown that blocking p21 (WAF1/Cip1) production may serve as a useful chemosensitization regimen for treating gliomas (Ruan et al, 1999).

The inducible nitric oxide synthetase (iNOS) gene is also the object of intensive research with the AdV tool, due to its ability to block apoptosis in endothelial cells (Tzeng et al, 1997; Kullo et al, 1997). As well, the pro-apoptotic effect of nitric oxide (NO) has been studied with AdVs (Chiche et al, 1998). Even if the mechanisms of NO-mediated inhibition of vascular smooth muscle (VSM) cell proliferation are still obscure, it is clear that cyclins A and E in association with cyclin-dependent kinase 2 (cdk2) serve as positive regulators for mammalian cell cycle progression through the G1/S checkpoint of the cell cycle and subsequent cell proliferation. The effect of AdV-mediated expression of the endothelial NOS gene on proliferation and expression of cell cycle regulatory molecules of coronary cells stimulated by platelet-derived growth factor (PDGF BB homodimer) was thus tested.

The results obtained suggested that inhibition of PDGF-stimulated expression of cyclin A, cyclin E, and PCNA is the target of NO action (Sharma et al, 1999).

As a preliminary work for a new strategy for cancer treatment, Ji et al (1999) showed that AdV overexpression of the fragile histidine triad (FHIT) gene induced apoptosis and inhibited tumor growth. Le et al (1998) similarly demonstrated that promyelocytic leukemia gene expression induced cell cycle arrest at the G1 phase and promoted apoptosis thus suppressing tumor growth.

Finally, AdVs were recently used to explore Alzheimer disease pathogenesis. The amyloid precursor protein gene was overexpressed and it was possible to demonstrate that this protein activates caspase 3, a principal cell death protease involved in neuron apoptosis (Uetsuki et al, 1999). Similar results were obtained with calcineurin showing that, AdV-mediated high level forced activity of calcineurin, induced cytochrome c/caspase-3-dependent apoptosis in neurons (Asai et al, 1999). It has also recently been shown with an AdV expressing the neuronal isoforms of human beta-amyloid precursor protein that AdVs are promising for protein trafficking studies in cultured neurons (Yuan et al, 1999). This could have important consequences for the study of the molecular pathogenesis of Alzheimer's disease. Ras and Gab1 pathways were also studied in such cells with AdV. It was shown that AdV-expressed constitutively activated Ras (RasV12) enhanced survival and the phosphorylation of Akt (protein kinase B) and MAP kinase (MAPK), two targets of Ras activity (Mazzoni et, 1999). It was also demonstrated that Gab1 mediates neurite outgrowth, DNA synthesis, and survival in PC12 cells (Korhonen et al, 1999).

5. AdV-mediated Gene Transfer in Germ Cells and Embryos

Although the vast majority of AdV-mediated gene transfers targeted newborn and adult somatic cells, a few studies have explored the potential of using AdVs for delivering genes into germ cells *in vitro* and embryonic cells both *in vitro* and *in utero* (Table 9). The main goals for those studies were in one instance to seek for improved methods for generating transgenic animals, and in other instances, for fetal gene therapy.

5.1. TRANSGENESIS

As mentioned previously, AdV integration occurs at fairly low frequency in somatic cells, and, as a result of their episomal status, transgenes delivered with AdVs are rapidly lost in actively proliferating cells. In keeping with this, the transduction of mouse, rat and cow zygotes at the 8-cell morula or blastocyst stages resulted in efficient, albeit transient expression of the Lac-Z reporter gene (Kubish et al., 1997). However, AdV-mediated gene transfer in zona-free mouse eggs at the pronucleus stage, allowed for efficient gene delivery in almost 100% of the one cell embryos with, surprisingly, stable integration of a single copy of the AdV genome in about 10% of the transduced pre-implantation embryos (Tsukui et al, 1996). Although this approach remains to be validated with other species, the use of AdV for transgenesis has two main benefits over standard methods, namely its relative simplicity and the ability to create single copy transgenics which reduces the occurrence of expression mosaics in the offspring.

This result also raises serious concerns about the potential of germ line transmission in somatic gene therapy. This was examined quite extensively in the study of Ye et al (1998) who concluded that even though both ovaries and testes contained vector DNA following high dose systemic administration of a second generation AdV in adult mice, no vector sequence was detected in 814 offspring, indicating that germ line transmission of AdV is an unlikely event under those conditions. Consistent with this report, Blanchard and Boekelheide (1997) observed that, following direct gene transfer to the testis, the expression of a lacZ reporter gene was only transient in Leydig cells and therefore mainly useful to study the short-term effects of specific genes on spermatogenesis *in vivo*. In another report, the direct exposure of boar spermatozoa to lacZ-expressing AdV resulted in acrosomal transfer in 22% of the cells. Seven percent of offspring obtained by artificial insemination with AdV-transduced sperm were positive. However, the mating of positive animals did not result in transmission of the transgene. Thus, AdV-mediated transgenesis through sperm cells was effective for the first generation only (Farre et al, 1999).

5.2. EMBRYO GENE TRANSFER

Although there is much controversy surrounding fetal gene therapy, the prospect of more efficient correction of inborn diseases constitutes a major incentive to move it forward (Caplan and Wilson, 2000; Walsh, 1999). As a result, many studies have explored this approach (Table 9). While approval of this type of therapeutic intervention has yet to come, interesting basic questions are being addressed in the field of embryology and developmental biology by embryo gene transfer. Several authors have addressed the feasibility of AdV-mediated gene transfer in embryos with reporter genes mainly in mice, although other species such as rat, cattle, sheep and chicks were also examined. For example, hippocampal neurons from 18 day rat embryos, which normally lack or show low amounts of GIRK proteins, were transduced *in vitro* to examine the inhibition of neuronal excitation by G protein-gated inward rectifier K+ channel subunits 1-4 (GIRK1-4) (Ehrengruber et al, 1997). In another study, Takahashi et al (1998a) examined the apoptotic elimination of neural crest cells *in ovo* by delivering an AdV expressing the Msx2 gene. Many other fetal tissues such as the heart, the musculature, the lung and the liver have also been targeted by AdV gene transfer *in utero* (see Table 9).

Some applications also examined the feasibility of correcting defects in neonatal organisms as a means of obtaining corrections for inborn genetic diseases (Takahashi et al, 1996). Thus, a genetic defect (citrullinemia) was corrected in neonatal calves using an AdV to show the feasibility of that approach for human gene therapy (Lee et al, 1999b). A similar work in embryos was performed in a mouse and sheep model of hemophilia (Schneider et al, 1999; Themis et al 1999). Finally, Yang et al (1999a) recently showed persistent postnatal transgene expression in muscle and liver after intramuscular delivery of AdV *in utero*, establishing the feasibility of fetal gene therapy for fatal genetic diseases.

TABLE 9- Embryo gene transfers with AdVs.

Species	Gene expressed	Tissue/ Organ targeted	References
Mouse	Reporter	Zone-free eggs	Tsukui et al, 1996
Pig	Reporter	Sperm	Farre et al, 1999
Rat	Reporter	Testis	Blanchart and Boekelheide, 1997
Mouse	Reporter	Systemic	Tsukui et al, 1995; 1996; Schachtner et al,1999; Turkay et al, 1999
Mouse	Reporter	Neurulating embryo	Hartig et al, 1998
Mouse	Reporter	Different stage embryos	Baldwin et al, 1997
Rat	GIRK 1-4	Embryonal neurons	Ehrengruber et al, 1997
Chicken	Msx2	Neural crests	Takahashi et al, 1998a
Mouse	Reporter	Heart	Woo et al, 1997
Mouse	Reporter	Heart	Christensen et al, 2000
Mouse	Reporter	Lung	Douar et al, 1997
Mouse, Rat, Cow	Reporter	Systemic	Kubisch et al, 1997
Human	Reporter	Isolated placental trophoblasts	MacCalman et al, 1996
Chicken	Reporter	Systemic	Yamagata et al, 1994
Chicken	Reporter	Chicken embryo fibroblasts	Adam et al, 1995
Sheep, Mouse	Reporter	Systemic	Holzinger et al, 1995
Mouse	Reporter	Musculature	Yang et al, 1999a
Mouse	Reporter	Embryonic stem cells	Mitani et al, 1995
Sheep	Factor IX	Systemic	Themis et al, 1999
Mouse	Factor IX	Systemic	Schneider et al, 1999
Mouse	Reporter	Peritoneum and hepatic parenchyma	Lipshutz and al, 1999a, b
Rabbit	Reporter	Hepatocytes	Wang et al, 1998e

6. Conclusion

As shown in this review, AdVs have extensively been used for functional studies in a wide variety of contexts. Indeed, in excess of 150 genes have been expressed to date with AdVs for functional study or gene therapy applications while several dozen viral antigens have been tested for vaccination. What's more, a literature survey indicates that there is currently an exponential growth of AdV use in functional genomics (see 1999-2000 references herein). In spite of their numerous applications, as explained in section 2, AdVs suffer from some intrinsic limitations like the transient nature of transgene expression and some toxicity at high MOIs. These limitations are currently being addressed in many ways, and, although the gene therapy applications are the main driving force behind this, it will also benefit the application in functional genomics. One emerging trend in vector development that has recently led to significant improvements is to combine the best features of different vectors to generate chimera such as Ad-retroviral or Ad-AAV vectors (Feng et al., 1997; Duisit et al., 1999; Lieber et al, 1999; Recchia et al., 1999). By allowing for long-term transgene expression, this type of hybrid vectors further expend the range of applications of Ad-based vectors and open avenues yet to be explored.

7. Acknowledgments

We thank Dick Mosser and Alaka Mullick for critical reading of the manuscript, and André Migneault for artwork. This work was supported by the National Research Council of Canada.

8. References

Abe, J., Wakimoto, H., Tsunoda, R., Okabe, S., Yoshida, Y., Aoyagi, M., Hirakawa, K. and Hamada, H. (1996) In vivo antitumor effect of cytotoxic T lymphocytes engineered to produce interferon-gamma by adenovirus-mediated genetic transduction. *Biochem.Biophys Res. Commun.***5**, 164-170.

Acsadi, G., Massie, B. and Jani., A. (1995) Adenovirus-mediated gene transfer into striated muscles. *J. Mol. Med.***73**, 165-180.

Adam, M., Oualikene, W., Le Cocq, H., Guittet, M. and Eloit, M. (1995) Replication-defective adenovirus type 5 as an in vitro and in vivo gene transfer vector in chickens. *J. Gen. Virol.***76**, 3153-3157.

Addison, C.L., Braciak, T., Ralston, R., Muller, W.J., Gauldie, J. and Graham, F.L. (1995) Intratumoral injection of an adenovirus expressing interleukin 2 induces regression and immunity in a murine breast cancer model. *Proc. Natl. Acad. Sci. U. S. A.***92**, 8522-8526.

Addison, C.L., Bramson, J.L., Hitt, M.M., Muller, W.J., Gauldie, J. and Graham, F.L. (1998) Intratumoral coinjection of adenoviral vectors expressing IL-2 and IL-12 results in enhanced frequency of regression of injected and untreated distal tumors. *Gene Ther.***5**, 1400-1409.

Adesanya, M.R., Redman, R.S., Baum, B.J. and O'Connell, B.C. (1996) Immediate inflammatory responses to adenovirus-mediated gene transfer in rat salivary glands. *Hum. Gene Ther.***7**, 1085-1093.

Agah, R., Kirshenbaum, L.A., Abdellatif, M., Truong,L.D., Chakraborty, S., Michael, L.H. and Schneider, M.D. (1997) Adenoviral delivery of E2F-1 directs cell cycle reentry and p53-independent apoptosis in postmitotic adult myocardium in vivo. *J. Clin. Invest.***100**, 2722-2728.

Agius, L., Peak, M., Newgard, C.B., Gomez-Foix, A.M. and Guinovart, J.J. (1996) Evidence for a role of glucose-induced translocation of glucokinase in the control of hepatic glycogen synthesis. *J. Biol. Chem.***271**, 30479-30486.

Ahmed, C.M., Sugarman, B.J., Johnson, D.E., Bookstein, R.E., Saha, D.P., Nagabhushan, T.L. and Wills, K.N. (1999) In vivo tumor suppression by adenovirus-mediated interferon alpha2b gene delivery. *Hum. Gene Ther.***10**, 77-84.

Aiston, S., Trinh, K.Y., Lange, A.J., Newgard, C.B. and Agius, L. (1999) Glucose-6-phosphatase overexpression lowers glucose 6-phosphate and inhibits glycogen synthesis and glycolysis in hepatocytes without affecting glucokinase translocation. Evidence against feedback inhibition of glucokinase. *J. Biol. Chem.***274**, 24559-24566.

Albrecht, J.H. and Hansen, L.K. (1999) Cyclin D1 promotes mitogen-independent cell cycle progression in hepatocytes. *Cell Growth Differ.***10**, 397-404.

Amalfitano, A. and Chamberlain, J.S. (1997) Isolation and characterization of packaging cell lines that coexpress the adenovirus E1, DNA polymerase, and preterminal proteins: implications for gene therapy. *Gene Ther.***4**, 258-263.

Amalfitano, A., Hauser, M.A., Hu, H., Serra, D., Begy, C.R. and Chamberlain, J.S. (1998) Production and characterization of improved adenovirus vectors with the E1, E2b, and E3 genes deleted. *J. Virol.***72**, 926-933.

Ambar, B.B., Frei, K., Malipiero, U., Morelli, A.E., Castro, M.G., Lowenstein, P.R. and Fontana, A. (1999) Treatment of experimental glioma by administration of adenoviral vectors expressing Fas ligand. *Hum. Gene Ther.***10**, 1641-1648.

Ameyar, M., Shatrov, V., Bouquet, C., Capoulade, C., Cai, Z., Stancou, R., Badie, C., Haddada, H. and Chouaib, S. (1999) Adenovirus-mediated transfer of wild-type p53 gene sensitizes TNF resistant MCF7 derivatives to the cytotoxic effect of this cytokine: relationship with c-myc and Rb. *Oncogene.***18**, 5464-5472.

Amin, R., Wilmott, R., Schwarz, Y., Trapnell, B. and Stark, J. (1995) Replication-deficient adenovirus induces expression of interleukin-8 by airway epithelial cell in vitro. *Hum. Gene Ther.***6**, 145-153.

Anderson, L.M., Swaminathan, S., Zackon, I., Tajuddin, A.K., Thimmapaya, B. and Weitzman, S.A. (1999) Adenovirus-mediated tissue-targeted expression of the HSVtk gene for the treatment of breast cancer. *Gene Ther.***6**, 854-864.

Antinozzi, P.A., Segall, L., Prentki, M., McGarry, J.D. and Newgard, C.B. (1998) Molecular or pharmacologic perturbation of the link between glucose and lipid metabolism is without effect on glucose-stimulated insulin secretion. A re-evaluation of the long-chain acyl-CoA hypothesis. *J. Biol. Chem.***273**, 16146-16154.

Antinozzi, P.A., Berman, H.K., O'Doherty, R.M. and Newgard, C.B. (1999) Metabolic engineering with recombinant adenoviruses. *Annu. Rev. Nutr.* **19**, 511-544.

Anton, M. and Graham, F.L. (1995) Site-specific recombination mediated by an adenovirus vector expressing the Cre recombinase protein: a molecular switch for control of gene expression. *J. Virol.* **69**, 4600-4606.

Apparailly, F., Verwaerde, C., Jacquet, C., Auriault, C., Sany, J. and Jorgensen, C. (1998) Adenovirus-mediated transfer of viral IL-10 gene inhibits murine collagen-induced arthritis. *J. Immunol.* **160**, 5213-5220.

Arai, Y., Kubo, T., Kobayashi, K., Ikeda, T., Takahashi, K., Takigawa, M., Imanishi, J. and Hirasawa, Y. (1999) Control of delivered gene expression in chondrocytes using heat shock protein 70B promoter. *J. Rheumatol.* **26**, 1769-1774.

Araki, T., Yamada, M., Ohnishi, H., Sano, S., Uetsuki, T. and Hatanaka, H. (2000) Shp-2 specifically regulates several tyrosine-phosphorylated proteins in brain-derived neurotrophic factor signaling in cultured cerebral cortical neurons. *J. Neurochem.* **74**, 659-668.

Arbuthnot, P.B., Bralet, M.P., Le Jossic, C., Dedieu, J.F., Perricaudet, M., Brechot, C. and Ferry, N. (1996) In vitro and in vivo hepatoma cell-specific expression of a gene transferred with an adenoviral vector. *Hum. Gene Ther.* **7**, 1503-1514.

Argaud, D., Lange, A.J., Becker, T.C., Okar, D.A., el-Maghrabi, M.R., Newgard, C.B. and Pilkis, S.J. (1995) Adenovirus-mediated overexpression of liver 6-phosphofructo-2-kinase/fructose-2,6-bisphosphatase in gluconeogenic rat hepatoma cells. Paradoxical effect on Fru-2,6-P2 levels. *J. Biol. Chem.* **270**, 24229-24236.

Armentano, D., Sookdeo, C.C., Hehir, K.M., Gregory, R.J., St.George, J.A., Prince, G.A., Wadsworth, S.C. and Smith, A.E. (1995) Characterization of an adenovirus gene transfer vector containing an E4 deletion. *Hum. Gene Ther.* **6**, 1343-1353.

Armentano, D., Smith, M.P., Sookdeo, C.C., Zabner, J., Perricone, M.A., St George, J.A.,Wadsworth, S.C. and Gregory, R.J. (1999) E4ORF3 requirement for achieving long-term transgene expression from the cytomegalovirus promoter in adenovirus vectors. *J.virol.* **73**, 7031-7034.

Asai, A., Qiu Jh, , Narita, Y., Chi, S., Saito, N., Shinoura, N., Hamada, H., Kuchino, Y. and Kirino, T. (1999) High level calcineurin activity predisposes neuronal cells to apoptosis. *J. Biol. Chem.* **274**, 34450-34458.

Baker, R.E., Dijkhuizen, P.A., Van Pelt, J. and Verhaagen, J. (1998) Growth of pyramidal, but not non-pyramidal, dendrites in long-term organotypic explants of neonatal rat neocortex chronically exposed to neurotrophin-3. *Eur. J. Neurosci.* **10**, 1037-1044.

Baque, S., Newgard, C.B., Gerard, R.D., Guinovart, J.J. and Gomez-Foix, A.M. (1994) Adenovirus-mediated delivery into myocytes of muscle glycogen phosphorylase, the enzyme deficient in patients with glycogen-storage disease type V. *Biochem. J.* **304**, 1009-1014.

Baque, S., Montell, E., Guinovart, J.J., Newgard, C.B. and Gomez-Foix, A.M. (1998) Expression of glucokinase in cultured human muscle cells confers insulin-independent and glucose concentration-dependent increases in glucose disposal and storage. *Diabetes* **47**, 1392-1398.

Barbara, G., Xing, Z., Hogaboam, C.M., Gauldie, J., and Collins, S.M. (2000) Interleukin 10 gene transfer prevents experimental colitis in rats. *Gut.* **46**, 344-349.

Baumgartner, B.J. and Shine, H.D. (1997) Targeted transduction of CNS neurons with adenoviral vectors carrying neurotrophic factor genes confers neuroprotection that exceeds the transduced population. *J. Neurosci.* **17**, 6504-6511.

Baumgartner, B.J. and Shine, H.D. (1998a) Neuroprotection of spinal motoneurons following targeted transduction with an adenoviral vector carrying the gene for glial cell line-derived neurotrophic factor. *Exp. Neurol.* **153**, 102-112.

Baumgartner, B.J. and Shine, H.D. (1998b) Permanent rescue of lesioned neonatal motoneurons and enhanced axonal regeneration by adenovirus-mediated expression of glial cell-line-derived neurotrophic factor. *J. Neurosci. Res.* **54**, 766-777.

Becker, T.C., BeltrandelRio, H., Noel, R.J., Johnson, J.H. and Newgard, C.B. (1994) Overexpression of hexokinase I in isolated islets of Langerhans via recombinant adenovirus. Enhancement of glucose metabolism and insulin secretion at basal but not stimulatory glucose levels. *J. Biol. Chem.* **269**, 21234-21238.

Bemelmans, A.P., Horellou, P., Pradier, L., Brunet, I., Colin, P. and Mallet, J. (1999) Brain-derived neurotrophic factor-mediated protection of striatal neurons in an excitotoxic rat model of Huntington's disease, as demonstrated by adenoviral gene transfer. *Hum. Gene Ther.* **10**, 2987-2997.

Bergelson, J.M., Cunningham, J.A., Droguett, G., Kurt-Jones, E.A., Krithivas, A., Hong, J.S., Horwitz, M.S., Crowell, R.L. and Finberg, R.W. (1997) Isolation of a common receptor for Coxsackie B viruses and adenoviruses 2 and 5. *Science* **275**, 1320-1323.

Berkner, K.L. (1988) Development of adenovirus vectors for the expression of heterologous genes. *BioTechniques* **6**, 616-629

Berman, H.K. and Newgard, C.B. (1998) Fundamental metabolic differences between hepatocytes and islet beta-cells revealed by glucokinase overexpression. *Biochemistry* **37**, 4543-4552.

Berman, H.K., O'Doherty, R.M., Anderson, P. and Newgard, C.B. (1998) Overexpression of protein targeting to glycogen (PTG) in rat hepatocytes causes profound activation of glycogen synthesis independent of normal hormone- and substrate-mediated regulatory mechanisms. *J. Biol. Chem.***273**, 26421-26425.

Bett, A.J., Prevec, L. and Graham, F.L. (1993) Packaging capacity and stability of human adenovirus type 5 vectors. *J. Virol.***67**, 5911-5921.

Bewley, M.C., Springer, K., Zhang, Y.B., Freimuth, P. and Flanagan, J.M. (1999) Structural analysis of the mechanism of adenovirus binding to its human cellular receptor, CAR. *Science* **286**, 1579-1583.

Bilbao, G., Contreras, J.L., Zhang, H.G., Pike, M.J., Overturf, K., Mikheeva, G., Krasnykh, V. and Curiel, D.T. (1999) Adenovirus-mediated gene expression in vivo is enhanced by the antiapoptotic bcl-2 gene. *J. Virol.***73**, 6992-7000.

Blanchard, K.T. and Boekelheide, K.(1997) Adenovirus-mediated gene transfer to rat testis in vivo. *Biol Reprod.***56**, 495-500.

Boehler, A., Chamberlain, D., Xing, Z., Slutsky, A.S., Jordana, M., Gauldie, J., Liu, M. and Keshavjee, S. (1998) Adenovirus-mediated interleukin-10 gene transfer inhibits post-transplant fibrous airway obliteration in an animal model of bronchiolitis obliterans. *Hum. Gene Ther.***9**, 541-551.

Bondeson, J., Foxwell, B., Brennan, F. and Feldmann, M. (1999) Defining therapeutic targets by using adenovirus: blocking NF-kappaB inhibits both inflammatory and destructive mechanisms in rheumatoid synovium but spares anti-inflammatory mediators. *Proc. Natl. Acad. Sci. U. S. A.***96**, 5668-5673.

Bondesson, M., Ohman, K., Manervik, M., Fan, S. and Akusjarvi, G. (1996) Adenovirus E4 open reading frame 4 protein autoregulates E4 transcription by inhibiting E1A transactivation of the E4 promoter. *J. Virol.***70**, 3844-3851.

Booz, G.W., Carl, L.L. and Baker, K.M. (1999) Amplification of angiotensin II signaling in cardiac myocytes by adenovirus-mediated overexpression of the AT1 receptor. *Ann. N. Y. Acad. Sci.***874**, 20-26.

Bouri, K., Feero, W.G., Myerburg, M.M., Wickham, T.J., Kovesdi, I., Hoffman, E.P. and Clemens, P.R. (1999) Polylysine modification of adenoviral fiber protein enhances muscle cell transduction. *Hum. Gene Ther.***10**, 1633-1640.

Bowman, L., Grossmann, M., Rill, D., Brown, M., Zhong, W.Y., Alexander, B., Leimig, T., Coustan-Smith, E., Campana, D., Jenkins, J., Woods, D., Kitchingman, G., Vanin, E. and Brenner, M. (1998) IL-2 adenovector-transduced autologous tumor cells induce antitumor immune responses in patients with neuroblastoma. *Blood* **92**, 1941-1949.

Braciak, T.A., Mittal, S.K., Graham, F.L., Richards, C.D. and Gauldie, J. (1993) Construction of recombinant human type 5 adenoviruses expressing rodent IL-6 genes. An approach to investigate in vivo cytokine function. *J. Immunol.***151**, 5145-5153.

Braciak, T.A., Bacon, K., Xing, Z., Torry, D.J., Graham, F.L., Schall, T.J., Richards, C.D., Croitoru, K. and Gauldie, J. (1996) Overexpression of RANTES using a recombinant adenovirus vector induces the tissue-directed recruitment of monocytes to the lung. *J. Immunol.***157**, 5076.

Bramson, J., Hitt, M., Gallichan, W.S., Rosenthal, K.L., Gauldie, J. and Graham, F.L. (1996a) Construction of a double recombinant adenovirus vector expressing a heterodimeric cytokine: in vitro and in vivo production of biologically active interleukin-12. *Hum. Gene Ther.***7**, 333-342.

Bramson, J.L., Hitt, M., Addison, C.L., Muller, W.J., Gauldie, J. and Graham, F.L. (1996b) Direct intratumoral injection of an adenovirus expressing interleukin-12 induces regression and long-lasting immunity that is associated with highly localized expression of interleukin-12. *Hum. Gene Ther.***7**, 1995-2002.

Brand, K., Loser, P., Arnold, W., Bartels, T. and Strauss, M. (1998) Tumor cell-specific transgene expression prevents liver toxicity of the adeno-HSVtk/GCV approach. *Gene Ther.***5**, 1363-1371.

Brand, K., Klocke, R., Possling, A., Paul, D. and Strauss, M. (1999) Iduction of apoptosis and G2/M arrest by infection with replication-deficient adenovirus at high multiplicity of infection.*Gene Ther.***6**, 1054-1063.

Brauner, R., Nonoyama, M., Laks, H., Drinkwater, D.C.,Jr., McCaffery, S., Drake, T., Berk, A.J., Sen, L. and Wu, L. (1997) Intracoronary adenovirus-mediated transfer of immunosuppressive cytokine genes prolongs allograft survival. J. Thorac. *Cardiovasc. Surg.***114**, 923-933.

Bridge, E. and Petterson, U. (1995) Nuclear organization of replication and gene expression in adenovirus-infected cells. The molecular repertoire of adenoviruses I: Virion structure and infection, *Current Topics in Microbiology and Immunology* **199**, Eds Doerfler W. and Böhm P., p.99-117.

Bruder, J.T., Appiah, A., Kirkman, W.M., Chen, P., Tian, J., Reddy, D., Brough, D.E., Lizonova, A. and Kovesdi, I. (2000) Improved production of adenovirus vectors expressing apoptotic transgenes. *Hum. Gene Ther.***11**, 139-149.

Bui, L.A., Butterfield, L.H., Kim, J.Y., Ribas, A., Seu, P., Lau, R., Glaspy, J.A., McBride, W.H. and Economou, J.S. (1997) In vivo therapy of hepatocellular carcinoma with a tumor-specific adenoviral vector expressing interleukin-2. Hum. *Gene Ther.***8**, 2173-2182.

Burcin, M.M., Schiedner, G., Kochanek, S., Tsai, S.Y. and O'Malley, B.W. (1999) Adenovirus-mediated regulable target gene expression in vivo. *Proc. Natl. Acad. Sci. U. S. A.***96**, 355-360.

Byrnes, A.P., Wood, M.J.A. and Charlton, H.M. (1996) Role of T cells in inflammation caused by adenovirus vectors in the brain. *Gene Ther.***3**, 644-651.

Caplan, L.A. and Wilson J.M. (2000) The ethical challenges of in utero gene therapy. *Nature genetics.***24**, 107-107

Caruso, M., Pham-Nguyen, K., Kwong, Y.L., Xu, B., Kosai, K.I., Finegold, M., Woo, S.L. and Chen, S.H. (1996) Adenovirus-mediated interleukin-12 gene therapy for metastatic colon carcinoma. *Proc. Natl. Acad. Sci. U. S. A.***93**, 11302-11306.

Cascallo, M., Mercade, E., Capella, G., Lluis, F., Fillat, C., Gomez-Foix, A.M. and Mazo, A. (1999) Genetic background determines the response to adenovirus-mediated wild-type p53 expression in pancreatic tumor cells. *Cancer Gene Ther.***6**, 428-436.

Cayouette, M., Behn, D., Sendtner, M., Lachapelle, P. and Gravel, C. (1998) Intraocular gene transfer of ciliary neurotrophic factor prevents death and increases responsiveness of rod photoreceptors in the retinal degeneration slow mouse. *J. Neurosci.***18**, 9282-9293.

Charron, F., Paradis, P., Bronchain, O., Nemer, G. and Nemer, M. (1999) Cooperative interaction between GATA-4 and GATA-6 regulates myocardial gene expression. *Mol. Cell Biol.***19**, 4355-4365.

Chen, D., Krasinski, K., Sylvester, A., Chen, J., Nisen, P.D. and Andres, V. (1997a) Downregulation of cyclin-dependent kinase 2 activity and cyclin A promoter activity in vascular smooth muscle cells by p27(KIP1), an inhibitor of neointima formation in the rat carotid artery. *J. Clin. Invest.***99**, 2334-2341

Chen, G., Koyama, K., Yuan, X., Lee, Y., Zhou, Y.T., O'Doherty, R., Newgard, C.B. and Unger, R.H. (1996) Disappearance of body fat in normal rats induced by adenovirus-mediated leptin gene therapy. *Proc. Natl. Acad. Sci. U. S. A.***93**, 14795-14799.

Chen, H.H., Mack, L.M., Kelly, R., Ontell, M., Kochanek, S. and Clemens, P.R. (1997b) Persistence in muscle of an adenoviral vector that lacks all viral genes. *Proc. Natl. Acad. Sci. U. S. A.***94**, 1645-1650.

Chen, J., Bezdek, T., Chang, J., Kherzai, A.W., Willingham, T., Azzara, M. and Nisen, P.D. (1998a) A glial-specific, repressible, adenovirus vector for brain tumor gene therapy. *Cancer Res.***58**, 3504-3507.

Chen, L., Chen, D., Block, E., O'Donnell, M., Kufe, D.W. and Clinton, S.K. (1997c) Eradication of murine bladder carcinoma by intratumor injection of a bicistronic adenoviral vector carrying cDNAs for the IL-12 heterodimer and its inhibition by the IL-12 p40 subunit homodimer. *J. Immunol.***159**, 351-359.

Chen, P., Tian, J., Kovesdi, I. and Bruder, J.T. (1998b) Interaction of the adenovirus 14.7-kDa protein with FLICE inhibits Fas ligand-induced apoptosis. *J. Biol. Chem.***6**, 5815-5820.

Chen P., Kovesdi I. and Bruder J.T. (2000) Effective repeat administration with adenovirus vectors to the muscle. *Gene Ther.***7**, 587-595.

Cheney, I.W., Neuteboom, S.T., Vaillancourt, M.T., Ramachandra, M. and Bookstein, R. (1999) Adenovirus-mediated gene transfer of MMAC1/PTEN to glioblastoma cells inhibits S phase entry by the recruitment of p27Kip1 into cyclin E/CDK2 complexes. *Cancer Res.***59**, 2318-2323.

Cheon, J., Ko, S.C., Gardner, T.A., Shirakawa, T., Gotoh, A., Kao, C. and Chung, L.W. (1997) Chemogene therapy: osteocalcin promoter-based suicide gene therapy in combination with methotrexate in a murine osteosarcoma model. *Cancer Gene Ther.***4**, 359-365.

Chiche, J.D., Schlutsmeyer, S.M., Bloch, D.B., de la Monte, S.M., Roberts, J.D.,Jr., Filippov, G., Janssens, S.P., Rosenzweig, A. and Bloch, K.D. (1998) Adenovirus-mediated gene transfer of cGMP-dependent protein kinase increases the sensitivity of cultured vascular smooth muscle cells to the antiproliferative and pro-apoptotic effects of nitric oxide/cGMP. *J. Biol. Chem.***273**, 34263-34271.

Chillon, M., Bosch, A., Zabner, J., Law, L., Armentano, D., Welsh, M.J. and Davidson, B.L. (1999) Group D adenoviruses infect primary central nervous system cells more efficiently than those from group C. *J.Virol.***73**, 2537-2540

Chinnadurai, G. (1998) Control of apoptosis by human Adenovirus genes. *Seminars in Virology*,**8**, 399-408

Chintala, S.K., Fueyo, J., Gomez-Manzano, C., Venkaiah, B., Bjerkvig, R., Yung, W.K., Sawaya, R., Kyritsis, A.P. and Rao, J.S. (1997) Adenovirus-mediated p16/CDKN2 gene transfer suppresses glioma invasion in vitro. *Oncogene* **15**, 2049-2057.

Choukroun, G., Hajjar, R., Kyriakis, J.M., Bonventre, J.V., Rosenzweig, A. and Force, T. (1998) Role of the stress-activated protein kinases in endothelin-induced cardiomyocyte hypertrophy. *J. Clin. Invest.***102**, 1311-1320.

Choukroun, G., Hajjar, R., Fry, S., del Monte, F., Haq, S., Guerrero, J.L., Picard, M., Rosenzweig, A. and Force, T. (1999) Regulation of cardiac hypertrophy in vivo by the stress-activated protein kinases/c-Jun NH(2)-terminal kinases. *J. Clin. Invest.***104**, 391-398.

Christ, M., Louis, B., Stoeckel, F., Dieterle, A., Grave, L., Dreyer, D., Kintz, J., Ali Hadji, D., Lusky, M. and Mehtali, M. (2000) Modulation of the inflammatory properties and hepatotoxicity of recombinant adenovirus vectors by the viral E4 gene products. *Hum. Gene Ther.***11**, 415-427.

Christensen, G., Minamisawa, S., Gruber, P.J., Wang, Y. and Chien, K.R. (2000) High-efficiency, long-term cardiac expression of foreign genes in living mouse embryos and neonates. *Circulation* **101**, 178-184.

Cichon, G. and Strauss, M. (1998) Transient immunosuppression with 15-deoxyspergualin prolongs reporter gene expression and reduces humoral immune response after adenoviral gene transfer. *Gene Ther.***5**, 85-90.

Cirielli, C., Inyaku, K,, Capogrossi, M.C., Yuan, X. and Williams, J.A. (1999) Adenovirus-mediated wild-type p53 expression induces apoptosis and suppresses tumorigenesis of experimental intracranial human malignant glioma. *J.Neurooncol.***43**, 99-108.

Clarke, M.F., Apel, I.J., Benedict, M.A., Eipers, P.G., Sumantran, V., Gonzalez-Garcia, M., Doedens, M., Fukunaga, N., Davidson, B. and Dick, J.E. (1995) A recombinant bcl-x s adenovirus selectively induces apoptosis in cancer cells but not in normal bone marrow cells. *Proc. Natl. Acad. Sci. U. S. A.***92**, 11024-11028.

Connelly, S., Gardner, J.M., McClelland, A. and Kaleko, M. (1996) High-level tissue-specific expression of functional human factor VIII in mice. *Hum. Gene Ther.***7**, 183-195.

Cordier, L., Duffour, M.T., Sabourin, J.C., Lee, M.G., Cabannes, J., Ragot, T., Perricaudet, M. and Haddada, H. (1995) Complete recovery of mice from a pre-established tumor by direct intratumoral delivery of an adenovirus vector harboring the murine IL-2 gene. *Gene Ther.***2**, 16-21.

Corti, O., Sanchez-Capelo, A., Colin, P., Hanoun, N., Hamon, M. and Mallet, J. (1999) Long-term doxycycline-controlled expression of human tyrosine hydroxylase after direct adenovirus-mediated gene transfer to a rat model of Parkinson's disease. *Proc. Natl. Acad. Sci. U. S. A.***96**, 12120-12125.

Craig, C., Kim, M., Ohri, E., Wersto, R., Katayose, D., Li, Z., Choi, Y.H., Mudahar, B., Srivastava, S., Seth, P. and Cowan. K. (1998) Effects of adenovirus-mediated p16INK4A expression on cell cycle arrest are determined by endogenous p16 and Rb status in human cancer cells. *Oncogene* **15**, 265-272.

Cregan, S.P., MacLaurin, J.G., Craig, C.G., Robertson, G.S., Nicholson, D.W., Park, D.S. and Slack, R.S. (1999) Bax-dependent caspase-3 activation is a key determinant in p53-induced apoptosis in neurons. *J. Neurosci.***19**, 7860-7869.

Dales, S. and Chardonnet, Y. (1973) Early events in the interaction of adenoviruses with HeLa cells. IV. Association with microtubules and the nuclear pore complex during vectorial movement of the inoculum. *Virology* **56**, 465-483.

Dedieu, J.F., Vigne, E., Torrent, C., Jullien, C., Mahfouz, I., Caillaud, J.M., Aubailly, N., Orsini, C., Perricaudet, M. and Yeh, P. (1997) Long-term gene delivery into the livers of immunocompetent mice with E1/E4-defective adenoviruses. *J. Virol.***71**, 4626-4637.

Deguchi, J., Namba, T., Hamada, H., Nakaoka, T., Abe, J., Sato, O., Miyata, T., Makuuchi, M., Kurokawa, K. and Takuwa, Y. (1999) Targeting endogenous platelet-derived growth factor B-chain by adenovirus-mediated gene transfer potently inhibits in vivo smooth muscle proliferation after arterial injury. *Gene Ther.***6**, 956-965.

DeMatteo, R.P., Markmann, J.F., Kozarsky, K.F., Barker, C.F. and Raper, S.E. (1996) Prolongation of adenoviral transgene expression in mouse liver by T lymphocyte subset depletion. *Gene Ther.***3,** 4-12.

Dematteo, R.P., McClane, S.J., Fisher, K., Yeh, H., Chu, G., Burke, C. and Raper, S.E. (1997) Engineering tissue-specific expression of a recombinant adenovirus: selective transgene transcription in the pancreas using the amylase promoter. *J. Surg. Res.***72**, 155-161.

De Nichilo, M.O., Katz, B.Z., O'Connell, B. and Yamada, K.M. (1999) De novo expression of pp125FAK in human macrophages regulates CSK distribution and MAP kinase activation but does not affect focal contact structure. *J. Cell Physiol.***178**, 164-172.

Descamps, V., Blumenfeld, N., Villeval, J.L., Vainchenker, W., Perricaudet, M. and Beuzard, Y. (1994) Erythropoietin gene transfer and expression in adult normal mice: use of an adenovirus vector. *Hum. Gene Ther.***5**, 979-985.

Di Polo, A., Aigner, L.J., Dunn, R.J., Bray, G.M. and Aguayo, A.J. (1998) Prolonged delivery of brain-derived neurotrophic factor by adenovirus-infected Muller cells temporarily rescues injured retinal ganglion cells. *Proc. Natl. Acad. Sci. U. S. A.***95**, 3978-3983.

Diehl, A.M., Johns, D.C., Yang, S., Lin, H., Yin, M., Matelis, L.A. and Lawrence, J.H. (1996) Adenovirus-mediated transfer of CCAAT/enhancer-binding protein-alpha identifies a dominant antiproliferative role for this isoform in hepatocytes. *J. Biol. Chem.***271**, 7343-7350.

Dijkhuizen, P.A., Hermens, W.T., Teunis, M.A. and Verhaagen, J. (1997) Adenoviral vector-directed expression of neurotrophin-3 in rat dorsal root ganglion explants results in a robust neurite outgrowth response. *J. Neurobiol.***33**, 172-184.

Dmitriev, I., Krasnykh, V., Miller, C.R., Wang, M., Kashentseva, E., Mikheeva, G., Belousova, N. and Curiel, D.T. (1998) An adenovirus vector with genetically modified fibers demonstrates expanded tropism via utilization of a coxsackievirus and adenovirus receptor-independent cell entry mechanism. *J. Virol.***72**, 9706-9713.

Doi, K., Itoh, H., Ikeda, T., Hosoda, K., Ogawa, Y., Igaki, T., Yamashita, J., Chun, T.H., Inoue, M., Masatsugu, K., Matsuda, K., Ohmori, K. and Nakao, K. (1997) Adenovirus-mediated gene transfer of C-type natriuretic peptide causes G1 growth inhibition of cultured vascular smooth muscle cells. Biochem. Biophys. *Res. Commun.***239**, 889-894.

Dole, M.G., Clarke, M.F., Holman, P., Benedict, M., Lu, J., Jasty, R., Eipers, P., Thompson, C.B., Rode, C., Bloch, C., Nunez, P. and Castle, V.P. (1996) Bcl-xS enhances adenoviral vector-induced apoptosis in neuroblastoma cells. *Cancer Res.***56**, 5734-5740.

Dong, Y.B., Yang, H.L., Jane, M., Liu, T.J., Stilwell, A., Atienza, C.-Jr. and McMasters, K.M. (1999) Adenovirus-mediated E2F-1 gene transfer efficiently induces apoptosis in melanoma cells. *Cancer* **86,** 2021-2033.

Dorai, T., Perlman, H., Walsh, K., Shabsigh, A., Goluboff, E.T., Olsson, C.A. and Buttyan, R. (1999) A recombinant defective adenoviral agent expressing anti-bcl-2 ribozyme promotes apoptosis of bcl-2-expressing human prostate cancer cells. *Int. J. Cancer* **82**, 846-852.

Douar, A.M., Adebakin, S., Themis, M., Pavirani, A., Cook, T. and Coutelle, C. (1997) Foetal gene delivery in mice by intra-amniotic administration of retroviral producer cells and adenovirus. *Gene Ther.***4**, 883-890.

Douglas, J.T., Rogers, B.E., Rosenfeld, M.E., Michael, S.I., Feng, M. and Curiel, D.T. (1996) Targeted gene delivery by tropism-modified adenoviral vectors. *Nat. Biotechnol.***14**, 1574-1578.

Douglas, J.T., Miller, C.R., Kim, M., Dmitriev, I., Mikheeva, G., Krasnykh, V. and Curiel, D.T. (1999) A system for the propagation of adenoviral vectors with genetically modified receptor specificities. *Nat. Biotechnol.***17**, 470-475.

Drazan, K.E., Wu, L., Olthoff, K.M., Jurim, O., Busuttil, R.W. and Shaked, A. (1995) Transduction of hepatic allografts achieves local levels of viral IL-10 which suppress alloreactivity in vitro. *J. Surg. Res.***59**, 219-223.

Drozdzik, M., Qian, C., Xie, X., Peng, D., Bilbao, R., Mazzolini, G. and Prieto, J. (2000) Combined gene therapy with suicide gene and interleukin-12 is more efficient than therapy with one gene alone in a murine model of hepatocellular carcinoma. *J. Hepatol.***32**, 279-286.

Duisit, G., Salvetti, A., Moullier, P. and Cosset, F.L. (1999) Functional characterization of adenoviral/retroviral chimeric vectors and their use for efficient screening of retroviral producer cell lines. *Hum. Gene Ther.***10**, 189-200.

Durham, H.D., Alonso-Vanegas, M.A., Sadikot, A.F., Zhu, L., Lochmuller. H., Massie, B., Nalbantoglu, J. and Karpati, G. (1997) The immunosuppressant FK506 prolongs transgene expression in brain following adenovirus-mediated gene transfer. *Neuroreport* **8**, 2111-2115.

Dusetti, N.J., Vasseur, S., Ortiz, E.M., Romeo, H., Dagorn, J.C., Burrone, O. and Iovanna, J.L. (1997) The pancreatitis-associated protein I promoter allows targeting to the pancreas of a foreign gene, whose expression is up-regulated during pancreatic inflammation. *J. Biol. Chem.***272**, 5800-5804.

Ealovega, M.W., McGinnis, P.K., Sumantran, V.N., Clarke, M.F. and Wicha, M.S. (1996) bcl-xs gene therapy induces apoptosis of human mammary tumors in nude mice. *Cancer Res.***56**, 1965-1969.

Ehrengruber, M.U., Doupnik, C.A., Xu, Y., Garvey, J., Jasek, M.C., Lester, H.A. and Davidson, N. (1997) Activation of heteromeric G protein-gated inward rectifier K+ channels overexpressed by adenovirus gene transfer inhibits the excitability of hippocampal neurons. *Proc. Natl. Acad. Sci. U. S. A.***94**, 7070-7075.

Einfeld, D.A., Brough, D.E., Roelvink, P.W., Kovesdi, I. and Wickham, T.J. (1999) Construction of a pseudoreceptor that mediates transduction by adenoviruses expressing a ligand in fiber or penton base. *J. Virol.***73**, 9130-9136.

Emtage, P.C., Wan, Y., Bramson, J.L., Graham, F.L. and Gauldie, J. (1998) A double recombinant adenovirus expressing the costimulatory molecule B7-1 (murine) and human IL-2 induces complete tumor regression in a murine breast adenocarcinoma model. *J. Immunol.***160**, 2531-2538.

Emtage, P.C., Wan, Y., Hitt, M., Graham, F.L., Muller, W.J., Zlotnik, A. and Gauldie, J. (1999) Adenoviral vectors expressing lymphotactin and interleukin 2 or lymphotactin and interleukin 12 synergize to facilitate tumor regression in murine breast cancer models. *Hum. Gene Ther.***10**, 697-709.

Endo, K., Sakatani, T., Watanabe, M., Yoshida, H., Nanba, E. and Ito, H. (1999) Wild-type p53 gene transfer resulted in cell cycle arrest, but not apoptosis of newly established human malignant fibrous histiocytoma cell line. *Int. J. Oncol.***15**, 935-942.

Engelhardt, J.F., Litzky, L. and Wilson, J.M. (1994a) Prolonged transgene expression in cotton rat lung with recombinant adenoviruses defective in E2a. *Hum. Gene Ther.***5**, 1217-1229.

Engelhardt, J.F., Ye, X., Doranz, B. and Wilson, J.M. (1994b) Ablation of E2A in recombinant adenoviruses improves transgene persistence and decreases inflammatory response in mouse liver. *Proc. Natl. Acad. Sci. U. S. A.***91**, 6196-6200.

Esandi, C.E., van Someren, G.D., Bout, A., Mulder, A.H., van Bekkum, D.W., Valerio, D. and Noteboom, J.L. (1998a) IL-1/IL-3 gene therapy of non-small cell lung cancer (NSCLC) in rats using 'cracked' adenoproducer cells. *Gene Ther.***5**, 778-788.

Esandi, C.E., van Someren, G.D., van der Velde, I., van Bekkum, D.W., Valerio, D., Noteboom, J.L. and Bout, A. (1998b) Cloning, biological characterization and high-level expression of rat interleukin-3 using recombinant adenovirus: description of a new splicing variant. *Gene* **211**, 151-158.

Etgen, G.J., Valasek, K.M., Broderick, C.L. and Miller, A.R. (1999) In vivo adenoviral delivery of recombinant human protein kinase C-zeta stimulates glucose transport activity in rat skeletal muscle. *J. Biol. Chem.***274**, 22139-22142.

Fallaux, F.J., Kranenburg, O., Cramer, S.J., Houweling, A., Van Ormondt, H., Hoeben, R.C. and Van Der Eb, A.J. (1996) Characterization of 911: a new helper cell line for the titration and propagation of early region 1-deleted adenoviral vectors. *Hum. Gene Ther.***7**, 215-222.

Fang, B., Eisensmith, R.C., Wang, H., Kay, M.A., Cross, R.E., Landen, C.N., Gordon, G., Bellinger, D.A., Read, M.S., Hu, P.C., Brinkhous, K.M. and Woo, S.L.C. (1995) Gene therapy for hemophilia B: host immunosuppression prolongs the therapeutic effect of adenovirus-mediated factor IX expression. *Hum. Gene Ther.***6**, 1039-1044.

Farre, L., Rigau, T., Mogas, T., Garcia-Rocha, M., Canal, M., Gomez-Foix, A.M. and Rodriguez-Gil, J.E. (1999) Adenovirus-mediated introduction of DNA into pig sperm and offspring. *Mol. Reprod. Dev.***53**,149-158.

Fathallah-Shaykh, H.M., Zhao, L.J., Kafrouni, A.I., Smith, G.M. and Forman, J. (2000) Gene transfer of IFN-gamma into established brain tumors represses growth by antiangiogenesis. *J. Immunol.***164**, 217-222.

Fechner, H., Haack, A., Wang, H., Wang, X., Eizema, K., Pauschinger, M., Schoemaker, R., Veghel, R., Houtsmuller, A., Schultheiss, H.P., Lamers, J. and Poller, W. (1999) Expression of coxsackie adenovirus receptor and alphav-integrin does not correlate with adenovector targeting in vivo indicating anatomical vector barriers. *Gene Ther.***6**, 1520-1535.

Feldman, E., Ahmed, T., Lutton, J.D., Farley, T., Tani, K., Freund, M., Asano, S. and Abraham, N.G. (1997) Adenovirus mediated alpha interferon (IFN-alpha) gene transfer into CD34+ cells and CML mononuclear cells. Stem. *Cells* **15**, 386-395.

Felzmann, T., Ramsey, W.J. and Blaese, R.M. (1997) Characterization of the antitumor immune response generated by treatment of murine tumors with recombinant adenoviruses expressing HSVtk, IL-2, IL-6 or B7-1. *Gene Ther.***4**, 1322-1329.

Feng, M., Jackson, W.H Jr., Goldman, C.K., Rancourt, C., Wang, M., Dusing, S.K., Siegal, G. and Curiel, D.T. (1997) Stable in vivo gene transduction via a novel adenoviral/retroviral chimeric vector. *Nat. Biotechnol.***15**, 866-870.

Fernandez, N.C., Levraud, J.P., Haddada, H., Perricaudet, M. and Kourilsky, P. (1999) High frequency of specific CD8+ T cells in the tumor and blood is associated with efficient local IL-12 gene therapy of cancer. *J. Immunol.***162**, 609-617.

Finke, S., Trojaneck, B., Lefterova, P., Csipai, M., Wagner, E., Kircheis, R., Neubauer, A., Huhn, D., Wittig, B. and Schmidt-Wolf, I.G. (1998) Increase of proliferation rate and enhancement of antitumor cytotoxicity of expanded human CD3+ CD56+ immunologic effector cells by receptor-mediated transfection with the interleukin-7 gene. *Gene Ther.***5**, 31-39.

Fisher, S.A., Siwik, E., Branellec, D., Walsh, K. and Watanabe, M. (1997) Forced expression of the homeodomain protein Gax inhibits cardiomyocyte proliferation and perturbs heart morphogenesis. *Development* **124**, 4405-4413.

Foger, B., Santamarina-Fojo, S., Shamburek, R.D., Parrot, C.L., Talley, G.D. and Brewer, H.B.,Jr. (1997) Plasma phospholipid transfer protein. Adenovirus-mediated overexpression in mice leads to decreased plasma high density lipoprotein (HDL) and enhanced hepatic uptake of phospholipids and cholesteryl esters from HDL. *J. Biol. Chem.***272**, 27393-27400.

Foretz, M., Guichard, C., Ferre, P. and Foufelle, F. (1999a) Sterol regulatory element binding protein-1c is a major mediator of insulin action on the hepatic expression of glucokinase and lipogenesis-related genes. *Proc. Natl. Acad. Sci. U. S. A.***96**, 12737-12742.

Foretz, M., Pacot, C., Dugail, I., Lemarchand, P., Guichard, C., Le Liepvre, X., Berthelier-Lubrano, C., Spiegelman, B., Kim, J.B., Ferre, P. and Foufelle, F. (1999b) ADD1/SREBP-1c is required in the activation of hepatic lipogenic gene expression by glucose. *Mol. Cell Biol.***19**, 3760-3768.

Franz, W.M., Rothmann, T., Frey, N. and Katus, H.A. (1997) Analysis of tissue-specific gene delivery by recombinant adenoviruses containing cardiac-specific promoters. *Cardiovasc. Res.***35**, 560-566.

Freimuth, P., Springer, K., Berard, C., Hainfeld, J., Bewley, M. and Flanagan, J. (1999) Coxsackievirus and adenovirus receptor amino-terminal immunoglobulin V-related domain binds adenovirus type 2 and fiber knob from adenovirus type 12. *J. Virol.***73**, 1392-1398.

Frevert, E.U. and Kahn, B.B. (1997) Differential effects of constitutively active phosphatidylinositol 3-kinase on glucose transport, glycogen synthase activity, and DNA synthesis in 3T3-L1 adipocytes. *Mol. Cell Biol.***17**, 190-198.

Frizelle, S.P., Grim, J., Zhou, J., Gupta, P., Curiel, D.T., Geradts, J. and Kratzke, R.A. (1998) Re-expression of p16INK4a in mesothelioma cells results in cell cycle arrest, cell death, tumor suppression and tumor regression. *Oncogene* **16**, 3087-3095.

Fueyo, J., Gomez-Manzano, C., Yung, W.K., Liu, T.J., Alemany, R., McDonnell, T.J., Shi, X., Rao, J.S., Levin, V.A. and Kyritsis, A.P.(1998) Overexpression of E2F-1 in glioma triggers apoptosis and suppresses tumor growth in vitro and in vivo. *Nat. Med.***4**, 685-690.

Fukazawa, T., Fujiwara, T., Morimoto, Y., Shao, J., Nishizaki, M., Kadowaki, Y., Hizuta, A., Owen-Schaub, L.B., Roth, J.A. and Tanaka, N. (1999) Differential involvement of the CD95 (Fas/APO-1) receptor/ligand system on apoptosis induced by the wild-type p53 gene transfer in human cancer cells. *Oncogene* **18**, 2189-2199

Gabaglia, C.R., Pedersen, B., Hitt, M., Burdin, N., Sercarz, E.E., Graham, F.L., Gauldie, J. and Braciak, T.A. (1999) A single intramuscular injection with an adenovirus-expressing IL-12 protects BALB/c mice against Leishmania major infection, while treatment with an IL-4-expressing vector increases disease susceptibility in B10.D2 mice. *J. Immunol.***162**, 753-760.

Gambotto, A., Tuting, T., McVey, D.L., Kovesdi, I., Tahara, H., Lotze, M.T. and Robbins, P.D. (1999) Induction of antitumor immunity by direct intratumoral injection of a recombinant adenovirus vector expressing interleukin-12. *Cancer Gene Ther.***6**, 45-53.

Garcia-Martinez, C., Opolon, P., Trochon, V., Chianale, C., Musset, K., Lu, H., Abitbol, M., Perricaudet, M. and Ragot, T. (1999) Angiogenesis induced in muscle by a recombinant adenovirus expressing functional isoforms of basic fibroblast growth factor. *Gene Ther.***6**, 1210-1221.

Garnier, A., Cote, J., Nadeau, I., Kamen, A. and Massie, B. (1994) Scale-up of the adenovirus expression system for the production of recombinant protein in human 293S cells. *Cytotechnology* **15**, 145-155.

Gasa, R., Trinh, K.Y., Yu, K., Wilkie, T.M. and Newgard, C.B. (1999) Overexpression of G11alpha and isoforms of phospholipase C in islet beta-cells reveals a lack of correlation between inositol phosphate accumulation and insulin secretion. *Diabetes* **48**, 1035-1044.

Ghivizzani, S.C., Lechman, E.R., Kang, R., Tio, C., Kolls, J., Evans, C.H. and Robbins, P.D. (1998) Direct adenovirus-mediated gene transfer of interleukin 1 and tumor necrosis factor alpha soluble receptors to rabbit knees with experimental arthritis has local and distal anti-arthritic effects. *Proc. Natl. Acad. Sci. U. S. A.***95**, 4613-4618.

Gibson, S.A., Pellenz, C., Hutchison, R.E., Davey, F.R. and Shillitoe, E.J. (2000) Induction of apoptosis in oral cancer cells by an anti-bcl-2 ribozyme delivered by an adenovirus vector. Clin. *Cancer Res.***6**, 213-222.

Gilgenkrantz, H., Duboc, D., Juillard, V., Couton, D., Pavirani, A., Guillet, J.G., Briand, P. and Kahn, A. (1995) Transient expression of genes transferred in vivo into heart using first-generation adenoviral vectors: role of the immune response. *Hum. Gene Ther.***6**, 1265-1274.

Gilly, F.N., Beaujard, A., Bienvenu, J., Trillet Lenoir, V., Glehen, O., Thouvenot, D., Malcus, C., Favrot, M., Dumontet, C., Lombard-Bohas, C., Garbit, F., Gueugniaud, P.Y., Vignal, J., Aymard, M., Touraine Moulin, F., Roos, M., Pavirani, A. and Courtney, M. (1999) Gene therapy with Adv-IL-2 in unresectable digestive cancer: phase I-II study, intermediate report. *Hepatogastroenterology* **46** Suppl 1, 1268-1273.

Gimenez, R., Revah, F., Pradier, L., Loquet, I., Mallet, J. and Privat, A. (1997) Prevention of motoneuron death by adenovirus-mediated neurotrophic factors. *J. Neurosci. Res.***48**, 281-285.

Goldman, C.K., Rogers, B.E., Douglas, J.T., Sosnowski, B.A., Ying, W., Siegal, G.P., Baird, A., Campain, J.A. and Curiel, D.T. (1997) Targeted gene delivery to Kaposi's sarcoma cells via the fibroblast growth factor receptor. *Cancer Res.***57**, 1447-1451.

Gomez-Foix, A.M., Coats, W.S., Baque, S., Alam, T., Gerard, R.D. and Newgard, C.B. (1992) Adenovirus-mediated transfer of the muscle glycogen phosphorylase gene into hepatocytes confers altered regulation of glycogen metabolism. *J. Biol. Chem.***267**, 25129-25134.

Gomez-Manzano, C., Fueyo, J., Kyritsis, A.P., McDonnell, T.J., Steck, P.A., Levin, V.A. and Yung, W.K. (1997) Characterization of p53 and p21 functional interactions in glioma cells en route to apoptosis. *J Natl Cancer Inst.***89**, 1036-1044.

Gomez-Manzano, C., Fueyo, J., Alameda, F., Kyritsis, A.P. and Yung, W.K. (1999) Gene therapy for gliomas: p53 and E2F-1 proteins and the target of apoptosis. *Int. J. Mol. Med.***3**, 81-85. Review.

Gorziglia, M.I., Kadan, M.J., Yei, S., Lim, J., Lee, G.M., Luthra, R. and Trapnell, B.C. (1996) Elimination of both E1 and E2 from adenovirus vectors further improves prospects for in vivo human gene therapy. *J. Virol.***70**, 4173-4178.

Gorziglia, M.I., Lapcevich, C., Roy, S., Kang, Q., Kadan, M., Wu, V., Pechan, P. and Kaleko, M. (1999) Generation of an adenovirus vector lacking E1, e2a, E3, and all of E4 except open reading frame 3. *J.Virol.***73**, 6048-6055.

Gossen, M., Bonin, A.L., Freundlieb, S. and Bujard, H. (1994) Inducible gene expression systems for higher eukaryotic cells. *Curr. Opin. Biotechnol.***5**, 516-520.

Gotoh, A., Kao, C., Ko, S.C., Hamada, K., Liu, T.J. and Chung, L.W. (1997) Cytotoxic effects of recombinant adenovirus p53 and cell cycle regulator genes (p21 WAF1/CIP1 and p16CDKN4) in human prostate cancers. *J. Urol.***158**, 636-641.

Gotoh, A., Ko, S.C., Shirakawa, T., Cheon, J., Kao, C., Miyamoto, T., Gardner, T.A., Ho, L.J., Cleutjens, C.B., Trapman, J., Graham, F.L. and Chung, L.W. (1998) Development of prostate-specific antigen promoter-based gene therapy for androgen-independent human prostate cancer. *J. Urol.***160**, 220-229.

Graham, F.L., Smiley, J., Russell, W.C. and Nairn, R. (1977) Characteristics of a human cell line transformed by DNA from human adenovirus type 5. *J. Gen. Virol.***36**, 59-74.

Graham, F.L. and Prevec, L.(1992) Adenovirus-based expression vectors and recombinant vaccines. In: Ellis, R.W. (ed.): Vaccines: New Approaches to Immunological Problems. *Butterworth-Heinemann*, Boston MA., pp 363-389.

Greber, U.F., Willetts, M., Webster, P. and Helenius, A. (1993) Stepwise dismantling of adenovirus 2 during entry into cells. *Cell* **75**, 477-486.

Greber, U.F., Webster, P., Weber, J. and Helenius, A. (1996) The role of the adenovirus protease on virus entry into cells. *EMBO J.***15**, 1766-1777.

Greenberger, M.J., Kunkel, S.L., Strieter, R.M., Lukacs, N.W., Bramson, J., Gauldie, J., Graham, F.L., Hitt, M., Danforth, J.M. and Standiford, T.J. (1996) IL-12 gene therapy protects mice in lethal Klebsiella pneumonia. *J. Immunol.***157**, 3006-3012.

Grey, S.T., Arvelo, M.B., Hasenkamp, W., Bach, F.H. and Ferran, C. (1999a) A20 inhibits cytokine-induced apoptosis and nuclear factor kappaB-dependent gene activation in islets. *J. Exp. Med.***190**, 1135-1146.

Grey, S.T., Arvelo, M.B., Hasenkamp, W.M., Bach, F.H. and Ferran, C. (1999b) Adenovirus-mediated gene transfer of the anti-apoptotic protein A20 in rodent islets inhibits IL-1 beta-induced NO release. *Transplant. Proc.***31**, 789.

Griscelli, F., Opolon, P., Chianale, C., Di Falco, N., Franz, W.M., Perricaudet, M. and Ragot, T. (1997) Expression from cardiomyocyte-specific promoter after adenovirus-mediated gene transfer in vitro and in vivo. *C. R. Acad. Sci. III* **320**, 103-112.

Griscelli, F., Gilardi-Hebenstreit, P., Hanania, N., Franz, W.M., Opolon, P., Perricaudet, M. and Ragot, T. (1998) Heart-specific targeting of beta-galactosidase by the ventricle-specific cardiac myosin light chain 2 promoter using adenovirus vectors. *Hum. Gene Ther.***9**, 1919-1928.

Gu, D.L., Gonzalez, A.M., Printz, M.A., Doukas, J., Ying, W., D'Andrea, M., Hoganson, D.K., Curiel, D.T., Douglas, J.T., Sosnowski, B.A., Baird, A., Aukerman, S.L. and Pierce, G.F. (1999) Fibroblast growth factor 2 retargeted adenovirus has redirected cellular tropism: evidence for reduced toxicity and enhanced antitumor activity in mice. *Cancer Res.***59**, 2608-2614.

Guibinga, G.H., Lochmuller, H., Massie, B., Nalbantoglu, J., Karpati, G. and Petrof, B.J. (1998) Combinatorial blockade of calcineurin and CD28 signaling facilitates primary and secondary therapeutic gene transfer by adenovirus vectors in dystrophic (mdx) mouse muscles. *J Virol.***72**, 4601-4609.

Guo, N., Faller, D.V. and Vaziri, C. (2000) A novel DNA damage checkpoint involving post-transcriptional regulation of cyclin A expression. *J. Biol. Chem.***275**, 1715-1722.

Guo, Z.S., Wang, L.H., Eisensmith, R.C. and Woo, S.L. (1996) Evaluation of promoter strength for hepatic gene expression in vivo following adenovirus-mediated gene transfer. *Gene Ther.***3**, 802-810.

Haase, G., Pettmann, B., Vigne, E., Castelnau-Ptakhine, L., Schmalbruch, H. and Kahn, A. (1998) Adenovirus-mediated transfer of the neurotrophin-3 gene into skeletal muscle of pmn mice: therapeutic effects and mechanisms of action. *J. Neurol. Sci.***160** Suppl 1, S97-105.

Haddada, H., Ragot, T., Cordier, L., Duffour, M.T. and Perricaudet, M. (1993) Adenoviral interleukin-2 gene transfer into P815 tumor cells abrogates tumorigenicity and induces antitumoral immunity in mice. *Hum. Gene Ther.***4**, 703-711.

Hahn, T.M., Copeland, K.C. and Woo, S.L. (1996) Phenotypic correction of dwarfism by constitutive expression of growth hormone. *Endocrinology* **137**, 4988-4993.

Haisma, H., Pinedo, H., Rijswijk, A., der Meulen-Muileman, I., Sosnowski, B., Ying, W., Beusechem, V., Tillman, B., Gerritsen, W. and Curiel, D. (1999) Tumor-specific gene transfer via an adenoviral vector targeted to the pan-carcinoma antigen EpCAM. *Gene Ther.***6**, 1469-1474.

Halbert, D.N., Cutt, J.R. and Shenk, T. (1985) Adenovirus early region 4 encodes functions required for efficient DNA replication, late gene expression, and host cell shutoff. *J. Virol.***56**, 250-257.

Hallenbeck, P.L., Chang, Y.N., Hay, C., Golightly, D., Stewart, D., Lin, J., Phipps, S. and Chiang, Y.L. (1999) A novel tumor-specific replication-restricted adenoviral vector for gene therapy of hepatocellular carcinoma. *Hum. Gene Ther.***10**, 1721-1733.

Hanna, A.K., Fox, J.C., Neschis, D.G., Safford, S.D., Swain, J.L. and Golden, M.A. (1997) Antisense basic fibroblast growth factor gene transfer reduces neointimal thickening after arterial injury. *J. Vasc. Surg.***25**, 320-325.

Harada, H., Nakagawa, K., Iwata, S., Saito, M., Kumon, Y., Sakaki, S., Sato, K. and Hamada, K. (1999) Restoration of wild-type p16 down-regulates vascular endothelial growth factor expression and inhibits angiogenesis in human gliomas. *Cancer Res.***59**, 3783-3789.

Harrod, K.S., Hermiston, T.W., Trapnell, B.C., Wold, W.S. and Whitsett, J.A. (1998) Lung-specific expression of adenovirus E3-14.7K in transgenic mice attenuates adenoviral vector-mediated lung inflammation and enhances transgene expression. *Hum. Gene Ther.***9**, 1885-1898.

Hartig, P.C. and Hunter, E.S. (1998) Gene delivery to the neurulating embryo during culture. *Teratology* **58**, 103-112.

Hashimoto, M., Aruga, J., Hosoya, Y., Kanegae, Y., Saito, I., and Mikoshiba, K. (1996) A neural cell-type-specific expression system using recombinant adenovirus vectors. *Hum. Gene Ther.*,**7**, 149-158.

Hashimoto, Y., Kohri, K., Akita, H., Mitani, K., Ikeda, K. and Nakanishi, M. (1997) Efficient transfer of genes into senescent cells by adenovirus vectors via highly expressed alpha v beta 5 integrin. *Biochem. Biophys. Res. Commun.***240**, 88-92.

He, X., Goldsmith, C.M., Marmary, Y., Wellner, R.B., Parlow, A.F., Nieman, L.K. and Baum, B.J. (1998) Systemic action of human growth hormone following adenovirus-mediated gene transfer to rat submandibular glands. *Gene Ther.***5**, 537-541.

Heeren, J., Steinwaerder, D.S., Schnieders, F., Cichon, G., Strauss, M. and Beisiegel, U. (1999) Nonphysiological overexpression of low-density lipoprotein receptors causes pathological intracellular lipid accumulation and the formation of cholesterol and cholesteryl ester crystals in vitro. *J. Mol. Med.***77**, 735-743.

Hegland, D.D., Sullivan, D.M., Rovira, I.I., Li, A., Kovesdi, I., Bruder, J.T. and Finkel, T. (1999) Regulation of endothelial cell adherens junctions by a Ras-dependent signal transduction pathway. *Biochem. Biophys. Res. Commun.***260**, 371-376.

Hehir, K.M., Armentano, D., Cardoza.L.M., Choquette,T.L., Berthelette,P.B., Whithe, G.A., Couture, L.A., Everton, M.B., Keegan, J., Martin, J.M., Pratt, D.A., Smith, M.P., Smith, A.E. and Wadsworth, S.C. (1996) Molecular charcterization of replication-competent variants of adenovirus vectors and genome modifications to prevent their occurence. *J.Virol.***70**, 8459-8467.

Heike, Y., Takahashi, M., Kanegae, Y., Sato, Y., Saito, I. and Saijo, N. (1997) Interleukin-2 gene transduction into freshly isolated lung adenocarcinoma cells with adenoviral vectors. *Hum. Gene Ther.***8**, 1-14.

Heldlund, T.E., Meech, S.J., Srikanth, S., Kraft, A.S., Miller, G.J., Schaack, J.B. and Duke, R.C. (1999) Adenovirus-mediated expression of Fas ligand induces apoptosis of human prostate cancer cells. *Cell Death Differ.***6**, 175-182.

Hemmi, S., Geertsen, R., Mezzacasa, A., Peter, I. and Dummer, R. (1998) The presence of human coxsackievirus and adenovirus receptor is associated with efficient adenovirus-mediated transgene expression in human melanoma cell cultures. *Hum. Gene Ther.***9**, 2363-2373.

Henke, P.K., DeBrunye, L.A., Strieter, R.M., Bromberg, J.S., Prince, M., Kadell, A.M., Sarkar, M., Londy, F. and Wakefield, T.W. (2000) Viral IL-10 gene transfer decreases inflammation and cell adhesion molecule expression in a rat model of venous thrombosis. *J. Immunol.***164**, 2131-2141.

Hirao, T., Sawada, H., Koyama, F., Watanabe, A., Yamada, Y., Sakaguchi, T., Tatsumi, M., Fujimoto, H., Emoto, K., Narikiyo, M., Oridate, N. and Nakano, H. (1999) Antisense epidermal growth factor receptor delivered by adenoviral vector blocks tumor growth in human gastric cancer. *Cancer Gene Ther.***6**, 423-427.

Hirschowitz, E.A. and Crystal, R.G. (1999a) Adenovirus-mediated expression of interleukin-12 induces natural killer cell activity and complements adenovirus-directed gp75 treatment of melanoma lung metastases. Am. J. Respir. *Cell Mol. Biol.***20**, 935-941.

Hirschowitz, E.A., Naama, H.A., Evoy, D., Lieberman, M.D., Daly, J. and Crystal, R.G. (1999b) Regional treatment of hepatic micrometastasis by adenovirus vector-mediated delivery of interleukin-2 and interleukin-12 cDNAs to the hepatic parenchyma. *Cancer Gene Ther.***6**, 491-498.

Hitt, M.M., Parks, R.J., and Graham, F.L. (1999) Structure and genetic organization of adenovirus vectors pp 61-86. In "The Development of Human Gene Therapy". Ed. *T. Friedman. Cold Spring Harbor Press*, Cold Spring Harbor, New York.

Hogaboam, C.M., Vallance, B.A., Kumar, A., Addison, C.L., Graham, F.L., Gauldie, J. and Collins, S.M. (1997) Therapeutic effects of interleukin-4 gene transfer in experimental inflammatory bowel disease. *J. Clin. Invest.***100**, 2766-2776.

Holt, J.R., Johns, D.C., Wang, S., Chen, Z.Y., Dunn, R.J., Marban, E. and Corey, D.P. (1999) Functional expression of exogenous proteins in mammalian sensory hair cells infected with adenoviral vectors. *J. Neurophysiol.***81**, 1881-1888.

Holzinger, A., Trapnell, B.C., Weaver, T.E., Whitsett, J.A. and Iwamoto, H.S. (1995) Intraamniotic administration of an adenoviral vector for gene transfer to fetal sheep and mouse tissues. *Pediatr. Res.***38**, 844-850.

Horwitz, M.S. (1990) Adenoviruses. In " Virology, 2nd edition " *Eds Field, B.N., Knipe, D.M. et al. Raven Press*, New York, 1723-1741.

Horwitz, M.S.: Adenoviruses. Fields Virology, Third Edition (1996) *Eds B.N.Fields, D.M. Knipe, P.M.Hawley et al., Lippincott-Raven Publishers*, Philadelphia. p2149-2171.

Hou, S.T., Callaghan, D., Hill, I., Kang, L., Massie, B., Morley, P., Murray, C., Rasquinha, I., Slack, R. and MacManus, J. (2000) Overexpression of E2F1 transcription factor in rat cortical cultures induces apoptosis of neurons. *Journal of Neuroscience*, in press

Howe, A.K., Gaillard, S., Bennett, J.S. and Rundell, K. (1998) Cell cycle progression in monkey cells expressing simian virus 40 small t antigen from adenovirus vectors. *J. Virol.***72**, 9637-9644.

Huang, H., Chen, S.H., Kosai, K., Finegold, M.J. and Woo, S.L. (1996) Gene therapy for hepatocellular carcinoma: long-term remission of primary and metastatic tumors in mice by interleukin-2 gene therapy in vivo. *Gene Ther.***3**, 980-987.

Huang, L., Sankar, S., Lin, C., Kontos, C.D., Schroff, A.D., Cha, E.H., Feng, S.M., Li, S.F., Yu, Z., Van Etten, R.L., Blanar, M.A. and Peters, K.G. (1999b) HCPTPA, a protein tyrosine phosphatase that regulates vascular endothelial growth factor receptor-mediated signal transduction and biological activity. *J. Biol. Chem.***274**, 38183-38188.

Huang, S., Reddy, V., Dasgupta, N. and Nemerow, G.R. (1999a) A single amino acid in the adenovirus type 37 fiber confers binding to human conjunctival cells. *J. Virol.***73**, 2798-2802.

Huang, T.G., Ip, S.M., Yeung, W.S. and Ngan, H.Y. (2000) Changes in p21WAF1, pRb, Mdm-2, Bax and Bcl-2 expression in cervical cancer cell lines transfected with a p53 expressing adenovirus. *Eur. J. Cancer.***36**, 249-256.

Huang, W. and Flint, S.J. (1998) The tripartite leader sequence of subgroup C adenovirus major late mRNAs can increase the efficiency of mRNA export. *J. Virol.***72**, 225-235.

Hunt, K.K., Deng, J., Liu, T.J., Wilson-Heiner, M., Swisher, S.G., Clayman, G. and Hung, M.C. (1997) Adenovirus-mediated overexpression of the transcription factor E2F-1 induces apoptosis in human breast and ovarian carcinoma cell lines and does not require p53. *Cancer Res.***57**, 4722-4726.

Igarashi, M., Wakasaki, H., Takahara, N., Ishii, H., Jiang, Z.Y., Yamauchi, T., Kuboki, K., Meier, M., Rhodes, C.J. and King, G.L. (1999) Glucose or diabetes activates p38 mitogen-activated protein kinase via different pathways. *J. Clin. Invest.***103**, 185-195.

Ilan, Y., Jona, V.K., Sengupta, K., Davidson, A., Horwitz, M.S., Roy-Chowdhury, N. and Roy-Chowdhury, J. (1997) Transient immunosuppression with FK506 permits long-term expression of therapeutic genes introduced into the liver using recombinant adenoviruses in the rat. *Hepatology* **26**, 949-956.

Im, S.A., Gomez-Manzano, C., Fueyo, J., Liu, T.J., Ke, L.D., Kim, J.S., Lee, H.Y., Steck, P.A., Kyritsis, A.P. and Yung, W.K. (1999) Antiangiogenesis treatment for gliomas: transfer of antisense-vascular endothelial growth factor inhibits tumor growth in vivo. *Cancer Res.***59**, 895-900.

Imler, J.L. (1995) Adenovirus vectors as recombinant viral vaccines. *Vaccine* **13**, 1143-1151. Review.

Imler, J.L., Chartier, C., Dreyer, D., Dieterle, A., Sainte-Marie, M., Faure, T., Pavirani, A. and Mehtali, M. (1996a) Novel complementation cell lines derived from human lung carcinoma A549 cells support the growth of E1-deleted adenovirus vectors. *Gene Ther.***3**, 75-84.

Imler, J.L., Dupuit, F., Chartier, C., Accart, N., Dieterle, A., Schultz, H., Puchelle, E. and Pavirani, A. (1996b) Targeting cell-specific gene expression with an adenovirus vector containing the lacZ gene under the control of the CFTR promoter. *Gene Ther.***3**, 49-58.

Inesi, G., Lewis, D., Sumbilla, C., Nandi, A., Strock, C., Huff, K.W., Rogers, T.B., Johns, D.C., Kessler, P.D. and Ordahl, C.P. (1998) Cell-specific promoter in adenovirus vector for transgenic expression of SERCA1 ATPase in cardiac myocytes. *Am. J. Physiol.***274**, C645-C653.

Isenmann, S., Cellerino, A., Gravel, C. and Bahr, M. (1999) Excess target-derived brain-derived neurotrophic factor preserves the transient uncrossed retinal projection to the superior colliculus. *Mol. Cell Neurosci.***14**, 52-65.

Jamieson, L., Carpenter, L., Biden, T.J. and Fields, A.P. (1999) Protein kinase Ciota activity is necessary for Bcr-Abl-mediated resistance to drug-induced apoptosis. *J. Biol. Chem.***274**, 3927-3930.

Jani, A., Menichella, D., Jiang, H., Chbihi, T., Acsadi, G., Kamholz, J. and Shy, M.E. (1999) Overcoming cellular immunity to prolong adenoviral-mediated gene expression in sciatic nerve. *Ann N .Y. Acad. Sci.***14**, 397-414.

Jasvinder, A.K., Massie, B., Miller, F.D. and Kaplan, D.R. (2000) The TrkB-Shc site signals neuronal survival and local axon growth via MEK and PI3-kinase. *Neuron*, in press

Ji, L., Fang, B., Yen, N., Fong, K., Minna, J.D. and Roth, J.A. (1999) Induction of apoptosis and inhibition of tumorigenicity and tumor growth by adenovirus vector-mediated fragile histidine triad (FHIT) gene overexpression. *Cancer Res.***59**, 3333-3339.

Jimenez-Chillaron, J.C., Newgard, C.B. and Gomez-Foix, A.M. (1999) Increased glucose disposal induced by adenovirus-mediated transfer of glucokinase to skeletal muscle in vivo. *FASEB J.***13**, 2153-2160.

Jin, X., Nguyen, D., Zhang, W.W., Kyritsis, A.P. and Roth, J.A. (1995) Cell cycle arrest and inhibition of tumor cell proliferation by the p16INK4 gene mediated by an adenovirus vector. *Cancer Res.***55**, 3250-3253.

Jobin, C., Holt, L., Bradham, C.A., Streetz, K., Brenner, D.A. and Sartor, R.B. (1999) TNF receptor-associated factor-2 is involved in both IL-1 beta and TNF-alpha signaling cascades leading to NF-kappa B activation and IL-8 expression in human intestinal epithelial cells. *J. Immunol.***162**, 4447-4454.

Johns, D.C., Marx, R., Mains, R.E., O'Rourke, B. and Marban, E. (1999) Inducible genetic suppression of neuronal excitability. *J. Neurosci.***19**, 1691-1697.

Johnson, M.D., Kinoshita, Y., Xiang, H., Ghatan, S. and Morrison, R.S. (1999) Contribution of p53 dependent caspase activation to neuronal cell death declines with neuronal maturation. *J. Neurosci.***19**, 2996-3006.

Jorgensen, C., Apparailly, F., Canovas, F., Verwaerde, C., Auriault, C., Jacquet, C. and Sany, J. (1999) Systemic viral interleukin-10 gene delivery prevents cartilage invasion by human rheumatoid synovial tissue engrafted in SCID mice. *Arthritis Rheum.***42**, 678-685.

Joshi, U.S., Chen, Y.Q., Kalemkerian, G.P., Adil, M.R., Kraut, M. and Sarkar, F.H. (1998) Inhibition of tumor cell growth by p21WAF1 adenoviral gene transfer in lung cancer. *Cancer Gene Ther.***5**, 183-191.

Ju, D.W., Wang, B.M. and Cao, X. (1998) Adenovirus-mediated combined suicide gene and interleukin-2 gene therapy for the treatment of established tumor and induction of antitumor immunity. *J. Cancer Res. Clin. Oncol.***124**, 683-689.

Kadowaki, Y., Fujiwara, T., Fukazawa, T., Shao, J., Yasuda, T., Itoshima, T., Kagawa, S., Hudson, L.G., Roth, J.A. and Tanaka, N. (1999) Induction of differentiation-dependent apoptosis in human esophageal squamous cell carcinoma by adenovirus-mediated p21sdi1 gene transfer. Clin. *Cancer Res.***5**, 4233-4241.

Kagawa, S., Fujiwara, T., Kadowaki, Y., Fukazawa, T., Sok-Joo, R., Roth, J.A. and Tanaka, N. (1999) Overexpression of the p21 sdi1 gene induces senescence-like state in human cancer cells: implication for senescence-directed molecular therapy for cancer. *Cell Death Differ.***6**, 765-772.

Kagawa, S., Pearson, S.A., Ji, L., Xu, K., McDonnell, T.J., Swisher, S.G., Roth, J.A. and Fang, B. (2000) A binary adenoviral vector system for expressing high levels of the proapoptotic gene bax. *Gene Ther.***7**, 75-79.

Kanai, F., Shiratori, Y., Yoshida, Y., Wakimoto, H., Hamada, H., Kanegae, Y., Saito, I., Nakabayashi, H., Tamaoki, T., Tanaka, T., Lan, K.H., Kato, N., Shiina, S. and Omata, M. (1996) Gene therapy for alpha-fetoprotein-producing human hepatoma cells by adenovirus-mediated transfer of the herpes simplex virus thymidine kinase gene. *Hepatology* **23**, 1359-1368.

Kaneko, S., Hallenbeck, P., Kotani, T., Nakabayashi, H., McGarrity, G., Tamaoki, T., Anderson, W.F. and Chiang, Y.L. (1995) Adenovirus-mediated gene therapy of hepatocellular carcinoma using cancer-specific gene expression. *Cancer Res.***55**, 5283-5287.

Kaner, R.J., Worgall, S., Leopold, P.L., Stolze, E., Milano, E., Hidaka, C., Ramalingam, R., Hackett, N.R., Singh, R., Bergelson, J., Finberg, R., Falck-Pedersen, E. and Crystal, R.G. (1999) Modification of the genetic program of human alveolar macrophages by adenovirus vectors in vitro is feasible but inefficient, limited in part by the low level of expression of the coxsackie/adenovirus receptor. Am. J. Respir. *Cell Mol. Biol.***20**, 361-370.

Kaplan, J.M., St George, J.A., Pennington, S.E., Keyes, L.D., Johnson, R.P., Wadsworth, S.C. and Smith, A.E. (1996) Humoral and cellular immune responses of nonhuman primates to long-term repeated lung exposure to Ad2/CFTR-2. *Gene Ther.***3**, 117-127.

Kaplan, J.M., Armentano, D., Scaria, A., Woodworth, L.A., Pennington, S.E., Wadsworth, S.C., Smith, A.E. and Gregory, R.J. (1999) Novel role for E4 region genes in protection of adenovirus vectors from lysis by cytotoxic T lymphocytes. *J.Virol.***73**, 4489-4492.

Kasono, K., Blackwell, J.L., Douglas, J.T., Dmitriev, I., Strong, T.V., Reynolds, P., Kropf, D.A., Carroll, W.R., Peters, G.E., Bucy, R.P., Curiel, D.T. and Krasnykh, V. (1999) Selective gene delivery to head and neck cancer cells via an integrin targeted adenoviral vector. Clin. *Cancer Res.***5**, 2571-2579.

Kass-Eisler, A., Leinwand, L., Gall, J., Bloom, B. and Falck-Pedersen, E. (1996) Circumventing the immune response to adenovirus-mediated gene therapy. *Gene Ther.***3**, 154-162.

Katabi, M.M., Chan, H.L., Karp, S.E. and Batist, G. (1999) Hexokinase type II: a novel tumor-specific promoter for gene-targeted therapy differentially expressed and regulated in human cancer cells. *Hum. Gene Ther.***10**, 155-164.

Katayose, D., Gudas, J., Nguyen, H., Srivastava, S., Cowan, K.H. and Seth, P. (1995) Cytotoxic effects of adenovirus-mediated wild-type p53 protein expression in normal and tumor mammary epithelial cells. *Clin. Cancer Res.***8**, 889-897.

Katayose, Y., Kim, M., Rakkar, A.N., Li, Z., Cowan, K.H. and Seth, P. (1997) Promoting apoptosis: a novel activity associated with the cyclin-dependent kinase inhibitor p27. *Cancer Res.***57**, 5441-5445.

Kay, M.A., Meuse, L., Gown, A.M., Linsley, P., Hollenbaugh, D., Aruffo, A., Ochs, H.D. and Wilson, C.B. (1997) Transient immunomodulation with anti-CD40 ligand antibody and CTLA4Ig enhances persistence and secondary adenovirus-mediated gene transfer into mouse liver. *Proc. Natl. Acad. Sci. U. S. A.***94**,4686-91

Kielian, T., Nagai, E., Ikubo, A., Rasmussen, C.A. and Suzuki, T. (1999) Granulocyte/macrophage-colony-stimulating factor released by adenovirally transduced CT26 cells leads to the local expression of macrophage inflammatory protein 1alpha and accumulation of dendritic cells at vaccination sites in vivo. *Cancer Immunol. Immunother.***48**, 123-131.

Kijima, T., Osaki, T., Nishino, K., Kumagai, T., Funakoshi, T., Goto, H., Tachibana, I., Tanio, Y. and Kishimoto, T. (1999) Application of the Cre recombinase/loxP system further enhances antitumor effects in cell type-specific gene therapy against carcinoembryonic antigen-producing cancer. *Cancer Res.***59**, 4906-4911.

Kim, K.N., Watanabe, S., Ma, Y., Thornton, S., Giannini, E.H. and Hirsch, R. (2000) Viral IL-10 and soluble TNF receptor act synergistically to inhibit collagen-induced arthritis following adenovirus-mediated gene transfer. *J. Immunol.***164**, 1576-1581.

Kim, M., Katayose, Y., Li, Q., Rakkar, A.N., Li, Z., Hwang, S.G., Katayose, D., Trepel, J., Cowan, K.H. and Seth, P. (1998) Recombinant adenovirus expressing Von Hippel-Lindau-mediated cell cycle arrest is associated with the induction of cyclin-dependent kinase inhibitor p27Kip1. Biochem. Biophys. *Res. Commun.***253**, 672-677.

Kirby, I., Davison, E., Beavil, A.J., Soh, C.P., Wickham, T.J., Roelvink, P.W., Kovesdi, I., Sutton, B.J. and Santis, G. (1999) Mutations in the DG loop of adenovirus type 5 fiber knob protein abolish high-affinity binding to its cellular receptor CAR. *J. Virol.***73**, 9508-9514.

Kirby, I., Davison, E., Beavil, A.J., Soh, C.P., Wickham, T.J., Roelvink, P.W., Kovesdi, I., Sutton, B.J. and Santis, G. (2000) Identification of Contact Residues and Definition of the CAR-Binding Site of Adenovirus Type 5 Fiber Protein. *J. Virol.***74**, 2804-2813.

Kircheis, R., Kupcu, Z., Wallner, G. and Wagner, E. (1998) Cytokine gene-modified tumor cells for prophylactic and therapeutic vaccination: IL-2, IFN-gamma, or combination IL-2 + IFN-gamma. Cytokines. *Cell Mol. Ther.***4**, 95-103.

Kitagawa, H., Sasaki, C., Sakai, K., Mori, A., Mitsumoto, Y., Mori, T., Fukuchi, Y., Setoguchi, Y. and Abe, K. (1999) Adenovirus-mediated gene transfer of glial cell line-derived neurotrophic factor prevents ischemic brain injury after transient middle cerebral artery occlusion in rats. J. Cereb. *Blood Flow Metab.***19**, 1336-1344.

Kiwaki, K., Kanegae, Y., Saito, I., Komaki, S., Nakamura, K., Miyazaki, J.I., Endo, F. and Matsuda, I. (1996) Correction of ornithine transcarbamylase deficiency in adult spf(ash) mice and in OTC-deficient human hepatocytes with recombinant adenoviruses bearing the CAG promoter. *Hum. Gene Ther.***7**, 821-830.

Ko, S.C., Cheon, J., Kao, C., Gotoh, A., Shirakawa, T., Sikes, R.A., Karsenty, G. and Chung, L.W. (1996) Osteocalcin promoter-based toxic gene therapy for the treatment of osteosarcoma in experimental models. *Cancer Res.***56**, 4614-4619.

Kobayashi, J., Applebaum-Bowden, D., Dugi, K.A., Brown, D.R., Kashyap, V.S., Parrott, C., Duarte, C., Maeda, N. and Santamarina-Fojo, S. (1996a) Analysis of protein structure-function in vivo. Adenovirus-mediated transfer of lipase lid mutants in hepatic lipase-deficient mice. *J. Biol. Chem.***271**, 26296-26301.

Kobayashi, K., Oka, K., Forte, T., Ishida, B., Teng, B., Ishimura-Oka, K., Nakamuta, M. and Chan, L. (1996b) Reversal of hypercholesterolemia in low density lipoprotein receptor knockout mice by adenovirus-mediated gene transfer of the very low density lipoprotein receptor. *J. Biol. Chem.***271**, 6852-6860.

Kobayashi, M., Nagata, S., Kita, Y., Nakatsu, N., Ihara, S., Kaibuchi, K., Kuroda, S., Ui, M., Iba, H., Konishi, H., Kikkawa, U., Saitoh, I. and Fukui, Y. (1997) Expression of a constitutively active phosphatidylinositol 3-kinase induces process formation in rat PC12 cells. Use of Cre/loxP recombination system. *J. Biol. Chem.***272**, 16089-16092.

Kolls, J.K., Lei, D., Odom, G., Nelson, S., Summer, W.R., Gerber, M.A., and Shellito, J.E. (1996) Use transient CD4 lymphocyte depletion to prolong transgene expression of E1-deleted adenoviral vectors. *Hum. Gene Ther.***7**, 489-497.

Konishi, H., Ochiya, T., Yasuda, Y., Sakamoto, H., Muto, T., Sugimura, T. and Terada, M. (1996) HST-1/FGF-4 stimulates proliferation of megakaryocyte progenitors synergistically and promotes megakaryocyte maturation. *Oncogene* **13**, 9-19.

Korhonen, J.M., Said, F.A., Wong, A.J. and Kaplan, D.R. (1999) Gab1 mediates neurite outgrowth, DNA synthesis, and survival in PC12 cells. *J. Biol. Chem.***274**, 37307-37314.

Kotani, K., Ogawa, W., Hino, Y., Kitamura, T., Ueno, H., Sano, W., Sutherland, C., Granner, D.K. and Kasuga, M. (1999) Dominant negative forms of Akt (protein kinase B) and atypical protein kinase Clambda do not prevent insulin inhibition of phosphoenolpyruvate carboxykinase gene transcription. *J. Biol. Chem.***274**, 21305-21312.

Kovesdi, I., Brough, D.E., Bruder, J.T., and Wickham, T.J. (1997) Adenoviral vectors for gene transfer. *Curr. Opin. in Biotech.***8**, 583-589.

Koyama, K., Chen, G., Wang, M.Y., Lee, Y., Shimabukuro, M., Newgard, C.B. and Unger, R.H. (1997) beta-cell function in normal rats made chronically hyperleptinemic by adenovirus-leptin gene therapy. *Diabetes* **46**, 1276-1280.

Koyama, K., Shimabukuro, M., Chen, G., Wang, M.Y., Lee, Y., Kalra, P.S., Dube, M.G., Kalra, S.P., Newgard, C.B. and Unger, R.H. (1998) Resistance to adenovirally induced hyperleptinemia in rats. Comparison of ventromedial hypothalamic lesions and mutated leptin receptors. *J. Clin. Invest.***102**, 728-733.

Krasnykh, V.N., Mikheeva, G.V., Douglas, J.T. and Curiel, D.T. (1996) Generation of recombinant adenovirus vectors with modified fibers for altering viral tropism. *J. Virol.***70**, 6839-6846.

Krasnykh, V., Dmitriev, I., Mikheeva, G., Miller, C.R., Belousova, N. and Curiel, D.T. (1998) Characterization of an adenovirus vector containing a heterologous peptide epitope in the HI loop of the fiber knob. *J. Virol.***72**, 1844-1852.

Krougliak, V. and Graham, F.L. (1995) Development of cell lines capable of complementing E1, E4, and protein IX defective adenovirus type 5 mutants. *Hum. Gene Ther.***6**, 1575-1586.

Kubisch, H.M., Larson, M.A., Eichen, P.A., Wilson, J.M. and Roberts, R.M. (1997) Adenovirus-mediated gene transfer by perivitelline microinjection of mouse, rat, and cow embryos. Biol. Reprod.,56, 119-124.

Kullo, I.J., Mozes, G., Schwartz, R.S., Gloviczki, P., Crotty, T.B., Barber, D.A., Katusic, Z.S., and O'Brien, T. (1997) Adventitial gene transfer of recombinant endothelial nitric oxide synthase to rabbit carotid arteries alters vascular reactivity. *Circulation* **96**, 2254-2261.

Kunisada, K., Tone, E., Fujio, Y., Matsui, H., Yamauchi-Takihara, K., and Kishimoto, T. (1998) Activation of gp130 transduces hypertrophic signals via STAT3 in cardiac myocytes. *Circulation* **98**, 346-352.

Kuriyama, S., Tominaga, K., Mitoro, A., Tsujinoue, H., Nakatani, T., Yamazaki, M., Nagao, S., Toyokawa, Y., Okamoto, S. and Fukui, H. (2000) Immunomodulation with FK506 around the time of intravenous re-administration of an adenoviral vector facilitates gene transfer into primed rat liver. *Int. J. Cancer* **85**, 839-844.

Kuzmin, A.I., Finegold, M.J. and Eisensmith, R.C. (1997) Macrophage depletion increases the safety, efficacy and persistence of adenovirus-mediated gene transfer in vivo. *Gene Ther.***4**, 309-3016.

Lamarche, N., Matton, G., Massie, B., Fontecave, M., Atta, M., Dumas, F., Gaudreau, P. and Langelier, Y. (1996) Production of herpes-simplex-virus ribonucleotide reductase R2 subunit with prokaryotic and eukaryotic expression systems: higher activity of R2 produced by eukaryotic cells related to higher iron binding capacity. *Biochemical Journal* **320**, 129-135.

Lan, K.H., Kanai, F., Shiratori, Y., Okabe, S., Yoshida, Y., Wakimoto, H., Hamada, H., Tanaka, T., Ohashi, M. and Omata, M. (1996) Tumor-specific gene expression in carcinoembryonic antigen--producing gastric cancer cells using adenovirus vectors. *Gastroenterology* **111**, 1241-1251.

Lang, F.F., Yung, W.K., Sawaya, R. and Tofilon, P.J. (1999) Adenovirus-mediated p53 gene therapy for human gliomas. *Neurosurgery* **45**, 1093-1104. Review.

Langelier, Y., Champoux, L., Hamel, M., Lamarche, N., Guilbault, C.,Gaudreau, P. and Massie, B. (1998) The R1 subunit of herpes simplex virus ribonucleotide reductase is a good substrate for host cell protein kinases but is not itself a protein kinase. *J. Biol. Chem.* **278**, 1435-1443.

Langelier, Y., Bergeron, S., Lippens, J., Gervais, A., Guilbault, C., Denis, S., Mosser, DD. and Massie, B.(2000) The R1 subunit of herpes simplex virus ribonucleotide reductase protects cells against apoptosis induced by TNFa and Fas L (submitted).

Larochelle, N., Lochmuller, H., Zhao, J., Jani, A., Hallauer, P., Hastings, K.E., Massie, B., Prescott, S., Petrof, B.J., Karpati, G. and Nalbantoglu, J. (1997) Efficient muscle-specific transgene expression after adenovirus-mediated gene transfer in mice using a 1.35 kb muscle creatine kinase promoter/enhancer. *Gene Ther.***4**, 465-472.

Lasarte, J.J., Corrales, F.J., Casares, N., Lopez-Diaz de Cerio, A., Qian, C., Xie, X., Borras-Cuesta, F. and Prieto, J. (1999) Different doses of adenoviral vector expressing IL-12 enhance or depress the immune response to a coadministered antigen: the role of nitric oxide. *J. Immunol.***162**, 5270-5277.

Latham, J.P., Searle, P.F., Mautner, V. and James, N.D. (2000) Prostate-specific antigen promoter/enhancer driven gene therapy for prostate cancer: construction and testing of a tissue-specific adenovirus vector. *Cancer Res.***60**, 334-341.

Lavoie, J.N., Nguyen, M., Marcellus, R.C., Branton, P.E. and Shore, G.C. (1998) E4orf4, a novel adenovirus death factor that induces p53-independent apoptosis by a pathway that is not inhibited by zVAD-fmk. *J. Cell Biol.***140**, 637-645.

Lazarous, D.F., Shou, M., Stiber, J.A., Hodge, E., Thirumurti, V., Goncalves, L. and Unger, E.F. (1999) Adenoviral-mediated gene transfer induces sustained pericardial VEGF expression in dogs: effect on myocardial angiogenesis. *Cardiovasc. Res.***44**, 294-302.

Lazennec, G., Alcorn, J.L. and Katzenellenbogen, B.S. (1999) Adenovirus-mediated delivery of a dominant negative estrogen receptor gene abrogates estrogen-stimulated gene expression and breast cancer cell proliferation. *Mol. Endocrinol.***13**, 969-980.

Le, X.F., Vallian, S., Mu, Z.M., Hung, M.C. and Chang, K.S. (1998) Recombinant PML adenovirus suppresses growth and tumorigenicity of human breast cancer cells by inducing G1 cell cycle arrest and apoptosis. *Oncogene* **16**, 1839-1849.

Leconte, I., Fox, J.C., Baldwin, H.S., Buck, C.A. and Swain, J.L. (1998) Adenoviral-mediated expression of antisense RNA to fibroblast growth factors disrupts murine vascular development. *Dev. Dyn.***213**, 421-430.

Lee, B., Dennis, J.A., Healy, P.J., Mull, B., Pastore, L., Yu, H., Aguilar-Cordova, E., O'Brien, W., Reeds, P. and Beaudet, A.L. (1999b) Hepatocyte gene therapy in a large animal: a neonatal bovine model of citrullinemia. *Proc. Natl. Acad. Sci. U. S. A.***96**, 3981-3986.

Lee, E.J., Anderson, L.M., Thimmapaya, B. and Jameson, J.L. (1999a) Targeted expression of toxic genes directed by pituitary hormone promoters: a potential strategy for adenovirus-mediated gene therapy of pituitary tumors. *J. Clin. Endocrinol. Metab.***84**, 786-794.

Lee, J.H., Lee, C.T., Yoo, C.G., Hong, Y.K., Kim, C.M., Han, S.K., Shim, Y.S., Carbone, D.P. and Kim, Y.W. (1998) The inhibitory effect of adenovirus-mediated p16INK4a gene transfer on the proliferation of lung cancer cell line. *Anticancer Res.***18**, 3257-3261.

Lee, L.Y., Patel, S.R., Hackett, N.R., Mack, C.A., Polce, D.R., El-Sawy, T., Hachamovitch, R., Zanzonico, P., Sanborn, T.A., Parikh, M., Isom, O.W., Crystal, R.G. and Rosengart, T.K. (2000) Focal angiogen therapy using intramyocardial delivery of an adenovirus vector coding for vascular endothelial growth factor 121. *Ann. Thorac. Surg.***69**,14-23.

Lee, Y.H., Sauer, B., Johnson, P.F. and Gonzalez, F.J. (1997) Disruption of the c/ebp alpha gene in adult mouse liver. *Mol. Cell Biol.***17**, 6014-6022.

Lei, D., Lehmann, M., Shellito, J.E., Nelson, S., Siegling, A., Volk, H.D. and Kolls, J.K.(1996) Nondepleting anti-CD4 antibody treatment prolongs lung-directed E1-deleted adenovirus-mediated gene expression inrats. *Hum. Gene Ther.***7**, 2273-2279.

Lei, X.F., Ohkawara, Y., Stampfli, M.R., Gauldie, J., Croitoru, K., Jordana, M. and Xing, Z. (1998) Compartmentalized transgene expression of granulocyte-macrophage colony-stimulating factor (GM-CSF) in mouse lung enhances allergic airways inflammation. *Clin. Exp. Immunol.***113**, 157-165.

Li, E., Stupack, D., Klemke, R., Cheresh, D.A. and Nemerow, G.R. (1998a) Adenovirus endocytosis via alpha(v) integrins requires phosphoinositide-3-OH kinase. *J. Virol.***72**, 2055-2061.

Li, H., Alonso-Vanegas, M., Colicos, M.A., Jung, S.S., Lochmuller, H., Sadikot, A.F. Snipes, G.J., Seth, P., Karpati, G. and Nalbantoglu, J. (1999b) Intracerebral adenovirus-mediated p53 tumor suppressor gene therapy for experimental human glioma. Clin. *Cancer Res.***5**, 637-642.

Li, H., Lochmuller, H., Yong, V.W., Karpati, G. and Nalbantoglu, J. (1997a) Adenovirus-mediated wild-type p53 gene transfer and overexpression induces apoptosis of human glioma cells independent of endogenous p53 status. *J. Neuropathol. Exp. Neurol.***56**, 872-878.

Li, H., Lu, H., Griscelli, F., Opolon, P., Sun, L.Q., Ragot, T., Legrand, Y., Belin, D., Soria, J., Soria, C., Perricaudet, M. and Yeh, P. (1998b) Adenovirus-mediated delivery of a uPA/uPAR antagonist suppresses angiogenesis-dependent tumor growth and dissemination in mice. *Gene Ther.***5**, 1105-1113.

Li, J.H., Li, P., Klamut, H. and Liu, F.F. (1997b) Cytotoxic effects of Ad5CMV-p53 expression in two human nasopharyngeal carcinoma cell lines. Clin. *Cancer Res.***3**, 507-514.

Li, Y.X., Xu, Y., Ju, D., Lester, H.A., Davidson, N. and Schuman, E.M. (1998c) Expression of a dominant negative TrkB receptor, T1, reveals a requirement for presynaptic signaling in BDNF-induced synaptic potentiation in cultured hippocampal neurons. *Proc. Natl. Acad. Sci. U. S. A.***95**, 10884-10889.

Li, Y., Pong, R.C., Bergelson, J.M., Hall, M.C., Sagalowsky, A.I., Tseng, C.P., Wang, Z. and Hsieh, J.T. (1999a) Loss of adenoviral receptor expression in human bladder cancer cells: a potential impact on the efficacy of gene therapy. *Cancer Res.***59**, 325-330.

Liao, W., Kobayashi, K. and Chan, L. (1999) Adenovirus-mediated overexpression of microsomal triglyceride transfer protein (MTP): mechanistic studies on the role of MTP in apolipoprotein B-100 biogenesis, by. *Biochemistry* **38**, 10215.

Lieber, A., He, C.Y., Kirillova, I. and Kay, M.A. (1996) Recombinant adenoviruses with large deletions generated by cre-mediated excision exhibit different biological properties compared with first-generation vectors in vitro and in vivo. *J. Virol.***70**, 8944-8960

Lieber, A., He, C.Y., Meuse, L., Himeda, C., Wilson, C. and Kay, M.A. (1998) Inhibition of NF-kappaB activation in combination with bcl-2 expression allows for persistence of first-generation adenovirus vectors in the mouse liver. *J. Virol.***72**, 9267-9277.

Lieber, A., Steinwaerder, D.S., Carlson, C.A. and Kay, M.A.(1999) Integrating adenovirus-adeno-associated virus hybrid vectors devoid of all viral genes. *J .Virol.***73**, 9314-24.

Liechty, K.W., Nesbit, M., Herlyn, M., Radu, A., Adzick, N.S. and Crombleholme, T.M. (1999) Adenoviral-mediated overexpression of platelet-derived growth factor-B corrects ischemic impaired wound healing. *J. Invest. Dermatol.***113**, 375-383.

Lim, R., Zaheer, A., Yorek, M.A., Darby, C.J. and Oberley, L.W. (2000) Activation of nuclear factor-kappaB in C6 rat glioma cells after transfection with glia maturation factor. *J. Neurochem.***74**, 596-602.

Lin, P., Buxton, J.A., Acheson, A., Radziejewski, C., Maisonpierre, P.C., Yancopoulos, G.D., Channon, K.M., Hale, L.P., Dewhirst, M.W., George, S.E. and Peters, K.G. (1998) Antiangiogenic gene therapy targeting the endothelium-specific receptor tyrosine kinase Tie2. *Proc. Natl. Acad. Sci. U. S. A.***95**, 8829-8834.

Lipshutz, G.S., Flebbe-Rehwaldt, L. and Gaensler, K.M. (1999a) Adenovirus-mediated gene transfer to the peritoneum and hepatic parenchyma of fetal mice in utero. *Surgery.***126**, 171-177.

Lipshutz, G.S., Sarkar, R., Flebbe-Rehwaldt, L., Kazazian, H. and Gaensler, K.M. (1999b) Short-term correction of factor VIII deficiency in a murine model of hemophilia A after delivery of adenovirus murine factor VIII in utero. *Proc. Natl. Acad. Sci. U. S. A.***96**, 13324-13329.

Liu, G., Excoffon, K.J., Wilson, J.E., McManus, B.M., Miao, L., Benoit, P., Duverger, N., Branellec, D., Denefle, P., Hayden, M.R. and Lewis, M.E. (1998b) Enhanced lipolysis in normal mice expressing liver-derived human lipoprotein lipase after adenoviral gene transfer. *Clin. Invest. Med.***21**, 172-185.

Liu, J.L., Grinberg, A., Westphal, H., Sauer, B., Accili, D., Karas, M. and LeRoith, D. (1998a) Insulin-like growth factor-I affects perinatal lethality and postnatal development in a gene dosage-dependent manner: manipulation using the Cre/loxP system in transgenic mice. *Mol. Endocrinol.***12**, 1452-1462.

Liu, T.J., Wang, M., Breau, R.L., Henderson, Y., El-Naggar, A.K., Steck, K.D., Sicard, M.W. and Clayman, G.L. (1999) Apoptosis induction by E2F-1 via adenoviral-mediated gene transfer results in growth suppression of head and neck squamous cell carcinoma cell lines. *Cancer Gene Ther.***6**, 163-171.

Lochmuller, H., Jani, A., Huard, J., Prescott, S., Simoneau, M., Massie, B., Karpati, G. and Acsadi, G. (1994) Emergence of early region 1-containing replication-competent adenovirus in stocks of replication-defective adenovirus recombinants (delta E1 + delta E3) during multiple passages in 293 cells. *Hum. Gene Ther.***5**, 1485-1491.

Lochmuller, H., Petrof, B.J., Allen, C., Prescott, S., Massie, B. and Karpati, G. (1995) Immunosuppression by FK506 markedly prolongs expression of adenovirus-delivered transgene in skeletal muscles of adult dystrophic [mdx] mice. *Biochem Biophys Res Commun.***213**,569-574.

Lochmuller, H., Petrof, B.J., Pari, G., Larochelle, N., Dodelet, V., Wang, Q., Allen, C., Prescott, S., Massie, B., Nalbantoglu, J. and Karpati, G. (1996) Transient immunosuppression by FK506 permits a sustained high-level dystrophin expression after adenovirus-mediated dystrophin minigene transfer to skeletal muscles of adult dystrophic (mdx) mice. *Gene Ther.***3**, 706-716.

Lu, W., Fidler, I.J. and Dong, Z. (1999) Eradication of primary murine fibrosarcomas and induction of systemic immunity by adenovirus-mediated interferon beta gene therapy. *Cancer Res.***59**, 5202-5208.

Lubberts, E., Joosten, L.A., van Den Bersselaar, L., Helsen, M.M., Bakker, A.C., van Meurs, J.B., Graham, F.L., Richards, C.D. and van Den Berg, W.B. (1999) Adenoviral vector-mediated overexpression of IL-4 in the knee joint of mice with collagen-induced arthritis prevents cartilage destruction. *J. Immunol.***163**, 4546-4556.

Lukacs, N.W., Addison, C.L., Gauldie, J., Graham, F., Simpson, K., Strieter, R.M., Warmington, K., Chensue, S.W. and Kunkel, S.L. (1997) Transgene-induced production of IL-4 alters the development and collagen expression of T helper cell 1-type pulmonary granulomas. *J. Immunol.***158**, 4478-4484.

Lundberg, M.S., Curto, K.A., Bilato, C., Monticone, R.E. and Crow, M.T. (1998) Regulation of vascular smooth muscle migration by mitogen-activated protein kinase and calcium/calmodulin-dependent protein kinase II signaling pathways. *J. Mol. Cell Cardiol.***30**, 2377-2389.

Lusky, M., Grave, L., Dieterle, A., Dreyer, D., Christ, M., Ziller, C., Furstenberger, P., Kintz, J., Hadji, D.A., Pavirani, A. and Mehtali, M. (1999) Regulation of adenovirus-mediated transgene expression by the viral E4 gene products: requirement for E4 ORF3. *J.Virol.***73**, 8308-8319.

Lynch, J., Suh, E.R., Silberg, D.G., Rulyak, S, Blanchard, N. and Traber, P.G. (2000) The caudal-related homeodomain protein Cdx1 inhibits proliferation of intestinal epithelial cells by down-regulation of D-type cyclins. *J. Biol. Chem.***275**, 4499-4506.

Ma, Y., Thornton, S., Duwel, L.E., Boivin, G.P., Giannini, E.H., Leiden, J.M., Bluestone, J.A. and Hirsch, R. (1998) Inhibition of collagen-induced arthritis in mice by viral IL-10 gene transfer. *J. Immunol.***161**, 1516-1524.

MacCalman, C.D., Furth, E.E., Omigbodun, A., Kozarsky, K.F., Coutifaris, C. and Strauss, J.F. (1996) Transduction of human trophoblast cells by recombinant adenoviruses is differentiation dependent. *Biol. Reprod.***54**, 682-691.

Macejak, D.G., Lin, H., Webb, S., Chase, J., Jensen, K., Jarvis, T.C., Leiden, J.M. and Couture, L. (1999) Adenovirus-mediated expression of a ribozyme to c-myb mRNA inhibits smooth muscle cell proliferation and neointima formation in vivo. *J. Virol.***73**, 7745-7751.

Mack, C.A., Song, W.R., Carpenter, H., Wickham, T. J., Kovesdi, I., Harvey, B.G., Magovern, C.J., Isom, O.W., Rosengart, T., Falck-Pedersen, E., Hackett, N.R., Crystal, R.G. and Mastrangeli, A. (1997) Circumvention of anti-adenovirus neutralizing immunity by administration of an adenoviral vector of an alternate serotype. *Hum. Gene Ther.***8**, 99-109.

Mack, C.A., Magovern, C.J., Budenbender, K.T., Patel, S.R., Schwarz, E.A., Zanzonico, P., Ferris, B., Sanborn, T., Isom, P., Isom, O.W., Crystal, R.G. and Rosengart, T.K. (1998) Salvage angiogenesis induced by adenovirus-mediated gene transfer of vascular endothelial growth factor protects against ischemic vascular occlusion. *J. Vasc. Surg.***27**, 699-709.

MacLachlan, T.K., Somasundaram, K., Sgagias, M., Shifman, Y., Muschel, R.J., Cowan, K.H. and El-Deiry, W.S. (2000) BRCA1 effects on the cell cycle and the DNA damage response are linked to altered gene expression. *J. Biol. Chem.***275**, 2777-2785.

Magovern, C.J., Mack, C.A., Zhang, J., Rosengart, T.K., Isom, O.W. and Crystal, R.G. (1997) Regional angiogenesis induced in nonischemic tissue by an adenoviral vector expressing vascular endothelial growth factor. *Hum. Gene Ther.***8**, 215-227.

Maillard, L. and Walsh, K. (1996) Growth-arrest homeobox gene Gax: a molecular strategy to prevent arterial restenosis. *Schweiz. Med. Wochenschr.***126**, 1721-1726.

Marcellus, R.C., Lavoie, J.N., Boivin, D., Shore, G.C., Ketner, G. and Branton, P.E. (1998) The early region 4 orf4 protein of human adenovirus type 5 induces p53-independent cell death by apoptosis. *J. Virol.***72**, 7144-7153.

Massie, B.: BMAdE1, a new complementing cell line for the large scale production of E1-deleted Adenovirus (Ad) vectors devoided of RCA (replication competent Ad). (filed on 26/04/96; issued 09/98) U.S. patent 5,891,6690.

Massie, B., Dionne, J., Lamarche, N., Fleurent, J. and Langelier, Y. (1995) Improved adenovirus vector provides herpes simplex virus ribonucleotide reductase R1 and R2 subunits very efficiently. *Biotechnology* **13**, 602-608.

Massie, B., Couture, F., Lamoureux, L., Mosser, D.D., Guilbault, C., Jolicoeur, P., Belanger, F. and Langelier, Y. (1998a) Inducible overexpression of a toxic protein by an adenovirus vector with a tetracycline-regulatable expression cassette. *J. Virol.***72**, 2289-2296.

Massie, B., Mosser, D.D., Koutromanis, M., Vitté-Mony, I., Lamoureux, L., Couture, F., Paquet, L., Guibault, C., Dionne, J., Chahla, D., Jolicoeur, P. and Langelier, Y. (1998b) New adenovirus vectors for protein production and gene transfer. *Cytotech.***28**, 53-64.

Mastrangeli, A., Harvey, B.-G., Yao, J., Wolff, G., Kovesdi, I., Crystal, R.G. and Falck-Pedersen, E. (1996) Sero-Switch Adenovirus-mediated in vivo gene transfer: circumvention of anti-adenovirus humoral immune defenses against repeat adenovirus serotype. *Hum. Gene Ther.***7**, 79-87.

Mathias, P., Wickham, T., Moore, M. and Nemerow G. (1994) Multiple adenovirus serotypes use alpha v integrins for infection. *J. Virol.***68**, 6811-6814.

Matsukawa, A., Lukacs, N.W., Standiford, T.J., Chensue, S.W. and Kunkel, S.L. (2000) Adenoviral-mediated overexpression of monocyte chemoattractant protein-1 differentially alters the development of Th1 and Th2 type responses in vivo. *J. Immunol.***164**, 1699-1704.

Matsuoka, N., Yukawa, H., Ishii, K., Hamada, H., Akimoto, M., Hashimoto, N. and Miyatake, S. (1999) Adenovirus-mediated gene transfer of Bcl-xL prevents cell death in primary neuronal culture of the rat. *Neurosci. Lett.***270**, 177-180.

Mazzolini, G., Qian, C., Xie, X., Sun, Y., Lasarte, J.J., Drozdzik, M. and Prieto, J. (1999) Regression of colon cancer and induction of antitumor immunity by intratumoral injection of adenovirus expressing interleukin-12. *Cancer Gene Ther.***6**, 514-522.

Mazzolini, G., Qian, C., Narvaiza, I., Barajas, M., Borras-Cuesta, F., Xie, X., Duarte, M., Melero, I. and Prieto, J. (2000) Adenoviral gene transfer of interleukin 12 into tumors synergizes with adoptive T cell therapy both at the induction and effector level. *Hum. Gene Ther.***11**, 113-125.

Mazzoni, I.E., Said, F.A., Aloyz, R., Miller, F.D. and Kaplan, D. (1999) Ras regulates sympathetic neuron survival by suppressing the p53-mediated cell death pathway. *J. Neurosci.***19**, 9716-9727.

McClaine, S.J., Hamilton, T.E., Burke, C.V. and Raper, S.E. (1997) Functional consequences of adenovirus-mediated murine pancreatic gene transfer. *Hum. Gene Ther.***8**, 739-746.

Mc Howell, J., Lochmüller, H., O'Hara, A., Fletcher, S., Kakulas, BA., Massie, B., Nalbantoglu, J. and Karpati, G. (1998) High level dystrophin expression after adenovirus-mediated dystrophin minigene transfer to skeletal muscle of dystrophic dogs: prolongation of expression with immunosuppression. *Hum. Gene. Ther.***9**, 629-634.

Medghalchi, S., Padmanabhan, R. and Ketner, G. (1997) Early region 4 modulates adenovirus DNA replication by two genetically separable mechanisms. *Virology* **236**, 8-17.

Meeker, T.C., Lay, L.T., Wroblewski, J.M., Turturro, F., Li, Z. and Seth, P. (1997) Adenoviral vectors efficiently target cell lines derived from selected lymphocytic malignancies, including anaplastic large cell lymphoma and Hodgkin's disease. *Clin. Cancer Res.***3**, 357-364.

Mehrara, B.J., Saadeh, P.B., Steinbrech, D.S., Dudziak, M., Spector, J.A., Greenwald, J.A., Gittes, G.K. and Longaker, M.T. (1999) Adenovirus-mediated gene therapy of osteoblasts in vitro and in vivo. *J. Bone Miner. Res.***14**, 1290-1301.

Menjo, M., Yamaguchi, S., Murata, Y., Hayashi, Y., Nagaya, T., Ohmori, S., Refetoff, S. and Seo, H. (1999) Responsiveness to thyroid hormone is enhanced in rat hepatocytes cultured as spheroids compared with that in monolayers: altered responsiveness to thyroid hormone possibly involves complex formed on thyroid hormone response elements. *Thyroid* **9**, 959-967.

Michael, S.J., Hong, J.S., Curiel, D.T. and Engler, J.A. (1995) Addition of a short peptide ligand to the adenovirus fiber protein. *Gene Ther.***2**, 660-668.

Milas, M., Yu, D., Lang, A., Ge, T., Feig, B., El-Naggar, A.K. and Pollock, R.E. (2000) Adenovirus-mediated p53 gene therapy inhibits human sarcoma tumorigenicity. *Cancer Gene Ther.***7**, 422-429.

Millecamps, S., Kiefer, H., Navarro, V., Geoffroy, M.C., Robert, J.J., Finiels, F., Mallet, J. and Barkats, M. (1999) Neuron-restrictive silencer elements mediate neuron specificity of adenoviral gene expression. *Nat. Biotechnol.***17**, 865-869.

Miller, C.R., Buchsbaum, D.J., Reynolds, P.N., Douglas, J.T., Gillespie, G.Y., Mayo, M.S., Raben, D. and Curiel, D.T. (1998) Differential susceptibility of primary and established human glioma cells to adenovirus infection: targeting via the epidermal growth factor receptor achieves fiber receptor-independent gene transfer. *Cancer Res.***58**, 5738-5748.

Miller, P.W., Sharma, S., Stolina, M., Butterfield, L.H., Luo, J., Lin, Y., Dohadwala, M., Batra, R.K., Wu, L., Economou, J.S. and Dubinett, S.M. (2000) Intratumoral administration of adenoviral interleukin 7 gene-modified dendritic cells augments specific antitumor immunity and achieves tumor eradication. *Hum. Gene Ther.***11**, 53-65.

Minaguchi, T., Mori, T., Kanamori, Y., Matsushima, M., Yoshikawa, H., Taketani, Y. and Nakamura, Y. (1999) Growth suppression of human ovarian cancer cells by adenovirus-mediated transfer of the PTEN gene. *Cancer Res.***59**, 6063-6067.

Mitani, K., Wakamiya, M., Hasty, P., Graham, F.L., Bradley, A. and Caskey, C.T. (1995) Gene targeting in mouse embryonic stem cells with an adenoviral vector. Somat. *Cell Mol. Genet.***21**, 221-231.

Mitry, R.R., Sarraf, C.E., Havlik, R. and Habib, N.A. (2000) Detection of adenovirus and initiation of apoptosis in hepatocellular carcinoma cells after Ad-p53 treatment. *Hepatology* **31**, 885-889.

Mittereder, N., March, K. and Trapnell, B.C. (1996) Evaluation of the concentration and bioactivity of adenovirus vectors for gene therapy. *J. Virol.***70**, 7498-7509.

Miyaguchi, K., Maeda, Y., Kojima, T., Setoguchi, Y. and Mori, N. (1999) Neuron-targeted gene transfer by adenovirus carrying neural-restrictive silencer element. *Neuroreport* **10**, 2349-2353.

Miyamoto, T., Leconte, I., Swain, J.L. and Fox, J.C. (1998) Autocrine FGF signaling is required for vascular smooth muscle cell survival in vitro. *J. Cell Physiol.***177**, 58-67.

Miyazaki, T., Katagiri, H., Kanegae, Y., Takayanagi, H., Sawada, Y., Yamamoto, A., Pando, M.P., Asano, T., Verma, I.M., Oda, H., Nakamura, K. and Tanaka, S. (2000a) Reciprocal role of ERK and NF-kappaB pathways in survival and activation of osteoclasts. *J. Cell Biol.***148**, 333-342.

Miyazaki, T., Takayanagi, H., Isshiki, M., Takahashi, T., Okada, M., Fukui, Y., Oda, H., Nakamura, K., Hirai, H., Kurokawa, T. and Tanaka, S. (2000b) In vitro and in vivo suppression of osteoclast function by adenovirus vector-induced csk gene. *J. Bone Miner. Res.***15**, 41-51.

Miyazawa, N., Leopold, P.L., Hackett, N.R., Ferris, B., Worgall, S., Falck-Pedersen, E. and Crystal, R.G. (1999) Fiber swap between adenovirus subgroups B and C alters intracellular trafficking of adenovirus gene transfer vectors. *J. Virol.***73**, 6056-6065.

Mizuguchi, H. and Kay, M.A.(1999) A simple method for constructing E1- and E1/E4-deleted recombinant adenoviral vectors. *Hum. Gene. Ther.***10**, 2013-2017.

Moore, M., Horikoshi, N. and Shenk,T. (1996) Oncogenic potential of the adenovirus E4orf6 protein. *Proc. Natl. Acad. Sci. U. S. A.***93**, 11295-11301.

Moorhead, J.W., Clayton, G.H., Smith, R.L. and Schaack, J. (1999) A replication-incompetent adenovirus vector with the preterminal protein gene deleted efficiently transduces mouse ears. *J. Virol.***73**, 1046-1053.

Morelli, A.E., Larregina, A.T., Smith-Arica, J., Dewey, R.A., Southgate, T.D., Ambar, B., Fontana, A., Castro, M.G. and Lowenstein, P.R. (1999) Neuronal and glial cell type-specific promoters within adenovirus recombinants restrict the expression of the apoptosis-inducing molecule Fas ligand to predetermined brain cell types, and abolish peripheral liver toxicity. *J. Gen. Virol.***80**, 571-583.

Morishita, K., Johnson, D.E. and Williams, L.T. (1995) A novel promoter for vascular endothelial growth factor receptor (flt-1) that confers endothelial-specific gene expression. *J. Biol. Chem.***270**, 27948-27953.

Morsy, M.A., Gu, M., Motzel, S., Zhao, J., Lin, J., Su, Q., Allen, H., Franlin, L., Parks, R.J., Graham, F.L., Kochanek, S., Bett, A.J. and Caskey, C.T. (1998) An adenoviral vector deleted for all viral coding sequences results in enhanced safety and extended expression of a leptin transgene. *Proc. Natl. Acad. Sci. U. S. A.***95**, 7866-7871.

Motoi, F., Sunamura, M., Ding, L., Duda, D.G., Yoshida, Y., Zhang, W., Matsuno, S. and Hamada, H. (2000) Effective gene therapy for pancreatic cancer by cytokines mediated by restricted replication-competent adenovirus. *Hum. Gene Ther.***11**, 223-235.

Mujoo, K., Maneval, D.C., Anderson, S.C. and Gutterman, J.U. (1996) Adenoviral-mediated p53 tumor suppressor gene therapy of human ovarian carcinoma. *Oncogene* **12**, 1617-1623.

Mukhopadhyay, T. and Roth, J.A. (1998) Superinduction of wild-type p53 protein after 2 methoxyestradiol treatment of Ad5p53-transduced cells induces tumor cell apoptosis. *Oncogene* **17**, 241-246.

Mulders, P., Tso, C.L., Pang, S., Kaboo, R., McBride, W.H., Hinkel, A., Gitlitz, B., Dannull, J., Figlin, R. and Belldegrun, A. (1998) Adenovirus-mediated interleukin-2 production by tumors induces growth of cytotoxic tumor-infiltrating lymphocytes against human renal cell carcinoma. *J. Immunother.***21**, 170-180.

Muzzin, P., Eisensmith, R.C., Copeland, K.C. and Woo, S.L. (1996) Correction of obesity and diabetes in genetically obese mice by leptin gene therapy. *Proc. Natl. Acad. Sci. U. S. A.***93**, 14804-14808.

Nagai, E., Ogawa, T., Kielian, T., Ikubo, A. and Suzuki, T. (1998) Irradiated tumor cells adenovirally engineered to secrete granulocyte/macrophage-colony-stimulating factor establish antitumor immunity and eliminate pre-existing tumors in syngeneic mice. *Cancer Immunol. Immunother.***47**, 72-80.

Nakamura, T., Akiyoshi, H., Shiota, G., Isono, M., Nakamura, K., Moriyama, M. and Sato, K. (1999) Hepatoprotective action of adenovirus-transferred HNF-3gamma gene in acute liver injury caused by CCl(4). *FEBS Lett.***459**, 1-4.

Nakaoka, T., Gonda, K., Ogita, T., Otawara-Hamamoto, Y., Okabe, F., Kira, Y., Harii, K., Miyazono, K., Takuwa, Y. and Fujita, T. (1997) Inhibition of rat vascular smooth muscle proliferation in vitro and in vivo by bone morphogenetic protein-2. *J. Clin. Invest.***100**, 2824-2832.

Nalbantoglu, J., Pari, G., Karpati, G. and Holland, P.C. (1999) Expression of the primary coxsackie and adenovirus receptor is downregulated during skeletal muscle maturation and limits the efficacy of adenovirus-mediated gene delivery to muscle cells. *Hum. Gene Ther.***10**, 1009-1019.

Narumi, K., Suzuki, M., Song, W., Moore, M.A. and Crystal, R.G. (1998) Intermittent, repetitive corticosteroid-induced upregulation of platelet levels after adenovirus-mediated transfer to the liver of a chimeric glucocorticoid-responsive promoter controlling the thrombopoietin cDNA. *Blood* **92**, 822-833.

Naruse, I., Heike, Y., Hama, S., Mori, M. and Saijo, N. (1998) High concentrations of recombinant adenovirus expressing p16 gene induces apoptosis in lung cancer cell lines. *Anticancer Res.***18**, 4275-4282.

Narvaiza, I., Mazzolini, G., Barajas, M., Duarte, M., Zaratiegui, M., Qian, C., Melero, I. and Prieto, J. (2000) Intratumoral coinjection of two adenoviruses, one encoding the chemokine IFN-gamma-inducible protein-10 and another encoding IL-12, results in marked antitumoral synergy. *J. Immunol.***164**, 3112-3122.

Nasu, Y., Bangma, C.H., Hull, G.W., Lee, H.M., Hu, J., Wang, J., McCurdy, M.A., Shimura, S., Yang, G., Timme, T.L. and Thompson, T.C. (1999) Adenovirus-mediated interleukin-12 gene therapy for prostate cancer: suppression of orthotopic tumor growth and pre-established lung metastases in an orthotopic model. *Gene Ther.***6**, 338-349.

Navarro, V., Millecamps, S., Geoffroy, M.C., Robert, J.J., Valin, A., Mallet, J. and Gal La Salle, G.L. (1999) Efficient gene transfer and long-term expression in neurons using a recombinant adenovirus with a neuron-specific promoter. *Gene Ther.***6**, 1884-1892.

Nesbit, M., Nesbit, H.K., Bennett, J., Andl, T., Hsu, M.Y., Dejesus, E., McBrian, M., Gupta, A.R., Eck, S.L. and Herlyn, M. (1999) Basic fibroblast growth factor induces a transformed phenotype in normal human melanocytes. *Oncogene* **18**,, 6469-6476.

Neschis, D.G., Safford, S.D., Hanna, A.K., Fox, J.C. and Golden, M.A. (1998) Antisense basic fibroblast growth factor gene transfer reduces early intimal thickening in a rabbit femoral artery balloon injury model. *J. Vasc. Surg.***27**, 126-134.

Nevins, J.R.: Adenovirus E1A: Transcription regulation and alteration of cell growth control. (1995) The molecular repertoire of adenoviruses III: Current Topics in Microbiology and Immunology 199, *Eds Doerfler W. and Böhm P.*, p.25-32.

Nishida, K., Kang, J.D., Gilbertson, L.G., Moon, S.H., Suh, J.K., Vogt, M.T., Robbins, P.D. and Evans, C.H. (1999) Modulation of the biologic activity of the rabbit intervertebral disc by gene therapy: an in vivo study of adenovirus-mediated transfer of the human transforming growth factor beta 1 encoding gene. *Spine* **24**, 2419-2425.

No, D., Yao, T.P. and Evans, R.M. (1996) Ecdysone-inducible gene expression in mammalian cells and transgenic mice. *Proc. Natl. Acad. Sci. U. S. A.***93**, 3346-3351.

Noel, R.J., Antinozzi, P.A., McGarry, J.D. and Newgard, C.B. (1997) Engineering of glycerol-stimulated insulin secretion in islet beta cells. Differential metabolic fates of glucose and glycerol provide insight into mechanisms of stimulus-secretion coupling. *J. Biol. Chem.***272**, 18621-18627.

Nuss, H.B., Marban, E. and Johns, D.C. (1999) Overexpression of a human potassium channel suppresses cardiac hyperexcitability in rabbit ventricular myocytes. J. *Clin. Invest.***103**, 889-896.

O'Doherty, R.M., Lehman, D.L., Seoane, J., Gomez-Foix, A.M., Guinovart, J.J. and Newgard, C.B. (1996) Differential metabolic effects of adenovirus-mediated glucokinase and hexokinase I overexpression in rat primary hepatocytes. *J. Biol. Chem.* **271**, 20524-20530.

O'Doherty, R.M., Lehman, D.L., Telemaque-Potts, S. and Newgard, C.B. (1999) Metabolic impact of glucokinase overexpression in liver: lowering of blood glucose in fed rats is accompanied by hyperlipidemia. *Diabetes* **48**, 2022-2027.

Oka, K., Kobayashi, K., Sullivan, M., Martinez, J., Teng, B.B., Ishimura-Oka, K. and Chan, L. (1997) Tissue-specific inhibition of apolipoprotein B mRNA editing in the liver by adenovirus-mediated transfer of a dominant negative mutant APOBEC-1 leads to increased low density lipoprotein in mice. *J. Biol. Chem.***272**, 1456-1460.

Okubo, Y., Bessho, K., Fujimura, K., Iizuka, T. and Miyatake, S. (1999) Expression of bone morphogenetic protein-2 via adenoviral vector in C2C12 myoblasts induces differentiation into the osteoblast lineage. *Biochem. Biophys. Res. Commun.***262**, 739-743.

Okubo, Y., Bessho, K., Fujimura, K., Iizuka, T. and Miyatake, S.I. (2000) Osteoinduction by bone morphogenetic protein-2 via adenoviral vector under transient immunosuppression. *Biochem. Biophys. Res. Commun.***267**, 382-387.

O'Malley, B.W.,Jr., Sewell, D.A., Li, D., Kosai, K., Chen, S.H., Woo, S.L. and Duan, L. (1997) The role of interleukin-2 in combination adenovirus gene therapy for head and neck cancer. *Mol. Endocrinol.***11**, 667-673.

O'Riordan C.R., Lachapelle A, Delgado C., Parkes V., Wadsworth S.C., Smith A.E. and Francis G.E. (1999) PEGylation of adenovirus with retention of infectivity and protection from neutralizing antibody in vitro and in vivo. *Hum. Gene Ther.***10**, 1349-1358.

Osada, S., Ebihara, I., Setoguchi, Y., Takahashi, H., Tomino, Y. and Koide, H. (1999) Gene therapy for renal anemia in mice with polycystic kidney using an adenovirus vector encoding the human erythropoietin gene. *Kidney Int.***55**, 1234-1240.

Osaki, T., Hashimoto, W., Gambotto, A., Okamura, H., Robbins, P.D., Kurimoto, M., Lotze, M.T. and Tahara, H. (1999) Potent antitumor effects mediated by local expression of the mature form of the interferon-gamma inducing factor, interleukin-18 (IL-18). *Gene Ther.***6**, 808-815.

Oualikene, W., Lamoureux, L., Weber, J.M. and Massie, B. (2000) Protease-deleted adenovirus vectors and complementing cell lines: Potential applications of single-round replication mutants for vaccination and gene therapy. *Hum. Gene Ther.***11**,1347-1353.

Overholt, S.M., Liu, T.J., Taylor, D.L., Wang, M., El-Naggar, A.K., Shillitoe, E., Adler-Storthz, K., John, L.S., Zhang, W.W., Roth, J.A. and Clayman, G.L. (1997) Head and neck squamous cell growth suppression using adenovirus-p53-FLAG: a potential marker for gene therapy trials. *Clin. Cancer Res.***3**, 185-191.

Oyama, T., Ran, S., Ishida, T., Nadaf, S., Kerr, L., Carbone, D.P. and Gabrilovich, D.I. (1998) Vascular endothelial growth factor affects dendritic cell maturation through the inhibition of nuclear factor-kappa B activation in hemopoietic progenitor cells. *J. Immunol.***160**, 1224-1232.

Ozawa, H., Ding, W., Torii, H., Hosoi, J., Seiffert, K., Campton, K., Hackett, N.R., Topf, N., Crystal, R.G. and Granstein, R.D. (1999) Granulocyte-macrophage colony-stimulating factor gene transfer to dendritic cells or epidermal cells augments their antigen-presenting function including induction of anti-tumor immunity. *J. Invest. Dermatol.***113**, 999-1005.

Palmer, K., Emtage, P.C., Strieter, R.M. and Gauldie, J. (1999) Transient gene transfer of non-ELR chemokines to rodent lung induces mononuclear cell accumulation and activation. *J. Interferon. Cytokine. Res.***19**, 1381-1390.

Park, M., Chae, H.D., Yun, J., Jung, M., Kim, Y.S., Kim, S.H., Han, M.H. and Shin, D.Y. (2000) Constitutive activation of cyclin B1-associated cdc2 kinase overrides p53-mediated G2-M arrest. *Cancer Res.***60**, 542-545.

Parks, R.J. and Graham, F.L. (1996) A helper-dependent system for adenovirus vector system: removal of helper virusby Cre-mediated excision of the viral packaging signal. *Proc. Natl. Acad. Sci. U. S. A.***93**, 13565-13570.

Parks, R.J. and Graham, F.L. (1997) A helper-dependent system for adenovirus vector production helps define a lower limit for efficient DNA packaging. *J.Virol.***71**, 3293-3298.

Parks, E., Strieter, R.M., Lukacs, N.W., Gauldie, J., Hitt, M., Graham, F.L. and Kunkel, S.L. (1998) Transient gene transfer of IL-12 regulates chemokine expression and disease severity in experimental arthritis. *J. Immunol.***160**, 4615-4619.

Parks, R.J., Bramson, J.L., Wan, Y., Addison, C.L. and Graham, F.L. (1999) Effects of stuffer DNA on transgene expression from helper-dependent adenovirus vectors. *J.Virol.***73**, 8027-8034.

Parr, M.J., Manome, Y., Tanaka, T., Wen, P., Kufe, D.W., Kaelin, W.G Jr. and Fine, H.A. (1997) Tumor-selective transgene expression in vivo mediated by an E2F-responsive adenoviral vector. *Nat. Med.***3**, 1145-1149.

Pastore, L., Morral, N., Zhou, H., Garcia, R., Parks, R.J., Kochanek, S., Graham, F.L., Lee, B. and Beaudet, A.L. (1999) Use of a liver-specific promoter reduces immune response to the transgene in adenoviral vectors. *Hum. Gene Ther.***10**, 1773-1781.

Pataer, A., Fang, B., Yu, R., Kagawa, S., Hunt, K.K., McDonnell, T.J., Roth, J.A. and Swisher, S.G. (2000) Adenoviral Bak overexpression mediates caspase-dependent tumor killing. *Cancer Res.***15**, 788-792.

Patel, S.R., Lee, L.Y., Mack, C.A., Polce, D.R., El-Sawy, T., Hackett, N.R., Ilercil, A., Jones, E.C., Hahn, R.T., Isom, O.W., Rosengart, T.K. and Crystal, R.G. (1999) Safety of direct myocardial administration of an adenovirus vector encoding vascular endothelial growth factor 121. *Hum. Gene Ther.***10**, 1331-1348.

Pearson, A.S., Spitz, F.R., Swisher, S.G., Kataoka, M., Sarkiss, M.G., Meyn, R.E., McDonnell, T.J., Cristiano, R.J. and Roth, J.A. (2000) Up-regulation of the proapoptotic mediators Bax and Bak after adenovirus-mediated p53 gene transfer in lung cancer cells. *Clin. Cancer Res.***6**, 887-890.

Perlman, H., Luo, Z., Krasinski, K., Le Roux, A., Mahfoudi, A., Smith, R.C., Branellec, D. and Walsh, K. (1999) Adenovirus-mediated delivery of the Gax transcription factor to rat carotid arteries inhibits smooth muscle proliferation and induces apoptosis. *Gene Ther.***6**, 758-763.

Petrof, B.J., Lochmüller, H., Massie, B., Yang, L., Macmillan, C., Zhao, J.E., Nalbantoglu, J. and Karpati, G. (1996) Impairment of force generation after adenovirus-mediated gene transfer to muscle is alliviated by adenoviral gene inactivation and host CD8+ Tcell deficiency. *Hum. Gene Ther.***7**, 1813-1826

Piché, A., Grim, J., Rancourt, C., Gomez-Navarro, J., Reed, J.C. and Curiel, D.T.(1998) Modulation of Bcl-2 protein levels by an intracellular anti-Bcl-2 single-chain antibody increases drug-induced cytotoxicity in the breast cancer cell line MCF-7. *Cancer Res.***15**, 2134-2140.

Pietersen, A.M., van der Eb, M.M., Rademaker, H.J, van den Wollenberg, D.J., Rabelink, M.J., Kuppen, P.J., van Dierendonck, J.H., van Ormondt, H., Masman, D., van de Velde, C.J., van der Eb, A.J., Hoeben, R.C. and Noteborn, M.H. (1999) Specific tumor-cell killing with adenovirus vectors containing the apoptin gene. *Gene Ther.***6**, 882-892.

Pili, R., Chang, J., Muhlhauser, J., Crystal, R.G., Capogrossi, M.C. and Passaniti, A. (1997) Adenovirus-mediated gene transfer of fibroblast growth factor-1: angiogenesis and tumorigenicity in nude mice. *Int. J. Cancer* **73**, 258-263.

Pinderski Oslund, L.J., Hedrick, C.C., Olvera, T., Hagenbaugh, A., Territo, M., Berliner, J.A. and Fyfe, A.I. (1999) Interleukin-10 blocks atherosclerotic events in vitro and in vivo. Arterioscler. *Thromb. Vasc. Biol.***19**, 2847-2853.

Piston, D.W., Knobel, S.M., Postic, C., Shelton, K.D. and Magnuson, M.A. (1999) Adenovirus-mediated knockout of a conditional glucokinase gene in isolated pancreatic islets reveals an essential role for proximal metabolic coupling events in glucose-stimulated insulin secretion. *J. Biol. Chem.***274**, 1000-1004.

Poliakova, L., Kovesdi, I., Wang, X., Capogrossi, M.C. and Talan, M. (1999) Vascular permeability effect of adenovirus-mediated vascular endothelial growth factor gene transfer to the rabbit and rat skeletal muscle. *J. Thorac. Cardiovasc. Surg.***118**, 339-347.

Poller, W., Schneider-Rasp, S., Liebert, U., Merklein, F., Thalheimer, P., Haack, A., Schwaab, R., Schmitt, C. and Brackmann, H.H. (1996) Stabilization of transgene expression by incorporation of E3 region genes into an adenoviral factor IX vector and by transient anti-CD4 treatment of the host. *Gene Ther.***3**, 521-530.

Prabhu, N.S., Blagosklonny, M.V., Zeng, Y.X., Wu, G.S., Waldman, T. and El-Deiry, W.S. (1996) Suppression of cancer cell growth by adenovirus expressing p21(WAF1/CIP1) deficient in PCNA interaction. *Clin. Cancer Res.***2**, 1221-1229.

Printz, M.A., Gonzalez, A.M., Cunningham, M., Gu, D.L., Ong, M., Pierce, G.F. and Aukerman, S.L. (2000) Fibroblast growth factor 2-retargeted adenoviral vectors exhibit a modified biolocalization pattern and display reduced toxicity relative to native adenoviral vectors. *Hum. Gene Ther.***11**, 191-204.

Putzer, B.M., Hitt, M., Muller, W.J., Emtage, P., Gauldie, J. and Graham, F.L. (1997) Interleukin 12 and B7-1 costimulatory molecule expressed by an adenovirus vector act synergistically to facilitate tumor regression. *Proc. Natl. Acad. Sci. U. S. A.***94**, 10889-10894.

Qi, Z., Atsuchi, N., Ooshima, A., Takeshita, A. and Ueno, H. (1999) Blockade of type beta transforming growth factor signaling prevents liver fibrosis and dysfunction in the rat. Proc. *Natl. Acad. Sci. U. S. A.***96**, 2345-2349.

Qian, J., Kaytor, E.N., Towle, H.C. and Olson, L.K. (1999) Upstream stimulatory factor regulates Pdx-1 gene expression in differentiated pancreatic beta-cells. *Biochem. J.***341**, 315-322.

Qiao, J., Chen, S.H., Pham-Nguyen, K.B., Mandeli, J. and Woo, S.L. (1999) Construction and characterization of a recombinant adenoviral vector expressing human interleukin-12. *Cancer Gene Ther.***6**, 373-379.

Qin, L., Ding, Y., Pahud, D.R., Robson, N.D., Shaked, A. and Bromberg, J.S. (1997) Adenovirus-mediated gene transfer of viral interleukin-10 inhibits the immune response to both alloantigen and adenoviral antigen. *Hum. Gene Ther.***8**, 1365-1374.

Qin, X.Q., Tao, N., Dergay, A., Moy, P., Fawell, S., Davis, A., Wilson, J.M. and Barsoum, J. (1998) Interferon-beta gene therapy inhibits tumor formation and causes regression of established tumors in immune-deficient mice. *Proc. Natl. Acad. Sci. U. S. A.***95**, 14411-14416.

Rakkar, A.N., Katayose, Y., Kim, M., Craig, C., Ohri, E., Li, Z., Cowan, K.H. and Seth, P. (1999) A novel adenoviral vector expressing human Fas/CD95/APO-1 enhances p53-mediated apoptosis. *Cell Death. Differ.***6**, 326-333.

Ramalingam, R., Rafii, S., Worgall, S., Hackett, N.R. and Crystal, R.G. (1999) Induction of endogenous genes following infection of human endothelial cells with an E1(-) E4(+) adenovirus gene transfer vector. *J. Virol.***73**, 10183-10190.

Ramondetta, L., Mills, G.B., Burke, T.W. and Wolf, J.K. (2000) Adenovirus-mediated expression of p53 or p21 in a papillary serous endometrial carcinoma cell line (SPEC-2) results in both growth inhibition and apoptotic cell death: potential application of gene therapy to endometrial cancer. *Clin. Cancer Res.***6**, 278-284.

Rancourt, C., Piche, A., Gomez-Navarro, J., Wang, M., Alvarez, R.D., Siegal, G.P., Fuller, G.M., Jones, S.A. and Curiel, D.T. (1999) Interleukin-6 modulated conditionally replicative adenovirus as an antitumor/cytotoxic agent for cancer therapy. *Clin. Cancer Res.***5**, 43-50.

Randrianarison-Jewtoukoff, V. and Perricaudet M. (1995) Recombinant adenoviruses as vaccines. *Biologicals* **23**, 145-157. Review.

Ravanti, L., Hakkinen, L., Larjava, H., Saarialho-Kere, U., Foschi, M., Han, J. and Kahari, V.M. (1999) Transforming growth factor-beta induces collagenase-3 expression by human gingival fibroblasts via p38 mitogen-activated protein kinase. *J. Biol. Chem.***274**, 37292-37300.

Recchia, A., Parks, R.J., Lamartina, S., Toniatti, C., Pieroni, L., Palombo, F., Ciliberto, G., Graham, F.L.,Cortese, R., La Monica, N. and Colloca, S. (1999) Site-specific integration mediated by a hybrid adenovirus/adeno-associated virus vector. *Proc. Natl. Acad. Sci. U. S. A.***16**, 2615-2620.

Reynolds, P., Dmitriev, I. and Curiel, D. (1999) Insertion of an RGD motif into the HI loop of adenovirus fiber protein alters the distribution of transgene expression of the systemically administered vector. *Gene Ther.***6**, 1336-1339.

Riew, K.D., Wright, N.M., Cheng, S., Avioli, L.V. and Lou, J. (1998) Induction of bone formation using a recombinant adenoviral vector carrying the human BMP-2 gene in a rabbit spinal fusion model. Calcif. *Tissue Int.***63**, 357-360.

Ring, C.J., Harris, J.D., Hurst, H.C. and Lemoine, N.R. (1996) Suicide gene expression induced in tumour cells transduced with recombinant adenoviral, retroviral and plasmid vectors containing the ERBB2 promoter. *Gene Ther.***3**, 1094-1103.

Rivard, A., Silver, M., Chen, D., Kearney, M., Magner, M., Annex, B., Peters, K. and Isner, J.M. (1999) Rescue of diabetes-related impairment of angiogenesis by intramuscular gene therapy with adeno-VEGF. *Am. J. Pathol.***154**, 355-363.

Rivera, V.M., Ye, X., Courage, N.L., Sachar, J., Cerasoli, F.,Jr., Wilson, J.M. and Gilman, M. (1999) Long-term regulated expression of growth hormone in mice after intramuscular gene transfer. *Proc. Natl. Acad. Sci. U. S. A.***96**, 8657-8662.

Rocco, J.W., Li, D., Liggett, W.H.,Jr., Duan, L., Saunders, J.K.,Jr., Sidransky, D. and O'Malley, B.W.,Jr. (1998) p16INK4A adenovirus-mediated gene therapy for human head and neck squamous cell cancer. *Clin. Cancer Res.***4**, 1697-1704.

Rodriguez, R., Schuur, E.R., Lim, H.Y., Henderson, G.A, Simons, J.W. and Henderson, D.R . (1997) Prostate attenuated replication competent adenovirus (ARCA) CN706: a selective cytotoxic for prostate-specific antigen-positive prostate cancer cells. *Cancer. Res.***57**, 2559-2563.

Roelvink, P.W., Lizonova, A., Lee, J.G., Li, Y., Bergelson, J.M., Finberg, R.W., Brough, D.E., Kovesdi, I. and Wickham, T.J. (1998) The coxsackievirus-adenovirus receptor protein can function as a cellular attachment protein for adenovirus serotypes from subgroups A, C, D, E, and F. *J. Virol.***72**, 7909-7915.

Roelvink, P.W., Mi Lee, G., Einfeld, D.A., Kovesdi, I. and Wickham, T.J. (1999) Identification of a conserved receptor-binding site on the fiber proteins of CAR-recognizing adenoviridae. *Science* **286**, 1568-1571.

Rogers, B.E., Douglas, J.T., Ahlem, C., Buchsbaum, D.J., Frincke, J. and Curiel, D.T. (1997) Use of a novel cross-linking method to modify adenovirus tropism. *Gene Ther.***4**, 1387-1392.

Romanczuk H., Galer C.D., Zabner J., Barsomian G., Wadsworth S.C. and O'Riordan C. (1999) Modification of an adenoviral vector with biologically selected peptides: a novel strategy for gene delivery to cells of choice. *Hum. Gene Ther.***10**, 2615-2626.

Rosenfeld, M.E., Rogers, B.E., Khazaeli, M.B., Mikheeva, G., Raben, D., Mayo, M.S., Curiel, D.T. and Buchsbaum, D.J. (1997) Adenoviral-mediated delivery of gastrin-releasing peptide receptor results in specific tumor localization of a bombesin analogue in vivo. *Clin. Cancer Res.***3**, 1187-1194.

Rosengart, T.K., Lee, L.Y., Patel, S.R., Kligfield, P.D., Okin, P.M., Hackett, N.R., Isom, O.W. and Crystal, R.G. (1999a) Six-month assessment of a phase I trial of angiogenic gene therapy for the treatment of coronary artery disease using direct intramyocardial administration of an adenovirus vector expressing the VEGF121 cDNA. *Ann. Surg.***230**, 466-70.

Rosengart, T.K., Lee, L.Y., Patel, S.R., Sanborn, T.A., Parikh, M., Bergman, G.W., Hachamovitch, R., Szulc, M., Kligfield, P.D., Okin, P.M., Hahn, R.T., Devereux, R.B., Post, M.R., Hackett, N.R., Foster, T., Grasso, T.M., Lesser, M.L., Isom, O.W. and Crystal, R.G. (1999b) Angiogenesis gene therapy: phase I assessment of direct intramyocardial administration of an adenovirus vector expressing VEGF121 cDNA to individuals with clinically significant severe coronary artery disease. *Circulation* **100**, 468-474.

Rothmann, T., Katus, H.A., Hartong, R., Perricaudet, M. and Franz, W.M. (1996) Heart muscle-specific gene expression using replication defective recombinant adenovirus. *Gene Ther.***3**, 919-926.

Rowe W.P., Huebner R.J., Gilmore L.K., Parrott R.H. and Ward T.G. (1953) Isolation of a cytopathic agent from human adenoids undergoing spontaneous degeneration in tissue culture. *Proc.Soc.Exp.Biol.Med.***84**, 570-573

Ruan, S., Okcu, M.F., Pong, R.C., Andreeff, M., Levin, V., Hsieh, J.T. and Zhang, W. (1999) Attenuation of WAF1/Cip1 expression by an antisense adenovirus expression vector sensitizes glioblastoma cells to apoptosis induced by chemotherapeutic agents 1,3-bis(2-chloroethyl)-1-nitrosourea and cisplatin. *Clin. Cancer Res.***5**, 197-202.

Russi, T.J., Hirschowitz, E.A. and Crystal, R.G. (1997) Delayed-type hypersensitivity response to high doses of adenoviral vectors. *Hum. Gene Ther.***8**, 323-330.

Saadeh, P.B., Mehrara, B.J., Steinbrech, D.S., Dudziak, M.E., Greenwald, J.A., Luchs, J.S., Spector, J.A., Ueno, H., Gittes, G.K. and Longaker, M.T. (1999) Transforming growth factor-beta1 modulates the expression of vascular endothelial growth factor by osteoblasts. *Am. J. Physiol.***277**, C628-C637.

Safi, J.,Jr., DiPaula, A.F.,Jr., Riccioni, T., Kajstura, J., Ambrosio, G., Becker, L.C., Anversa, P. and Capogrossi, M.C. (1999) Adenovirus-mediated acidic fibroblast growth factor gene transfer induces angiogenesis in the nonischemic rabbit heart. Microvasc. *Res.***58**, 238-249.

Sakai, K., Mitani, K. and Miyazaki, J. (1995) Efficient regulation of gene expression by adenovirus vector-mediated delivery of the CRE recombinase. *Biochem. Biophys. Res. Commun.***217**, 393-401.

Sanchez-Capelo, A., Corti, O. and Mallet, J. (1999) Adenovirus-mediated over-expression of TGFbeta1 in the striatum decreases dopaminergic cell survival in embryonic nigral grafts. *Neuroreport.***10**, 2169-2173.

Sandig, V., Loser, P., Lieber, A., Kay, M.A. and Strauss, M. (1996) HBV-derived promoters direct liver-specific expression of an adenovirally transduced LDL receptor gene. *Gene Ther.***3**, 1002-1009.

Sandig, V., Brand, K., Herwig, S., Lukas, J., Bartek, J. and Strauss, M. (1997) Adenovirally transferred p16INK4/CDKN2 and p53 genes cooperate to induce apoptotic tumor cell death. *Nat. Med.***3**, 313-319.

Sandig, V., Youil, R., Bett, A.J., Franlin, L.L., Oshima, M., Maione, D., Wang, F., Metzker, M.L., Savino, R. and Caskey, C.T. (2000) Optimization of the helper-dependent adenovirus system for production and potency in vivo. *Proc. Natl. Acad. Sci. U. S. A.***97**, 1002-1007.

Sandy, P., Gostissa, M., Fogal, V., Cecco, L.D., Szalay, K., Rooney, R.J., Schneider, C. and Del Sal, G. (2000) p53 is involved in the p120E4F-mediated growth arrest. *Oncogene* **19**, 188-199.

Sato, Y., Tanaka, K., Lee, G., Kanegae, Y., Sakai, Y., Kaneko, S., Nakabayashi, H., Tamaoki, T. and Saito, I. (1998) Enhanced and specific gene expression via tissue-specific production of Cre recombinase using adenovirus vector. *Biochem. Biophys. Res. Commun.***244**, 455-462.

Sauter, E.R., Nesbit, M., Litwin, S., Klein-Szanto, A.J., Cheffetz, S. and Herlyn, M. (1999) Antisense cyclin D1 induces apoptosis and tumor shrinkage in human squamous carcinomas. *Cancer Res.***59**, 4876-4881.

Sawchuk, S.J., Boivin, G.P., Duwel, L.E., Ball, W., Bove, K., Trapnell, B. and Hirsch, R.(1996) Anti-T- cell receptor monoclonal antibody prolongs transgene expression following adenovirus-mediated in vivo gene transfer to mouse synovium. *Hum. Gene Ther.***7**, 499-506.

Schaack, J., Guo, X. and Langer, S.J. (1996) Characterization of a replication-incompetent adenovirus type 5 mutant deleted for the preterminal protein gene. *Proc. Natl. Acad. Sci. U. S. A.***93**, 14686-14691.

Schachtner, S., Buck, C., Bergelson, J. and Baldwin, H. (1999) Temporally regulated expression patterns following in utero adenovirus-mediated gene transfer. *Gene Ther.***6**, 1249-1257.

Scheinman, M., Ascher, E., Kallakuri, S., Hingorani, A., Gade, P., Sherman, M., Seth, P. and Jacob, T. (1999) P53 gene transfer to the injured rat carotid artery promotes apoptosis. *Surgery* **126**, 863-868.

Schiedner, G., Morral, N., Parks, R.J., Wu, Y., Koopmans, S.C., Langston, C., Graham, F.L., Beaudet, A.L. and Kochanek, S. (1998) Genomic DNA transfer with a high-capacity adenovirus vector results in improved in vivo gene expression and decreased toxicity. *Nat. Genet.***18**, 180-183.

Schneider, H., Adebakin, S., Themis, M., Cook, T., Douar, A.M., Pavirani, A. and Coutelle, C. (1999) Therapeutic plasma concentrations of human factor IX in mice after gene delivery into the amniotic cavity: a model for the prenatal treatment of haemophilia B. *J. Gene Med.***1**, 424-432.

Schreiber, M., Muller, W.J., Singh, G. and Graham, F.L. (1999) Comparison of the effectiveness of adenovirus vectors expressing cyclin kinase inhibitors p16INK4A, p18INK4C, p19INK4D, p21(WAF1/CIP1) and p27KIP1 in inducing cell cycle arrest, apoptosis and inhibition of tumorigenicity. *Oncogene* **18**, 1663-1676.

Schrump, D.S., Chen, G.A., Consuli, U., Jin, X. and Roth, J.A. (1996) Inhibition of esophageal cancer proliferation by adenovirally mediated delivery of p16INK4. *Cancer Gene Ther.***3**, 357-364.

Schwartz, M.L., Katagi, C., Bruce, J. and Schlaepfer, W.W. (1994) Brain-specific enhancement of the mouse neurofilament heavy gene promoter in vitro. *J. Biol. Chem.***269**, 13444-13450.

Schwarzenberger, P., La Russa, V., Miller, A., Ye, P., Huang, W., Zieske, A., Nelson, S., Bagby, G.J., Stoltz, D., Mynatt, R.L., Spriggs, M. and Kolls, J.K. (1998) IL-17 stimulates granulopoiesis in mice: use of an alternate, novel gene therapy-derived method for in vivo evaluation of cytokines. *J. Immunol.***161**, 6383-6389.

Scott, D.K., O'Doherty, R.M., Stafford, J.M., Newgard, C.B. and Granner, D.K. (1998) The repression of hormone-activated PEPCK gene expression by glucose is insulin-independent but requires glucose metabolism. *J. Biol. Chem.***273**, 24145-24151.

Seijffers, R., Ben-David, O., Cohen, Y., Karasik, A., Berezin, M., Newgard, C.B. and Ferber, S. (1999) Increase in PDX-1 levels suppresses insulin gene expression in RIN 1046-38 cells. *Endocrinology* **140**, 3311-3317.

Seoane, J., Trinh, K., O'Doherty, R.M., Gomez-Foix, A.M., Lange, A.J., Newgard, C.B. and Guinovart, J.J. (1997) Metabolic impact of adenovirus-mediated overexpression of the glucose-6-phosphatase catalytic subunit in hepatocytes. *J. Biol. Chem.***272**, 26972-26977.

Seoane, J., Barbera, A., Telemaque-Potts, S., Newgard, C.B. and Guinovart, J.J. (1999) Glucokinase overexpression restores glucose utilization and storage in cultured hepatocytes from male Zucker diabetic fatty rats. *J. Biol. Chem.***274**, 31833-31838.

Setoguchi, Y., Danel, C. and Crystal, R.G. (1994) Stimulation of erythropoiesis by in vivo gene therapy: physiologic consequences of transfer of the human erythropoietin gene to experimental animals using an adenovirus vector. *Blood* **84**, 2946-2953.

Shao, J., Fujiwara, T., Kadowaki, Y., Fukazawa, T., Waku, T., Itoshima, T., Yamatsuji, T., Nishizaki, M., Roth, J.A. and Tanaka, N. (2000) Overexpression of the wild-type p53 gene inhibits NF-kappaB activity and synergizes with aspirin to induce apoptosis in human colon cancer cells. *Oncogene* **10**, 726-736.

Sharma, P.M., Egawa, K., Huang, Y., Martin, J.L., Huvar, I., Boss, G.R. and Olefsky, J.M. (1998) Inhibition of phosphatidylinositol 3-kinase activity by adenovirus-mediated gene transfer and its effect on insulin action. *J. Biol. Chem.* **273,** 18528-18537.

Sharma, R.V., Tan, E., Fang, S., Gurjar, M.V. and Bhalla, R.C. (1999) NOS gene transfer inhibits expression of cell cycle regulatory molecules in vascular smooth muscle cells. *Am. J. Physiol.***276,** H1450-H1459.

Shatrov, V.A., Ameyar, M., Bouquet, C., Cai, Z., Stancou, R., Haddada, H. and Chouaib, S. (2000) Adenovirus-mediated wild-type-p53-gene expression sensitizes TNF-resistant tumor cells to TNF-induced cytotoxicity by altering the cellular redox state. *Int. J. Cancer.***85,** 93-97.

Shaw, A.R. and Ziff, E.B. (1980) Transcripts from the adenovirus-2 major late promoter yield a single early family of 3' coterminal mRNAs and five late families. *Cell* **22,** 905-916.

Shenk, T. Adenoviridea: The virus and Their replication. Fields Virology, Third Edition (1996) Eds B.N.Fields, D.M. Knipe, P.M.Hawley et al., *Lippincott-Raven Publishers*, Philadelphia, p211-2148.

Shi, Q., Wang, Y. and Worton, R. (1997) Modulation of the specificity and activity of a cellular promoter in an adenoviral vector. *Hum. Gene Ther.***8,** 403-410.

Shibata, H., Toyama, K., Shioya, H., Ito, M., Hirota, M., Hasegawa, S., Matsumoto, H., Takano, H., Akiyama, T., Toyoshima, K., Kanamaru, R., Kanegae, Y., Saito, I., Nakamura, Y., Shiba, K. and Noda, T. (1997) Rapid colorectal adenoma formation initiated by conditional targeting of the Apc gene. *Science* **278,** 120-123.

Shinoura, N., Yoshida, Y., Sadata, A., Hanada, K.I., Yamamoto, S., Kirino, T., Asai, A. and Hamada, H. (1998) Apoptosis by retrovirus- and adenovirus-mediated gene transfer of Fas ligand to glioma cells: implications for gene therapy. *Hum. Gene Ther.***9,** 1983-1993.

Shinoura, N., Yoshida, Y., Nishimura, M., Muramatsu, Y., Asai, A., Kirino, T. and Hamada, H. (1999a) Expression level of Bcl-2 determines anti- or proapoptotic function. *Cancer Res.***59,** 4119-4128.

Shinoura, N., Muramatsu, Y., Nishimura, M., Yoshida, Y., Saito, A., Yokoyama, T., Furukawa, T., Horii, A., Hashimoto, M., Asai, A., Kirino, T. and Hamada, H. (1999b) Adenovirus-mediated transfer of p33ING1 with p53 drastically augments apoptosis in gliomas. *Cancer Res.***1,** 5521-5528.

Shinoura, N., Satou, R., Yoshida, Y., Asai, A., Kirino, T. and Hamada, H. (2000) Adenovirus-Mediated Transfer of Bcl-X(L) Protects Neuronal Cells from Bax-Induced Apoptosis. Exp. *Cell Res.***254,** 221-231.

Shinozaki, K., Yahata, H., Tanji, H., Sakaguchi, T., Ito, H. and Dohi, K. (1999) Allograft transduction of IL-10 prolongs survival following orthotopic liver transplantation. *Gene Ther.***6,** 816-822.

Shiratori, Y., Kanai, F., Hikiba, Y., Moriyama, H., Hamada, H., Matsumura, M., Tanaka, T., Lan, K.H., Ohashi, M., Okano, K., Naito, M. and Omata, M. (1998) Gene therapy for hepatic micrometastasis of murine colon carcinoma. *J. Hepatol.***28,** 886-895.

Shisler, J.L. and Gooding, L.R. (1998) Adenoviral inhibitors of the apoptotic cascade. *Trends in Microbiology* **6,** 337-339.

Shtrichman, R. and Kleinberger, T. (1998) Adenovirus type 5 E4 open reading frame 4 protein induces apoptosis in transformed cells. *J. Virol.***72,** 2975-2982.

Siders, W.M., Halloran, P.J. and Fenton, R.G. (1996) Transcriptional targeting of recombinant adenoviruses to human and murine melanoma cells. *Cancer Res.***56,** 5638-5646.

Siders, W.M., Wright, P.W., Hixon, J.A., Alvord, W.G., Back, T.C., Wiltrout, R.H. and Fenton, R.G. (1998) T cell- and NK cell-independent inhibition of hepatic metastases by systemic administration of an IL-12-expressing recombinant adenovirus. *J. Immunol.***160,** 5465-5474.

Sime, P.J., Xing, Z., Graham, F.L., Csaky, K.G. and Gauldie, J. (1997) Adenovector-mediated gene transfer of active transforming growth factor-beta1 induces prolonged severe fibrosis in rat lung. *J. Clin. Invest.***100,** 768-776.

Simons, M., Beinroth, S., Gleichmann, M., Liston, P., Korneluk, R.G., MacKenzie, A.E., Bahr, M., Klockgether, T., Robertson, G.S., Weller, M. and Schulz, J.B. (1999) Adenovirus-mediated gene transfer of inhibitors of apoptosis protein delays apoptosis in cerebellar granule neurons. *J. Neurochem.***72,** 292-301.

Smith, D.K., Korbutt, G.S., Suarez-Pinzon, W.L., Kao, D., Rajotte, R.V. and Elliott, J.F. (1997a) Interleukin-4 or interleukin-10 expressed from adenovirus-transduced syngeneic islet grafts fails to prevent beta cell destruction in diabetic NOD mice. *Transplantation* **64,** 1040-1049.

Smith, G. M. and Romero, M. I. (1999b) Adenoviral-mediated gene transfer to enhance neuronal survival, growth, and regeneration. J. Neurosc. Res.**55,** 147-157.

Smith, J.S., Keller, J.R., Lohrey, N.C., McCauslin, C.S., Ortiz, M., Cowan, K. and Spence, S.E. (1999a) Redirected infection of directly biotinylated recombinant adenovirus vectors through cell surface receptors and antigens. *Proc. Natl. Acad. Sci. U. S. A.***96**, 8855-8860.

Smith, R.C., Branellec, D., Gorski, D.H., Guo, K., Perlman, H., Dedieu, J.F., Pastore, C., Mahfoudi, A., Denefle, P., Isner, J.M. and Walsh, K. (1997b) p21CIP1-mediated inhibition of cell proliferation by overexpression of the gax homeodomain gene. *Genes Dev.***11**, 1674-1689.

Smith, R.C., Wills, K.N., Antelman, D., Perlman, H., Truong, L.N., Krasinski, K. and Walsh, K. (1997c) Adenoviral constructs encoding phosphorylation-competent full-length and truncated forms of the human retinoblastoma protein inhibit myocyte proliferation and neointima formation. *Circulation* **16**, 1899-1905.

Somers, J.P., DeLoia, J.A. and Zeleznik, A.J. (1999) Adenovirus-directed expression of a nonphosphorylatable mutant of CREB (cAMP response element-binding protein) adversely affects the survival, but not the differentiation, of rat granulosa cells. *Mol. Endocrinol.***13**, 1364-1372.

Stampfli, M.R., Cwiartka, M., Gajewska, B.U., Alvarez, D., Ritz, S.A., Inman, M.D., Xing, Z. and Jordana, M. (1999) Interleukin-10 gene transfer to the airway regulates allergic mucosal sensitization in mice. Am. J. Respir. *Cell Mol. Biol.***21**, 586-596.

Stec, D.E., Davisson, R.L., Haskell, R.E., Davidson, B.L. and Sigmund, C.D. (1999) Efficient liver-specific deletion of a floxed human angiotensinogen transgene by adenoviral delivery of Cre recombinase in vivo. *J. Biol. Chem.***274**, 21285-21290.

Steegenga, W.T., Riteco, N., Jochemsen, A.G., Fallaux, F.J. and Bos, J.L. (1998) The large E1B protein together with the E4orf6 protein target p53 for active degradation in adenovirus infected cells. *Oncogene* **16**, 349-357.

Steegenga, W.T., Shvarts, A., Riteco, N., Bos, J.L. and Jochemsen, A.G. (1999) Distinct regulation of p53 and p73 activity by adenovirus E1A, E1B, and E4orf6 proteins. *Mol. Cell Biol.***19**, 3885-3894.

Steiner, M.S., Zhang, Y., Carraher, J. and Lu, Y. (1999) In vivo expression of prostate-specific adenoviral vectors in a canine model. *Cancer Gene Ther.***6**, 456-464.

Stempien-Otero, A., Karsan, A., Cornejo, C.J., Xiang, H., Eunson, T., Morrison, R.S., Kay, M., Winn, R. and Harlan, J. (1999) Mechanisms of hypoxia-induced endothelial cell death. Role of p53 in apoptosis. *J. Biol. Chem.***274**, 8039-8045.

Stephan, D., San, H., Yang, Z.Y., Gordon, D., Goelz, S., Nabel, G.J. and Nabel, E.G. (1997) Inhibition of vascular smooth muscle cell proliferation and intimal hyperplasia by gene transfer of beta-interferon. *Mol. Med.***3**, 593-599.

Stevenson, S.C., Rollence, M., Marshall-Neff, J. and McClelland, A. (1997) Selective targeting of human cells by a chimeric adenovirus vector containing a modified fiber protein. *J.Virol.***71**, 4782-4790

Stewart, A.K., Lassam, N.J., Graham, F.L., Gauldie, J., Addison, C.L., Bailey, D.J., Dessureault, S., Dube, I.D., Gallenger, S., Krajden, M., Rotstein, L.E., Quirt, I.C. and Moen, R. (1997) A phase I study of adenovirus mediated gene transfer of interleukin 2 cDNA into metastatic breast cancer or melanoma. *Hum. Gene Ther.***8**, 1403-1414.

Stewart, A.K., Lassam, N.J., Quirt, I.C., Bailey, D.J., Rotstein, L.E., Krajden, M., Dessureault, S., Gallinger, S., Cappe, D., Wan, Y., Addison, C.L., Moen, R.C., Gauldie, J. and Graham, F.L. (1999) Adenovector-mediated gene delivery of interleukin-2 in metastatic breast cancer and melanoma: results of a phase 1 clinical trial. *Gene Ther.***6**, 350-363.

Stoeckle, M.Y., Falck-Pederson, E., Rubin, B.Y., Anderson, S.L. and Murray, H.W. (1996) Delivery of human interferon-gamma via gene transfer in vitro: prolonged expression and induction of macrophage antimicrobial activity. J. Interferon. Cytokine. *Res.***16**, 1015-1019.

Svensson, E.C., Black, H.B., Dugger, D.L., Tripathy, S.K., Goldwasser, E., Hao, Z., Chu, L. and Leiden, J.M. (1997) Long-term erythropoietin expression in rodents and non-human primates following intramuscular injection of a replication-defective adenoviral vector. *Hum. Gene Ther.***8**, 1797-1806.

Swaminathan, S. and Thimmapaya, B. (1995) Regulation of adenovirus E2 transcription unit. In The molecular repertoire of adenoviruses III: Current Topics in Microbiology and Immunology 199, *Eds Doerfler W. and Böhm P.*, p.177-194.

Taigen, T., De Windt, L.J., Lim, H.W. and Molkentin, J.D. (2000) Targeted inhibition of calcineurin prevents agonist-induced cardiomyocyte hypertrophy. *Proc. Natl. Acad. Sci. U. S. A.***97**, 1196-1201.

Takahashi, M., Ilan, Y., Chowdhury, N.R., Guida, J., Horwitz, M. and Chowdhury, J.R. (1996) Long term correction of bilirubin-UDP-glucuronosyltransferase deficiency in Gunn rats by administration of a recombinant adenovirus during the neonatal period. *J. Biol. Chem.***271**, 26536-26542.

Takahashi, K., Nuckolls, G.H., Tanaka, O., Semba, I., Takahashi, I., Dashner, R., Shum, L. and Slavkin, H.C. (1998a) Adenovirus-mediated ectopic expression of Msx2 in even-numbered rhombomeres induces apoptotic elimination of cranial neural crest cells in ovo. *Development* **125**, 1627-1635.

Takahashi, T., Yamada, K., Tanaka, T., Kumano, K., Kurokawa, M., Hirano, N., Honda, H., Chiba, S., Tsuji, K., Yazaki, Y., Nakahata, T. and Hirai, H. (1998b) A potential molecular approach to ex vivo hematopoietic expansion with recombinant epidermal growth factor receptor-expressing adenovirus vector. *Blood* **91**, 4509-4515.

Takayama, K., Ueno, H., Pei, X.H., Nakanishi, Y., Yatsunami, J. and Hara, N. (1998) The levels of integrin alpha v beta 5 may predict the susceptibility to adenovirus-mediated gene transfer in human lung cancer cells. *Gene Ther.***5**, 361-368.

Takayanagi, H., Juji, T., Miyazaki, T., Iizuka, H., Takahashi, T., Isshiki, M., Okada, M., Tanaka, Y., Koshihara, Y., Oda, H., Kurokawa, T., Nakamura, K. and Tanaka, S. (1999) Suppression of arthritic bone destruction by adenovirus-mediated csk gene transfer to synoviocytes and osteoclasts. *J. Clin. Invest.***104**, 137-146.

Tamamori, M., Ito, H., Hiroe, M., Terada, Y., Marumo, F. and Ikeda, M.A. (1998) Essential roles for G1 cyclin-dependent kinase activity in development of cardiomyocyte hypertrophy. *Am. J. Physiol.***275**, H2036-H2040.

Tanaka, T., Kanai, F., Okabe, S., Yoshida, Y., Wakimoto, H., Hamada, H., Shiratori, Y., Lan, K., Ishitobi, M. and Omata, M. (1996) Adenovirus-mediated prodrug gene therapy for carcinoembryonic antigen-producing human gastric carcinoma cells in vitro. *Cancer Res.***56**, 1341-1345.

Tanaka, T., Kanai, F., Lan, K.H., Ohashi, M., Shiratori, Y., Yoshida, Y., Hamada, H. and Omata, M. (1997a) Adenovirus-mediated gene therapy of gastric carcinoma using cancer-specific gene expression in vivo. *Biochem. Biophys. Res. Commun.***231**, 775-779.

Tanaka, T., Manome, Y., Wen, P., Kufe, D.W. and Fine, H.A. (1997b) Viral vector-mediated transduction of a modified platelet factor 4 cDNA inhibits angiogenesis and tumor growth. *Nat. Med.***3**, 437-442.

Tanaka, S., Takahashi, T., Takayanagi, H., Miyazaki, T., Oda, H., Nakamura, K., Hirai, H. and Kurokawa, T. (1998) Modulation of osteoclast function by adenovirus vector-induced epidermal growth factor receptor. *J. Bone Miner. Res.***13**, 1714-1720.

Taniguchi, K., Kohsaka, H., Inoue, N., Terada, Y., Ito, H., Hirokawa, K. and Miyasaka, N. (1999) Induction of the p16INK4a senescence gene as a new therapeutic strategy for the treatment of rheumatoid arthritis. *Nat. Med.***5**, 760-767.

Tanizawa, Y., Ohta, Y., Nomiyama, J., Matsuda, K., Tanabe, K., Inoue, H., Matsutani, A., Okuya, S. and Oka, Y. (1999) Overexpression of dominant negative mutant hepatocyte nuclear factor (HNF)-1alpha inhibits arginine-induced insulin secretion in MIN6 cells. *Diabetologia* **42**, 887-891.

Tatebe, S., Matsuura, T., Endo, K., Teramachi, K., Nakamura, T., Sato, K. and Ito, H. (1999) Adenovirus-mediated transfer of wild-type p53 gene results in apoptosis or growth arrest in human cultured gastric carcinoma cells. *Int. J. Oncol.***15**, 229-235.

Teodoro, J.G. and Branton, P.E. (1997) Regulation of apoptosis by viral gene products. *J.Virol.***71**, 1739-1746.

Terada, K., Nakako, T., Yang, X.L., Iida, M., Aiba, N., Minamiya, Y., Nakai, M., Sakaki, T., Miura, N. and Sugiyama, T. (1998) Restoration of holoceruloplasmin synthesis in LEC rat after infusion of recombinant adenovirus bearing WND cDNA. *J. Biol. Chem.***273**, 1815-1820.

Terasaki, M., Kato, S., Sugita, Y., Shigemori, M. and Morimatsu, M. (1999) Adenovirus-mediated over expression of p21Waf-1 differentially affects the proliferation, telomerase activity, and expression of glial fibrillary acidic protein in human glioma cells. *Int. J. Oncol.***14**, 63-69.

Themis, M., Schneider, H., Kiserud, T., Cook,T., Adebakin, S., Jezzard, S., Forbes, S., Hanson, M., Pavirani, A., Rodeck, C. and Coutelle, C. (1999) Successful expression of beta-galactosidase and factor IX transgenes in fetal and neonatal sheep after ultrasound-guided percutaneous adenovirus vector administration into the umbilical vein. *Gene Ther.***6**, 1239-1248.

Thibault, V., Terlain, B., Graham, F.L. and Gauldie, J. (1997) Construction and characterization of a replication-deficient adenovirus expressing rat-soluble interleukin-6 receptor. *Mol. Med.***3**, 519-529.

Tillman, B.W., de Gruijl, T.D., Luykx-de Bakker, S.A., Scheper, R.J., Pinedo, H.M., Curiel, T.J., Gerritsen, W.R. and Curiel, D.T. (1999) Maturation of dendritic cells accompanies high-efficiency gene transfer by a CD40-targeted adenoviral vector. *J. Immunol.* **162**, 6378-6383.

Todd, M.C., Sclafani, R.A., and Langan, T.A. (2000) Ovarian cancer cells that coexpress endogenous Rb and p16 are insensitive to overexpression of functional p16 protein. *Oncogene* **19**, 258-264.

Tollefson, A.E., Ryerse, J.S., Scaria, A., Hermiston, T.W. and Wold, W.S. (1996a) The E3-11.6-kDa adenovirus death protein (ADP) is required for efficient cell death: characterization of cells infected with adp mutants. *Virology* **220**, 152-162.

Tollefson, A.E., Scaria, A., Hermiston, T.W., Ryerse, J.S., Wold, L.J. and Wold, W.S. (1996b) The adenovirus death protein (E3-11.6K) is required at very late stages of infection for efficient cell lysis and release of adenovirus from infected cells. *J. Virol.* **70**, 2296-2306.

Toloza, E.M., Hunt, K., Swisher, S., McBride, W., Lau, R., Pang, S., Rhoades, K., Drake, T., Belldegrun, A., Glaspy, J. and Economou, J.S. (1996) In vivo cancer gene therapy with a recombinant interleukin-2 adenovirus vector. *Cancer Gene Ther.* **3**, 11-17.

Tomanin, R., Bett, A.J., Picci, L., Scarpa, M. and Graham, F.L. (1997) Development and characterization of a binary gene expression system based on bacteriophage T7 components in adenovirus vectors. *Gene* **193**, 129-140.

Tomko, R.P., Johansson, C.B., Totrov, M., Abagyan, R., Frisen, J. and Philipson, L. (2000) Expression of the Adenovirus Receptor and Its Interaction with the Fiber Knob. *Exp. Cell Res.* **255**, 47-55.

Trapnell, B.C. and Gorzilia, M. (1994) Gene therapy using adenoviral vectors. *Curr. Opinion in Biotech.* **5**, 617-625.

Trinh, K., Minassian, C., Lange, A.J., O'Doherty, R.M. and Newgard, C.B. (1997) Adenovirus-mediated expression of the catalytic subunit of glucose-6-phosphatase in INS-1 cells. Effects on glucose cycling, glucose usage, and insulin secretion. *J. Biol. Chem.* **272**, 24837-24842.

Tripathy, S.K., Goldwasser, E., Lu, M.M., Barr, E. and Leiden, J.M. (1994) Stable delivery of physiologic levels of recombinant erythropoietin to the systemic circulation by intramuscular injection of replication-defective adenovirus. *Proc. Natl. Acad. Sci. U. S. A.* **91**, 11557-11561.

Tsao, Y.P., Huang, S.J., Chang, J.L., Hsieh, J.T., Pong, R.C. and Chen, S.L. (1999) Adenovirus-mediated p21((WAF1/SDII/CIP1)) gene transfer induces apoptosis of human cervical cancer cell lines. *J. Virol.* **73**, 4983-4990.

Tsukui, T., Miyake, S., Azuma, S., Ichise, H., Saito, I. and Toyoda, Y. (1995) Gene transfer and expression in mouse preimplantation embryos by recombinant adenovirus vector. *Mol. Reprod. Dev.* **42**, 291-297.

Tsukui, T., Kanegae, Y., Saito, I. and Toyoda, Y. (1996) Transgenesis by adenovirus-mediated gene transfer into mouse zona-free eggs. *Nat. Biotechnol.* **14**, 982-985.

Turkay, A., Saunders, T. and Kurachi, K. (1999) Intrauterine gene transfer: gestational stage-specific gene delivery in mice. *Gene Ther.* **6**, 1685-1694.

Turturro, F., Seth, P. and Link, C.J-Jr. (2000) In vitro adenoviral vector p53-mediated transduction and killing correlates with expression of coxsackie-adenovirus receptor and alpha(nu)beta5 integrin in SUDHL-1 cells derived from anaplastic large-cell lymphoma. Clin. *Cancer Res.* **6**, 185-192.

Tzeng, E., Kim, Y.M., Pitt, B.R., Lizonova, A., Kovesdi, I. and Billiar, T.R. (1997) Adenoviral transfer of the inducible nitric oxide synthase gene blocks endothelial cell apoptosis. *Surgery* **122**, 255-263.

Ueno, H., Li, J.J., Masuda, S., Qi, Z., Yamamoto, H. and Takeshita, A. (1997) Adenovirus-mediated expression of the secreted form of basic fibroblast growth factor (FGF-2) induces cellular proliferation and angiogenesis in vivo. *Arterioscler. Thromb. Vasc. Biol.* **17**, 2453-2460.

Uetsuki, T., Takemoto, K., Nishimura, I., Okamoto, M., Niinobe, M., Momoi, T., Miura, M. and Yoshikawa, K. (1999) Activation of neuronal caspase-3 by intracellular accumulation of wild-type Alzheimer amyloid precursor protein. *J. Neurosci.* **19**, 6955-6964.

Uhlmann, E.J., Subramanian, T., Vater, C.A., Lutz, R. and Chinnadurai, G. (1998) A potent cell death activity associated with transient high level expression of BCL-2. *J. Biol. Chem.* **273**, 17926-17932.

Vaillant, A.R., Mazzoni, I., Tudan, C., Boudreau, M., Kaplan, D.R. and Miller, F.D. (1999) Depolarization and neurotrophins converge on the phosphatidylinositol 3-kinase-Akt pathway to synergistically regulate neuronal survival. *J. Cell Biol.* **146**, 955-966.

Vanderkwaak, T.J., Wang, M., Gomez-Navarro, J., Rancourt, C., Dmitriev, I., Krasnykh, V., Barnes, M., Siegal, G.P., Alvarez, R. and Curiel, D.T. (1999) An advanced generation of adenoviral vectors selectively enhances gene transfer for ovarian cancer gene therapy approaches. *Gynecol. Oncol.***74**, 227-234.

Vassaux, G., Hurst, H. and Lemoine, N. (1999) Insulation of a conditionally expressed transgene in an adenoviral vector. *Gene Ther.***6**, 1192-1197.

Vigne, E., Mahfouz, I., Dedieu, J.F., Perricaudet, M. and Yeh, P.(1999) RGD inclusion in the hexon monomer provides adenovirus type 5-based vectors with a fiber knob-independent pathway for infection. *J. Virol.***73**, 5156-5161.

Vilquin, J.T., Guerette, B., Kinoshita, I., Roy, B., Goulet, M., Gravel, C., Roy, R. and Tremblay, J.P. (1995) FK506 immunosuppression to control the immune reactions triggered by first-generation adenovirus-mediated gene transfer. *Hum. Gene Ther.***6**, 1391-1401.

Volloch, V., Mosser, D.D., Massie, B. and Sherman, M.Y. (1998) Reduced thermotolerance in aged cells results from a loss of an hsp72-mediated control of JNK signaling pathway. *Cell stress and chaperones* **3**, 265-271.

Von Harsdorf, R., Hauck, L., Mehrhof, F., Wegenka, U., Cardoso, M.C. and Dietz, R. (1999) E2F-1 overexpression in cardiomyocytes induces downregulation of p21CIP1 and p27KIP1 and release of active cyclin-dependent kinases in the presence of insulin-like growth factor I. *Circ. Res.***85**, 128-136.

Von Seggern, D.J., Huang, S., Fleck, S.K., Stevenson, S.C. and Nemerow, G.R. (2000) Adenovirus vector pseudotyping in fiber-expressing cell lines : improved transduction of Epstein-Barr virus-transformed B cells. *J. Virol.***74**, 354-362.

Waheed, I., Guo, Z.S., Chen, G.A., Weiser, T.S., Nguyen, D.M. and Schrump, D.S. (1999) Antisense to SV40 early gene region induces growth arrest and apoptosis in T-antigen-positive human pleural mesothelioma cells. *Cancer Res.***15**, 6068-6073.

Walsh, C.E. (1999) Fetal gene therapy. *Hum. Gene ther.***6**, 1200-1201.

Walton, T., Wang, J.L., Ribas, A., Barsky, S.H., Economou, J. and Nguyen, M. (1998) Endothelium-specific expression of an E-selectin promoter recombinant adenoviral vector. *Anticancer Res.***18**, 1357-1360.

Wang, G., Williamson, R., Mueller, G., Thomas, P., Davidson, B.L. and McCray, P.B Jr. (1998e) Ultrasound-guided gene transfer to hepatocytes in utero. *Fetal Diagn. Ther.***13**, 197-205.

Wang, J., Palmer, K., Lotvall, J., Milan, S., Lei, X.F., Matthaei, K.I., Gauldie, J., Inman, M.D., Jordana, M. and Xing, Z. (1998d) Circulating, but not local lung, IL-5 is required for the development of antigen-induced airways eosinophilia. *J. Clin. Invest.***102**, 1132-1141.

Wang, M.Y., Koyama, K., Shimabukuro, M., Mangelsdorf, D., Newgard, C.B. and Unger, R.H. (1998a) Overexpression of leptin receptors in pancreatic islets of Zucker diabetic fatty rats restores GLUT-2, glucokinase, and glucose-stimulated insulin secretion. *Proc. Natl. Acad. Sci. U. S. A.***95**, 11921-11926.

Wang, M.Y., Koyama, K., Shimabukuro, M., Newgard, C.B. and Unger, R.H. (1998b) OB-Rb gene transfer to leptin-resistant islets reverses diabetogenic phenotype. *Proc. Natl. Acad. Sci. U. S. A.***95**, 714-718.

Wang, M.Y., Koyama, K., Shimabukuro, M., Mangelsdorf, D., Newgard, C.B. and Unger, R.H. (1998c) Overexpression of leptin receptors in pancreatic islets of Zucker diabetic fatty rats restores GLUT-2, glucokinase, and glucose-stimulated insulin secretion. *Proc. Natl. Acad. Sci .U. S. A.***29**, 11921-11926.

Wang, M.Y., Shimabukuro, M., Lee, Y., Trinh, K.Y., Chen, J.L., Newgard, C.B. and Unger, R.H. (1999a) Adenovirus-mediated overexpression of uncoupling protein-2 in pancreatic islets of Zucker diabetic rats increases oxidative activity and improves beta-cell function. *Diabetes* **48**, 1020-1025.

Wang, N., Verna, L., Hardy, S., Forsayeth, J., Zhu, Y. and Stemerman, M.B. (1999b) Adenovirus-mediated overexpression of c-Jun and c-Fos induces intercellular adhesion molecule-1 and monocyte chemoattractant protein-1 in human endothelial cells. *Arterioscler. Thromb. Vasc. Biol.***19**, 2078-2084.

Wang, N., Verna, L., Hardy, S., Zhu, Y., Ma, K.S., Birrer, M.J. and Stemerman, M.B. (1999c) c-Jun triggers apoptosis in human vascular endothelial cells. *Circ. Res.***85**, 387-393.

Wang, Y., Krushel, L.A. and Edelman, G.M. (1996) Targeted DNA recombination in vivo using an adenovirus carrying the cre recombinase gene. *Proc. Natl. Acad. Sci. U. S. A.***93**, 3932-3936.

Weber, K.S., Draude, G., Erl, W., de Martin, R. and Weber, C. (1999) Monocyte arrest and transmigration on inflamed endothelium in shear flow is inhibited by adenovirus-mediated gene transfer of IkappaB-alpha. *Blood* **93**, 3685-3693.

Wei, M.X., Li, F., Ono, Y., Gauldie, J. and Chiocca, E.A. (1998) Effects on brain tumor cell proliferation by an adenovirus vector that bears the interleukin-4 gene. *J. Neurovirol.***4**, 237-241.

Wersto, R.P., Rosenthal, E.R., Seth, P.K., Eissa, N.T. and Donahue, R.E. (1998) Recombinant, replication-defective adenovirus gene transfer vectors induce cell cycle dysregulation and inappropriate expression of cyclin proteins. *J. Virol.***72**, 9491-9502.

Whalen, J.D., Lechman, E.L., Carlos, C.A., Weiss, K., Kovesdi, I., Glorioso, J.C., Robbins, P.D. and Evans, C.H. (1999) Adenoviral transfer of the viral IL-10 gene periarticularly to mouse paws suppresses development of collagen-induced arthritis in both injected and uninjected paws. *J. Immunol.***162**, 3625-3632.

White, E. (1998) Regulation of apoptosis by E1A and E1B oncogenes. *Seminars in Virology* **8**, 505-513.

Wickham, T.J., Mathias, P., Cheresh, D.A. and Nemerow, G.R. (1993) Integrins alpha v beta 3 and alpha v beta 5 promote adenovirus internalization but not virus attachment. *Cell* **73**, 309-319.

Wickham, T.J., Carrion, M.E. and Kovesdi, I. (1995) Targeting of adenovirus penton base to new receptors through replacement of its RGD motif with other receptor-specific peptide motifs. *Gene Ther.***2**, 750-756.

Wickham, T.J., Roelvink, P.W., Brough, D.E. and Kovesdi, I. (1996) Adenovirus targeted to heparan-containing receptors increases its gene delivery efficiency to multiple cell types. *Nat. Biotechnol.***14**, 1570-1573.

Wickham, T.J., Lee, G.M., Titus, J.A., Sconocchia, G., Bakacs, T., Kovesdi, I. and Segal, D.M. (1997a) Targeted adenovirus-mediated gene delivery to T cells via CD3. *J. Virol.***71**, 7663-7669.

Wickham, T.J., Tzeng, E., Shears, L.L., Roelvink, P.W., Li, Y., Lee, G.M., Brough, D.E., Lizonova, A. and Kovesdi, I. (1997b) Increased in vitro and in vivo gene transfer by adenovirus vectors containing chimeric fiber proteins. *J. Virol.***71**, 8221-8229.

Willimsky, G. and Blankenstein, T. (2000) Interleukin-7/B7.1-encoding adenoviruses induce rejection of transplanted but not nontransplanted tumors. *Cancer Res.***60**, 685-692.

Wirtz, S., Galle, P.R. and Neurath, M.F. (1999) Efficient gene delivery to the inflamed colon by local administration of recombinant adenoviruses with normal or modified fibre structure. *Gut* **44**, 800-807.

Wivel, N.A., Gao, G.-P. and Wilson, J.M. (1999) Adenovirus vectors pp 87-110. In "The Development of Human Gene Therapy". *Ed. T. Friedman. Cold Spring Harbor Press*, Cold Spring Harbor, New York.

Wold, W.S.M. and Tollefson, A.E. (1998) Adenovirus E3 proteins: 14.7,RID, and gp19K inhibit immune-induced cell death; Adenovirus death proteins promotes cell death. *Seminars in Virology* **8**, 515-523.

Wolf, J.K., Mills, G.B., Bazzet, L., Bast, R.C-Jr, Roth, J.A. and Gershenson, D.M. (1999a) Adenovirus-mediated p53 growth inhibition of ovarian cancer cells is independent of endogenous p53 status. *Gynecol. Oncol.***75**, 261-266.

Wolf, J.K., Kim, T.E., Fightmaster, D., Bodurka, D., Gershenson, D.M., Mills, G. and Wharton, J.T. (1999b) Growth suppression of human ovarian cancer cell lines by the introduction of a p16 gene via a recombinant adenovirus. *Gynecol. Oncol.***73**, 27-34.

Wolff, G., Worgall, S., Van Rooijen, N., Song, W.R., Harvey, B.G. and Crystal, R.G. (1997) Enhancement of in vivo adenovirus-mediated gene tranfer and expression by prior depletion of tissue macrophages in the target organ. *J.Virol.*,**71**, 624-629.

Woo, Y.J., Raju, G.P., Swain, J.L., Richmond, M.E., Gardner, T.J. and Balice-Gordon, R.J. (1997) In utero cardiac gene transfer via intraplacental delivery of recombinant adenovirus. *Circulation* **96**, 3561-3569.

Woods, J.M., Tokuhira, M., Berry, J.C., Katschke, K.J.,Jr., Kurata, H., Damergis, J.A.,Jr., Arai, K. and Koch, A.E. (1999) Interleukin-4 adenoviral gene therapy reduces production of inflammatory cytokines and prostaglandin E2 by rheumatoid arthritis synovium ex vivo. *J. Investig. Med.***47**, 285-292.

Worgall, S., Wolf, G., Falck-Pederson, E. and Crystal, R.G. (1997) Innate immune mechanisms dominate elimination of adenoviral vectors following in vitro administration. *Hum. Gene Ther.***8**, 37-44.

Wrighton, C.J., Hofer-Warbinek, R., Moll, T., Eytner, R., Bach, F.H. and de Martin, R. (1996) Inhibition of endothelial cell activation by adenovirus-mediated expression of I kappa B alpha, an inhibitor of the transcription factor NF-kappa B. *J. Exp. Med.***183**, 1013-1022.

Xiang, J., Gomez-Navarro, J., Arafat, W., Liu, B., Barker, S.D., Alvarez, R.D., Siegal, G.P. and Curiel, D.T. (2000) Pro-apoptotic treatment with an adenovirus encoding Bax enhances the effect of chemotherapy in ovarian cancer. *J. Gene Med.***2**, 97-106.

Xing, Z., Braciak, T., Jordana, M., Croitoru, K., Graham, F.L. and Gauldie, J. (1994) Adenovirus-mediated cytokine gene transfer at tissue sites. Overexpression of IL-6 induces lymphocytic hyperplasia in the lung. *J. Immunol.***153**, 4059-4069.

Xing, Z., Ohkawara, Y., Jordana, M., Graham, F.L. and Gauldie, J. (1997) Adenoviral vector-mediated interleukin-10 expression in vivo: intramuscular gene transfer inhibits cytokine responses in endotoxemia. *Gene Ther.***4**, 140-149.

Xing, Z., Braciak, T., Chong, D., Feng, X., Schroeder, J.A. and Gauldie, J. (1999) Adenoviral-mediated gene transfer of interleukin-6 in rat lung enhances antiviral immunoglobulin A and G responses in distinct tissue compartments. *Biochem. Biophys. Res. Commun.***258**, 332-335.

Xu, X., Dai, Y., Heidenreich, O. and Nerenberg, M.I. (1996) Adenovirus-mediated interferon-gamma transfer inhibits growth of transplanted HTLV-1 Tax tumors in mice. *Hum. Gene Ther.***7**, 471-477.

Xu, R., Ying, B., Zhao, S., Li, C., and Wang, Y. (1997) Construction and identification of a recombinant adenovirus which expresses human interferon-gamma. Chin. *J. Biotechnol.***13**, 1-8.

Yamagata, M., Jaye, D.L. and Sanes, J.R. (1994) Gene transfer to avian embryos with a recombinant adenovirus. *Dev. Biol.***166**, 355-359.

Yamamoto, H., Ueno, H., Ooshima, A. and Takeshita, A. (1996) Adenovirus-mediated transfer of a truncated transforming growth factor-beta (TGF-beta) type II receptor completely and specifically abolishes diverse signaling by TGF-beta in vascular wall cells in primary culture. *J. Biol. Chem.***271**, 16253-16259.

Yamochi, T., Kaneita, Y., Akiyama, T., Mori, S. and Moriyama, M. (1999) Adenovirus-mediated high expression of BCL-6 in CV-1 cells induces apoptotic cell death accompanied by down-regulation of BCL-2 and BCL-X(L). *Oncogene* **18**, 487-494.

Yang, Y., Nunes, F. A., Berencsi, K., Furth , E.E., Gönczöl, E., Engelhardt, J. and Wilson, J.M. (1994) Cellular immunity to viral antigens limits E1-deleted adenoviruses for gene therapy. *Proc. Natl. Acad. Sci. U. S. A.***91**, 4407-4411.

Yang, Y., Li, Q., Ertl, H.C. and Wilson, J.M. (1995) Cellular and humoral immune responses to viral antigens create barriers to lung-directed gene therapy with recombinant adenoviruses. *J. Virol.***69**, 2004-2015.

Yang, Y., Su, Q., Grewal, I.S., Schilz, R., Flavell, R.A. and Wilson, J.M. (1996a) Transient subversion of CD40 ligand function diminishes immune responses to adenovirus vectors in mouse liver and lung tissues. *J. Virol.***70**, 6370-6377.

Yang, Z.Y., Simari, R.D., Perkins, N.D., San, H., Gordon, D., Nabel, G.J. and Nabel, E.G. (1996b) Role of the p21 cyclin-dependent kinase inhibitor in limiting intimal cell proliferation in response to arterial injury. *Proc. Natl. Acad. Sci. U. S. A.***93**, 7905-7910.

Yang, E.Y., Kim, H.B., Shaaban, A.F., Milner, R., Adzick, N.S. and Flake, A.W. (1999a) Persistent postnatal transgene expression in both muscle and liver after fetal injection of recombinant adenovirus. *J. Pediatr. Surg.***34**, 766-772.

Yang, H.L., Dong, Y.B., Elliott, M.J., Liu, T.J., Atienza, C.,Jr., Stilwell, A. and McMasters, K.M. (1999b) Adenovirus-mediated E2F-1 gene transfer inhibits MDM2 expression and efficiently induces apoptosis in MDM2-overexpressing tumor cells. *Clin. Cancer Res.***5**, 2242-2250.

Yang, H.L., Dong, Y.B., Elliott, M.J., Liu, T.J. and McMasters, K.M. (2000) Caspase activation and changes in Bcl-2 family member protein expression associated with E2F-1-mediated apoptosis in human esophageal cancer cells. *Clin. Cancer Res.***6**, 1579-1589.

Yasuda, H., Nagata, M., Arisawa, K., Yoshida, R., Fujihira, K., Okamoto, N., Moriyama, H., Miki, M., Saito, I., Hamada, H., Yokono, K. and Kasuga, M. (1998) Local expression of immunoregulatory IL-12p40 gene prolonged syngeneic islet graft survival in diabetic NOD mice. *J. Clin. Invest.***102**, 1807-1814.

Ye, X., Gao, G.-P., Pabin C., Raper S.E. and Wilson J.M. (1998) Evaluating the potential of germ line transmission after intravenous administration of recombinant adenovirus in the C3H mouse. *Hum. Gene Ther.***9**, 2135-2142.

Yew, N.S., Marshall, J., Przybylska, M., Wysokenski, D.M., Ziegler, R.J., Rafter, P.W., Li, C., Armentano, D. and Cheng, S.H. (1999) Increased duration of transgene expression in the lung with plasmid DNA vectors harboring adenovirus E4 open reading frame 3. *Hum. Gene. Ther.***10**, 1833-1843.

Yoshida, Y., Sadata, A., Zhang, W., Saito, K., Shinoura, N. and Hamada, H. (1998) Generation of fiber-mutant recombinant adenoviruses for gene therapy of malignant glioma. *Hum.Gene Ther.***9**, 2503-2515.

Yu, D.C., Sakamoto, G.T. and Henderson, D.R. (1999) Identification of the transcriptional regulatory sequences of human kallikrein 2 and their use in the construction of calydon virus 764, an attenuated replication competent adenovirus for prostate cancer therapy. *Cancer Res.***59**, 1498-1504.

Yuan, H., Zhai, P., Anderson, L.M., Pan, J., Thimmapaya, B., Koo, E.H. and Marquez-Sterling, N.R. (1999) Recombinant adenovirus is an appropriate vector for endocytotic protein trafficking studies in cultured neurons. *J. Neurosci. Methods* **88**, 45-54.

Zabner, J., Chillon, M., Grunst, T., Moninger, T.O., Davidson, B.L., Gregory, R. and Armentano, D. (1999) A chimeric type 2 adenovirus vector with a type 17 fiber enhances gene transfer to human airway epithelia. *J. Virol.***73**, 8689-8695

Zaheer, A., Weiss, J.L., Goyal, P. and Lim, R. (1999) Enhanced expression of neurotrophic factors by C6 rat glioma cells after transfection with glia maturation factor. *Neurosci. Lett.***265**, 203-206.

Zen, K., Karsan, A., Stempien-Otero, A., Yee, E., Tupper, J., Li, X., Eunson, T., Kay, M.A., Wilson, C.B., Winn, R.K. and Harlan, J.M. (1999) NF-kappaB activation is required for human endothelial survival during exposure to tumor necrosis factor-alpha but not to interleukin-1beta or lipopolysaccharide. *J. Biol. Chem.***274**, 28808-28815.

Zhang, J.F., Hu, C., Geng, Y., Selm, J., Klein, S.B., Orazi, A. and Taylor, M.W. (1996) Treatment of a human breast cancer xenograft with an adenovirus vector containing an interferon gene results in rapid regression due to viral oncolysis and gene therapy. *Proc. Natl. Acad. Sci. U. S. A.***93**, 4513-4518.

Zhao, J., Sime, P.J., Bringas, P.,Jr., Gauldie, J. and Warburton, D. (1998) Epithelium-specific adenoviral transfer of a dominant-negative mutant TGF-beta type II receptor stimulates embryonic lung branching morphogenesis in culture and potentiates EGF and PDGF-AA. *Mech. Dev.***72**, 89-100.

Zhao, J., Sime, P.J., Bringas, P.,Jr., Gauldie, J. and Warburton, D. (1999) Adenovirus-mediated decorin gene transfer prevents TGF-beta-induced inhibition of lung morphogenesis. *Am. J. Physiol.***277**, L412-L422.

Zhou, H., O'Neal, W., Morral, N. and Beaudet, A.L. (1996) Development of a complementing cell line and a system for construction of adenovirus vectors with E1 and E2a deleted. *J. Virol.***70**, 7030-7038

Zhou, Y.T., Wang, Z.W., Higa, M., Newgard, C.B. and Unger, R.H. (1999) Reversing adipocyte differentiation: implications for treatment of obesity. *Proc. Natl. Acad. Sci. U. S. A.* **96**, 2391-2395.

Zhu, J., Grace, M., Casale, J., Chang, A.T., Musco, M.L., Bordens, R., Greenberg, R., Schaefer, E. and Indelicato, S.R. (1999) Characterization of replication-competent adenovirus isolates from large-scale production of a recombinant adenoviral vector. *Hum. Gene Ther.***10**, 113-121.

Zsengeller, Z.K., Boivin, G.P., Sawchuck, S.S., Trapnell, B.C., Witsett, J.A. and Hirsch, R. (1997). Anti-T cell receptor antibody prologs transgene expression and reduces lung inflammation after adenovirus-mediated gene transfer. *Hum. Gene Ther.***8**, 935-941.

Zsengeller, Z.K., Reed, J.A., Bachurski, C.J., LeVine, A.M., Forry-Schaudies, S., Hirsch, R. and Whitsett, J.A. (1998) Adenovirus-mediated granulocyte-macrophage colony-stimulating factor improves lung pathology of pulmonary alveolar proteinosis in granulocyte-macrophage colony-stimulating factor-deficient mice. *Hum. Gene Ther.***9**, 2101-2109.

ADENO-ASSOCIATED VIRUS: A PROMISING TOOL FOR GENE DELIVERY

A.M. DOUAR
Genethon III – CNRS URA 1923
1 bis, rue de l'Internationale
BP 60 - 91002 Evry Cedex, France.
E-mail: douar@genethon.fr

Abstract

Among viral vectors for gene therapy, adeno-associated virus (AAV) has been shown to be very attractive, due to its ability to transduce numerous cell types, both dividing and quiescent, and its lack of pathogenicity. Its potential to integrate into the host genome and the demonstration of persistence *in vivo* make it suitable for long-term correction. The ability to reach therapeutic levels in pre-clinical animal models and the absence of adverse effects related to AAV vector administration have opened the route to clinical trials in humans. Although the status of the immune response has been well evaluated in animal models, issue concerning AAV immunity in humans following AAV administration is poorly documented. Recent improvements in AAV vector manufacturing processes have led to the production of highly purified concentrated stocks, although large-scale production of recombinant AAV is still hampered by insufficient technology. Given the large amount of data concerning these different issues, AAV is retaining attention as a likely very efficient vehicle for gene therapy.

1 Biology of adeno-associated virus

1.1 MOLECULAR BIOLOGICAL FEATURES OF AAV

Adeno-associated virus (AAV) is a small non-enveloped single stranded (ss) DNA virus with a diameter of 18 to 25 nm and a density of 1.41 g/cm^3. It belongs to the parvoviridae family, and is classified in the Dependovirus genus. The icosahedral capsid contains only one viral genome. At least six strains have been isolated in humans and primates, and many other strains can be found in other mammals (Berns, 1996; Rutledge, Halbert and Russell, 1998). AAV is very resistant to extreme conditions of pH, detergent and temperature, making it easy to manipulate. In this review, we will focus on the AAV-2 strain, which is the best characterised and the most studied, and from which the vast majority of vectors have been derived so far (Samulski *et al.*, 1982; Samulski, Chang and Shenk, 1987; Tratschin *et al.*, 1985). AAV tropism is large and it

M. Al-Rubeai (ed.), Cell Engineering, 155–181.

is able to infect both dividing and non-dividing cells. Lastly, AAV has the capability to integrate at a specific site into the human genome.

The 4679 bp genome of AAV-2 (Genebank AF043303) contains two open reading frames (ORF) encoding four regulatory and three structural proteins (Srivastava, Lusby and Berns, 1983; Tratschin, Miller and Carter, 1984). The compacted genome of AAV is framed by two inverted terminal repeats (ITRs). The *rep* gene encodes four proteins from two different promoters p5 and p19 combined with an alternative splice site (figure 1). The two major Rep proteins, Rep78 and Rep68, are involved in viral genome excision, rescue and replication. Both proteins possess a helicase activity, necessary for genome elongation, and a nickase activity required for the genome resolution step. They bind to the ITR via a consensus binding site and mediate site-specific integration in the human genome (Weitzman *et al.*, 1994). They also regulate gene expression from AAV and heterologous promoters (Horer *et al.*, 1995; Pereira, McCarty and Muzyczka, 1997). The two minor Rep proteins, Rep52 and Rep40, are involved in replicated ssDNA genome accumulation and packaging. The second ORF is initiated at the p40 promoter and encodes the three structural capsid proteins VP1, VP2 and VP3. Their stoichiometry in an assembled capsid is 1:1:10. All the transcripts share the same polyadenylation signal. Equal amounts of virions are found containing strands of plus or minus polarity.

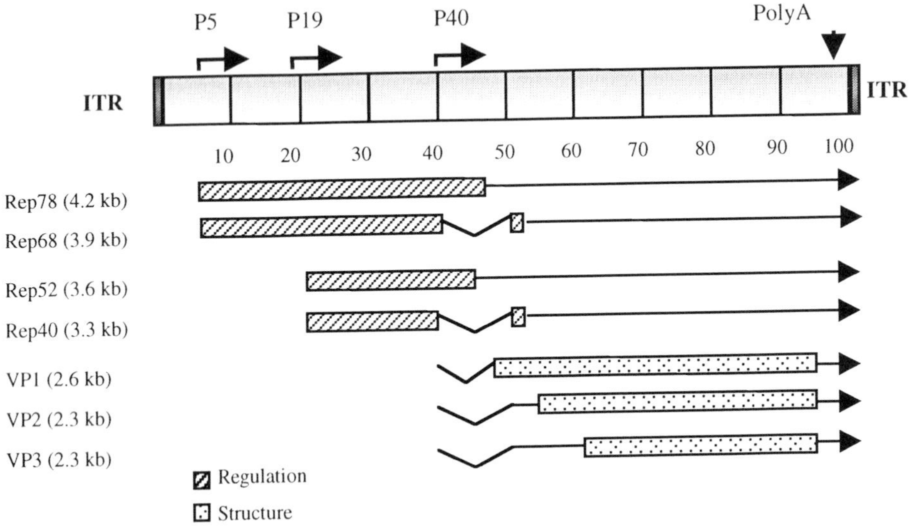

Figure 1. Wild type AAV-2 genome organisation (arbitrary map units).
Dark boxes indicate the ITRs. The arrows indicate the three promoters The solid lines indicate the transcripts; the introns are shown by the broken lines. The arrowheads represent the poly-adenylation signal (position 96) common to all transcripts. The first ORF encodes the four regulatory proteins (shaded boxes) arising from the promoters p5 and p 19 and alternative splicing. The second ORF encodes the three capsid proteins (dotted boxes) from two transcripts. VP-2 and VP-3 are initiated at two different codon sites.

The ITRs are base-paired hairpin structures of 145 nucleotide length framing the viral genome. The distal 125 bases form the hairpin structure and comprise three internal palindromes. The 20 proximal bases of the ITR are not involved in the palindromic sequence (Bohenzky, LeFebvre and Berns, 1988). The ITRs contain the only necessary regulatory cis acting sequences, namely the origin of replication of the genome and both packaging and integration signals.

1.2 CLINICAL FEATURES OF ADENO-ASSOCIATED VIRUS INFECTION

Most of the human population (70-80 %) has been exposed to AAV and a majority of AAV-positive individuals possesses antibodies to more than one serotype. Most AAV infection events appear to occur in early childhood (Blacklow, Hoggan and Rowe, 1968; Erles, Sebokova and Schlehofer, 1999). The presence of AAV sequences has also been detected by PCR in various human clinical samples (Tobiasch *et al.*, 1998). A recent study indicates that there is no specific geographical prevalence of AAV infections (Erles, Sebokova and Schlehofer, 1999). *In utero* infection has been shown to occur and results in partial immunotolerance to AAV. Re-infection or reactivation of latently established virus has been demonstrated after the age of 30, as well as during pregnancy. Interestingly, the immune response to AAV infection does not seem to protect from re-infection or reactivation (Erles, Sebokova and Schlehofer, 1999).

No known adverse clinical consequences are associated with AAV infection or latency in humans. Moreover, several studies have suggested that AAV infection might provide an anti-neoplastic effect. Two types of observations have led to this conclusion. First, it has been observed that the development of tumours induced *in vivo*, irrespective to the induction method (either viral or transformed cell transplantation), could be inhibited by infection of the animal with AAV. Secondly, sero-epidemiological investigations in human populations have shown that antibodies against AAV are less frequently found in cancer patients than in healthy individuals (reviewed in (Schlehofer, 1994)).

1.3 AAV LIFE CYCLE

AAV was initially discovered as a contaminant of adenovirus preparations (Hoggan, Blacklow and Rowe, 1966). This reflects the main feature of this virus, which lies in its dependency on a helper virus to achieve a productive cycle and gave the genus name of Dependovirus. Several viruses have been shown to have the ability to ensure the helper functions: adenovirus was is the first, but herpes simplex virus and vaccinia virus (Schlehofer, Ehrbar and zur Hausen, 1986; Weindler and Heilbronn, 1991) as well as cytomegalovirus (CMV) (McPherson, Rosenthal and Rose, 1985) may also provide helper functions.

In the course of infection, AAV enters the cell and once in the nucleus, undergoes a step in which its ss genome is converted to a replicative double stranded (ds) form suitable for gene expression. In the absence of infection with a helper virus, the AAV genome integrates into the host genome and latently persists as a provirus (figure 2). If superinfection by a helper virus occurs, the AAV genome is rescued by excision from the host genome, and actively replicated. In a co-infection situation, the AAV genome is directly replicated. Capsid proteins are synthesised and DNA packaging occurs in the nucleus (Wistuba *et al.*, 1997). AAV does not possess a lytic capability by itself, and in the natural infection process, the liberation of AAV virions relies on the lytic effect of the helper virus.

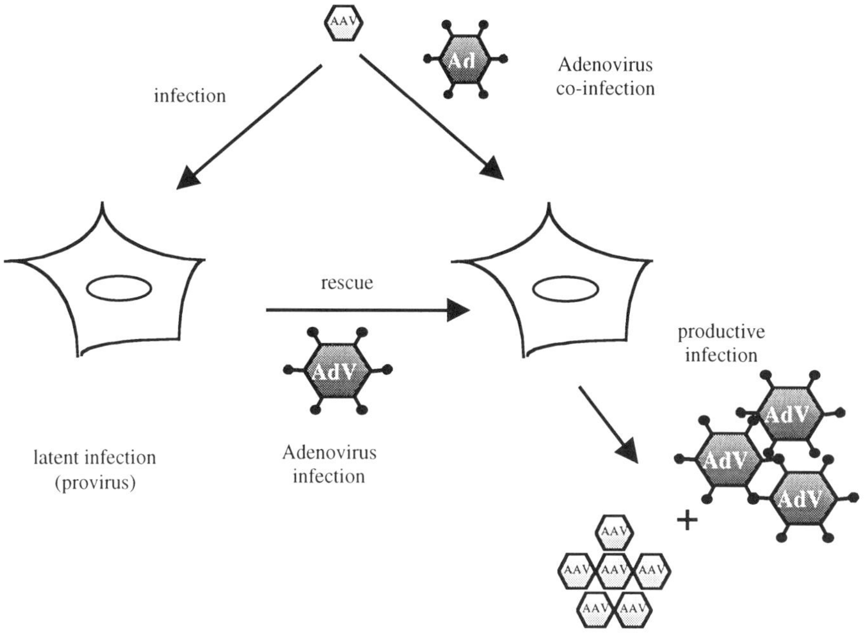

Figure 2. The AAV life cycle.
Following infection of the cell, AAV integrates into the host genome (in a specific site on chromosome 19 in humans) and remains latent until a helper virus infects the cell. The proviral genome is then rescued and actively replicated. In the presence of helper virus, AAV infection leads directly to a productive infection.

1.4 AAV DNA REPLICATION

AAV requires cellular and/or helper virus proteins for its own replication. The Rep proteins do not possess polymerase activity and none of the adenoviral polymerases are required for AAV DNA polymerisation. Rather, helper virus functions are thought to induce AAV and cellular genes that are necessary for AAV replication. Moreover, the repeated observations that AAV can replicate in the absence of helper virus support the involvement of cellular polymerases in the replication process (Ni *et al.*, 1998). Adenovirus helper functions have been well characterised. The set of genes that are required, E1a, E1b (Richardson and Westphal, 1984), E2a, E4, and VAI RNA (West *et al.*, 1987), are all early adenovirus functions. In the presence of adenovirus, E1a induces both p5 and p19 promoters and the p5 Rep product induces RNA synthesis from all three AAV promoters (Berns, 1990). Moreover, E1A is necessary to induce the early functions of adenovirus itself. The role of the other helper sequences is only partially understood (reviewed in (Berns, 1990)).

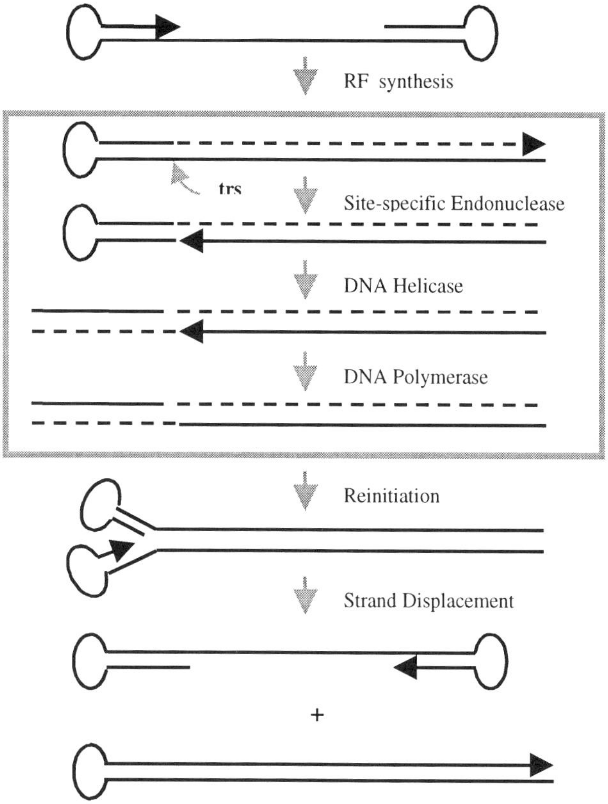

Figure 3. Mechanism of AAV DNA replication.
The major steps of AAV DNA replication are represented. RF: replicative form. The polymerisation and resolution steps are framed (see text for details).

AAV genome replication follows a leading-strand DNA synthesis process (figure 3), and both strands of AAV are used as templates. The first step of viral DNA replication is the conversion of the single stranded genome to a duplex form. The palindromic structure of the 3' terminal repeat provides a structural primer from which DNA replication is initiated. The helicase activity necessary to unfold the T-shaped terminal repeat is attributed to Rep78 and Rep68 (Walker, Wonderling and Owens, 1997). In the next step, Rep78 and Rep68 endonuclease activity allows to resolve the covalent terminal hairpin by nicking the parental strand at the terminal resolution site (trs). After unfolding of the terminal T-shaped structure, polymerisation of DNA is completed (Berns, 1996; Hong, Ward and Berns, 1992; Ni *et al.*, 1994). The ends of the two molecules can fold back to a hairpin structure, providing a 3' OH primer and allowing strand displacement. The resulting ss DNA molecules can either be packaged into preformed capsids or undergo a new replication step. It should be noted that the nicking step during the replicative process leads to an exchange of the 125 3'-terminal bases between the parental and the progeny strands, which are therefore inverted in the progeny strand (figure 3). As a consequence, the AAV terminal repeat exists in two orientations called flip and flop (Berns, 1996).

1.5 CELLULAR FACTORS INTERFERING WITH AAV

Before the replication process is completed in the cell, synthesis of the first strand appears to be critical for both replication and gene expression. It has been shown that the crucial and rate-limiting step for AAV replication as well as gene transduction by recombinant AAV (rAAV) is the conversion of the single stranded genome to the replicative form suitable for gene transcription (Ferrari *et al.*, 1996; Fisher *et al.*, 1996). Several factors can enhance this step (figure 4). This is the main role of the helper virus through the action of the E4ORF6 region of adenovirus (Ferrari *et al.*, 1996).

Chemicals (such as etoposid or hydroxyurea) and physical agents (X-rays, gamma rays, UV) can also lead to increased gene transduction by promoting genome conversion (Russell, Alexander and Miller, 1995). The mechanism is thought to be mediated by the induction of genotoxic shocks leading to the activation of specific cellular responses such as cell cycle arrest and DNA repair activation, although the precise nature of the involved pathways remains to be elucidated. Preliminary indications have suggested that rAAV genome replication is induced by the presence of AdE4ORF6 expression, whereas UV irradiation favours the formation of circular intermediates reminiscent of the head-to-tail concatemerized structures observed in latent-phase persistence of rAAV (Sanlioglu, Duan and Engelhardt, 1999).

Cellular factors have been shown to interact with the AAV genome. Only a small subset of cellular proteins seems to be required for full ss viral genome replication (Ni *et al.*, 1998). A cellular protein called the single strand D sequence Binding Protein (ssD-BP) was shown to interact with the D sequence of the 3' ITR of AAV. The dephosphorylation status of the ssD-BP correlates with the replication of the ss AAV genome (Qing *et al.*, 1998). The authors showed that the higher the ratio of dephosphorylated/phosphorylated forms, the better the transduction by rAAV. It was further demonstrated that the phosphorylation of the ssD-BP is mediated by the protein kinase activity of the EGF receptor (Mah *et al.*, 1998).

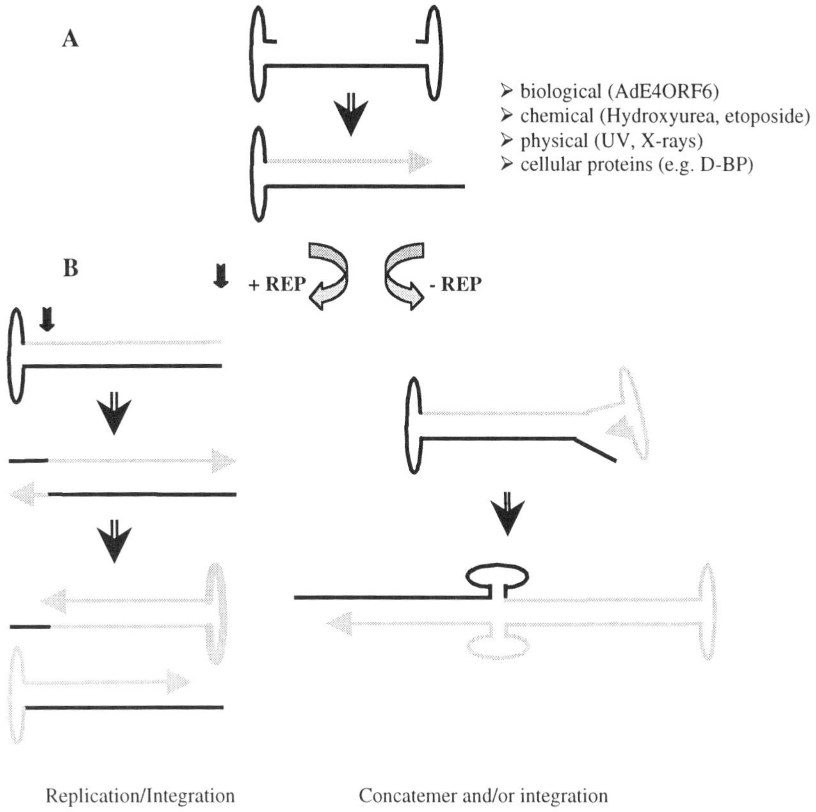

Figure 4. Conversion and fate of the AAV genome.
A: Several chemical, physical or biological factors can induce the conversion of the genome to a replicative form. B: The fate of the AAV genome is dependent on the Rep78/68 proteins. In their presence, the wt AAV genome is replicated and undergoes an integration process (site-specific in humans). If Rep78/68 are not expressed (the case of rAAV), the genome is converted into multimeric forms that integrate less efficiently without site-specificity.

1.6 AAV RECEPTORS

AAV possesses a wide range of tropisms. It is known to infect many different cell types from various species, leading to the idea that ubiquitous cell surface receptors were likely to account for this broad host range tropism. Elucidation of the primary steps of virus infection is recent. Heparan sulfate proteoglycans (HSPG) were shown to be involved in virus attachment to the cellular membrane and cell infection by acting as a primary receptor for AAV. HSPGs are associated with the extracellular matrix as well as plasma membranes. They are widely distributed in most tissues and may be in part responsible for the large host range of AAV (Summerford and Samulski, 1998).

Secondary receptors have been described, suggesting a two-step process for AAV infection. AAV appears to bind secondarily but directly to the β5 subunit of the αVβ5 integrin (Summerford, Bartlett and Samulski, 1999). Moreover, this work showed that the αVβ5 integrin facilitates AAV internalisation. Another co-receptor, the FGF receptor 1, was shown to be co-expressed along with HSPGs on permissive cell surfaces, allowing successful AAV binding and entry into the cell (Qing *et al.*, 1999).

Interestingly, HSPGs play a role in herpes simplex infection where they serve as the primary attachment receptors (Shieh *et al.*, 1992). Similarly, αVβ5 integrin is the secondary receptor for adenovirus allowing its internalisation and cell membrane permeabilisation (Wickham *et al.*, 1994; Wickham *et al.*, 1993). This can reflect an evolutionary pressure that led AAV to exploit the same mode of entry and/or the same tropism as its helper viruses.

Transport to the nucleus appears to be very fast, the virus being delivered to the nucleus within few hours (Berns, 1996). However, the exact trafficking mode of AAV is still essentially not understood, although it seems to be predominantly internalised through the clathrin dependent endocytic pathway (Duan *et al.*, 1999).

1.7 INTEGRATION OF AAV IN THE HOST GENOME

In vitro studies have shown that during a wild type (wt) infection, in the absence of helper virus infection, the AAV genome undergoes stable integration into the human host genome of most of the infected cells, therefore establishing a latent infection state. The integration is site-specific, although not absolute (70%) and takes place at a preferential site called AAVS1 (Kotin *et al.*, 1991; Samulski *et al.*, 1991). AAVS1 has been characterised and is located on chromosome 19q13.3-ter (Kotin, Linden and Berns, 1992; Kotin *et al.*, 1991). In addition, integration was also shown to occur in non-dividing cells (Russell, Miller and Alexander, 1994).

Integrated AAV genomes are found mostly as head to tail tandems comprising up to ten copies of the viral genome. Integration is thought to occur through a non-homologous recombination process (Linden *et al.*, 1996). The Rep78/68 proteins play a major role in this process (figure 4), by mediating a complex formation between the AAVS1 site and viral ITRs. Rep78/68 recognise a dodecameric sequence $(GCTC)_3$ that is found both in the ITRs and in AAVS1 (Linden, Winocour and Berns, 1996). An *in vitro* system has been developed to elucidate the process of site-specific integration that clearly showed that Rep mediated the formation of covalent junctions between AAVS1 and AAV (Dyall, Szabo and Berns, 1999).

The mechanism by which AAV vectors persist is different from its wild-type counterpart (figure 4). Recombinant AAVs (rAAV), because they lack the *rep* genes, do not retain the site-specificity of integration (Kearns *et al.*, 1996; Ponnazhagan *et al.*, 1997). The ability to integrate into the host genome is lowered and random (Yang *et al.*, 1997). Only vectors retaining the Rep proteins maintain site-specificity of integration (Surosky *et al.*, 1997).

The *in vivo* integration state following rAAV transduction is less documented. Many studies indicate that following *in vivo* gene transfer, the rAAV genome is maintained for a long period (up to a year and a half) as high molecular weight head to tail concatemers (Fisher *et al.*, 1997; Snyder *et al.*, 1997b; Xiao, Li and Samulski, 1996). These concatemers were shown to form circular intermediates, which not only allow episomal persistence but also serve as a pre-integrating structure in muscle (Duan *et al.*, 1998).

More recently, fluorescent *in situ* hybridisation and pulse-field gel electrophoresis analysis have shown that random integration occurs in mice following rAAV gene transfer (Miao *et al.*, 1998). These data were further confirmed by the isolation of vector-cellular DNA junctions from transduced mouse liver (Nakai *et al.*, 1999). Nevertheless, it is still unclear how frequently integration of rAAV occurs into the host genome *in vivo*. The recent development of transgenic rodent models for AAVS1 should lead to further elucidation of the different events leading to site specific integration of AAV in humans (Rizzuto *et al.*, 1999).

Finally, the ability of AAV to site-specifically integrate in chromosome 19 through non-homologous recombination has been exploited to promote targeted gene transfer of plasmid vectors, providing that they express the Rep protein (Balague, Kalla and Zhang, 1997; Lamartina *et al.*, 1998; Pieroni *et al.*, 1998). In addition, AAV vectors were used *in vitro* to introduce specific gene modifications by homologous recombination at higher frequency than using the traditional method (Russell and Hirata, 1998).

2 AAV-based vectors

Recombinant AAVs are derived by replacing the whole viral genome between the ITRs by the gene of interest and the desired regulatory sequences. Due to the small size of the AAV genome, the packaging capacity cannot exceed 119 % of the wt size (Hermonat *et al.*, 1997). This constitutes the major drawback of AAV, which therefore has a maximal capacity for exogenous sequences of about 4.6 kb. Nevertheless, in some cases, mini-cDNAs derived from large size cDNAs have been shown to remain fully active and have allowed vectorisation into recombinant AAV (Flotte *et al.*, 1993b; Wang *et al.*, 1999; Zhang *et al.*, 1998).

2.1 rAAV PRODUCTION

2.1.1 Transient transfection-based systems

AAV vector manufacturing of large batches of high quality virus is a crucial requirement for clinical use and GMP procedures remain to be established. Classical methods of production are laborious and time consuming. These procedures rely on trans-complementation and transitory transfection. Several rounds of purification are necessary, with poor yields and unsatisfying purity grade preparations.

The conventional protocol is based on the co-transfection of human 293 cells (an adenovirus E1A-transformed embryonic kidney cell line) with a vector and helper AAV plasmids (figure 5), followed by infection with E1 deleted adenovirus. The vector plasmid bears the expression cassette framed by the ITRs. The helper plasmid encodes the *rep* and *cap* genes necessary for recombinant genome excision, replication and encapsidation. This method led to reasonably high titers of 10^{11}-10^{12} physical particles (pp) but also to substantial levels of adenovirus contamination. This is highly undesirable because of the immunity associated with adenovirus administration, and therefore an additional purification step is required.

A recent improvement was achieved by replacing the adenoviral infection step by a mini-Ad plasmid encoding the adenoviral helper functions (Matsushita *et al.*, 1998;

Salvetti *et al.*, 1998; Xiao, Li and Samulski, 1998). This tri-transfection method eliminates adenovirus contamination (figure 5), but a single cell has to be hit three times in order to produce rAAV. However, in both methods, several important limitations still persist. First, the preparations are constantly contaminated with pseudo-wt AAV arising from non-homologous recombination between vector and helper AAV plasmids (Allen *et al.*, 1997; Salvetti *et al.*, 1998). Secondly, to achieve high titer production, it is necessary to perform large-scale transfection that might turn out to be difficult to scale up.

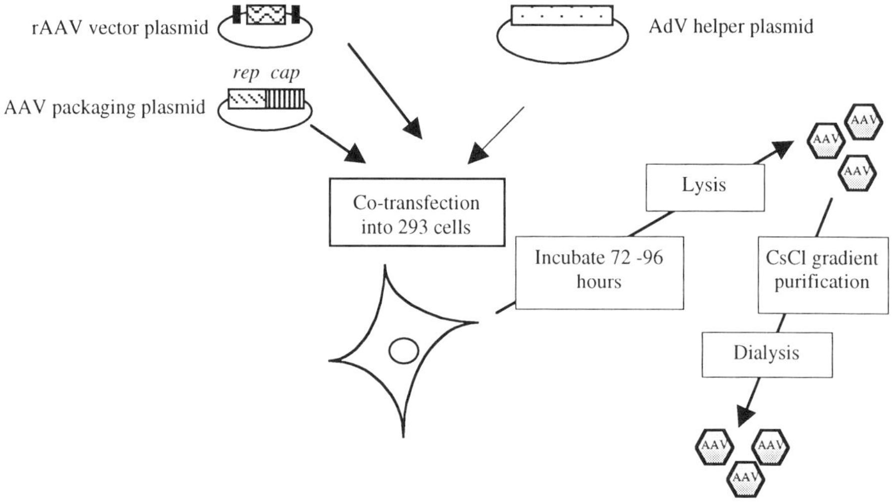

Figure 5. rAAV adenovirus-free production strategy. The modified conventional protocol is based on the use of a non-infectious mini-Ad helper plasmid. Further steps of purification remain unchanged and include virus fractionation, two to three rounds of CsCl density gradient ultracentrifugation and extensive dialysis (see text for details).

These studies have highlighted the tight regulation that is necessary between *rep* and *cap* genes. It has been shown that for optimum virion production per cell, the *rep* gene must be weakly expressed, whereas the *cap* gene must be over-expressed, capsid protein representing the limiting factor for virion accumulation (Li, Samulski and Xiao, 1997).

A two-component packaging plasmid (pDG) was generated that contains the AAV *rep* and *cap* genes and Ad helper functions. This method only requires a double hit transfection efficiency of cells with the hybrid helper and AAV vector plasmids, but results in 10-fold higher titers than those achieved using conventional protocols. Again, the replacement of the AAV-2 p5 promoter by the weaker MMTV-LTR promoter led to reduced expression of *rep* genes and increased expression of the *cap* gene (Grimm *et al.*, 1998). Nevertheless, the same authors, in a recent study, indicate that this is only true until a threshold of assembled capsids is reached, and that beyond this limit, efficient encapsidation is more likely to be rate limiting than capsid protein expression (Grimm *et al.*, 1999).

In transfection methods, the amount of plasmid successfully targeted to the nucleus is low, compared to infection efficiencies. Hybrid recombinant adenoviruses were constructed that contain the rAAV genome in place of the E1 region (Gao *et al.*, 1998; Liu, Clark and Johnson, 1999). This approach still requires adenoviral co-infection, either by wt or E1 deleted adenovirus, depending on the cell line used. With this system, the question of adenovirus contamination persists and further heat inactivation is required with the inherent associated problems of rAAV titer loss and incomplete adenovirus elimination. Similarly, an amplicon based on HSV-1 expressing *rep* and *cap* was developed, leading to an increased virion production per cell (up to 380 transducing units per cell). Combined with the packaging system containing the vector sequence (a proviral cell line), superinfection with such an amplicon might be a scaleable alternative for rAAV production, providing it can be made rHSV-1 free (Conway *et al.*, 1999).

In conclusion, these approaches give titers in a range of $10^{12\text{-}13}$ physical particles per ml and $10^{9\text{-}12}$ infectious particles per ml. These titers are still insufficient for clinical application in humans, and further efforts are still required to gain at least one log increase in the titers.

2.1.2 AAV packaging systems

An alternative to multiple hit transfection would be to develop stable packaging cell lines, on a model based on the existing and well established system for retroviral vector production. This has been the subject of several attempts. The initial packaging systems established were cell lines bearing the *rep* and *cap* genes either under the control of their own or exogenous promoters (Clark, Voulgaropoulou and Johnson, 1996). Improvements in these cell lines were made by establishing cell lines containing not only integrated copies of the *rep* and *cap* genes but also of the rAAV vector sequence (Clark *et al.*, 1995). As observed in the transitory producer systems, these initial attempts showed the necessity of a very tight regulation between the *rep* and *cap* genes in order to achieve good yields of rAAV ((Li, Samulski and Xiao, 1997; Vincent, Piraino and Wadsworth, 1997). Packaging cell lines have been developed in which the AAV ITRs are inducible upon wt adenovirus infection (Clark, Voulgaropoulou and Johnson, 1996; Tamayose, Hirai and Shimada, 1996) or using exogenous inducible promoters (Holscher *et al.*, 1994).

Titers of rAAV produced by most of these methods are far below those obtained in wt production. This has been suggested to be due to the inability of the *rep-cap* plasmid to replicate inside the transfected or packaging cell (Fan and Dong, 1997). Packaging cell lines were designed that possess integrated copies of both helper and vector constructs containing an SV40 origin of replication. These cell lines also express the SV40 T antigen regulatable (Tet-ON) by tetracycline (Inoue and Russell, 1998), leading to a production of up to 10^4 virions per cell, similar to that of the wt.

TABLE 1 : Overview of the main approaches for rAAV production and purification.
Comparison between protocols must be cautious as the initial number of cells varies between studies. However, they are good examples of both purification and production methods which have been evaluated so far.

Production system	**Purification method**	**References**
Pack. cell line / Ad infection	Affinity HPLC	(Clark *et al.*, 1999) (Clark *et al.*, 1995)
Tri-transfection (mini-Ad plasmid)	CsCl	(Salvetti *et al.*, 1998) (Xiao, Li and Samulski, 1998) (Matsushita *et al.*, 1998)
Cotransfection (mini-Ad plasmid-rep-cap : pDG)	Iodixanol / centricon	(Hermens *et al.*, 1999)
Pack. cell line / Ad. infection	Chromatography	(Tamayose, Hirai and Shimada, 1996)
Cotransfection (pDG)	Chromatography	(Grimm *et al.*, 1998)
Transfection / infection	CsCl	(Snyder, 1996)
Cotranstfection (pDG)	Iodixanol/affinity HPLC	(Zolotukhin *et al.*, 1999)
Pack. Cell line /Hybrid Ad shuttle vector infection	CsCl	(Gao *et al.*, 1998)
Inducible gene amplification pack. cell lines/ Ad. infection	CsCl	(Inoue and Russell, 1998)
Inducible pack. cell lines	-	(Holscher *et al.*, 1994)
Pack. cell lines/ Ad. infection	CsCl	(Clark *et al.*, 1995)
rHSV(rep-cap)	CsCl	(Conway *et al.*, 1999)

2.2 AAV VECTOR PURIFICATION

Downstream processing of AAV production, *i.e.* concentration and purification, is still cumbersome and constitutes the second challenge for high quality vector preparations. Once harvested, the cells are lysed and cellular and plasmidic DNAs are eliminated by DNAse treatment and high-speed centrifugation. The resulting supernatant is referred to as the "crude lysate". In order to reduce the volume of the lysate, the virus is precipitated by fractionation with saturated ammonium sulfate and resuspended in few ml of a buffer solution. At this stage, the viral preparation is essentially unpurified and further processing is necessary.

2.2.1 Cesium gradient

Development of a purification protocol based on cesium chloride (CsCl) density centrifugation has led to satisfactory stock purity for basic research (Snyder, 1996). In this procedure, rAAV is purified from fractionated crude lysate by two to three rounds of ultracentrifugation in a CsCl isopicnic gradient. The step is time consuming, 6 to 24 h per round of centrifugation depending on the rotor used. Most importantly, in order to perform *in vivo* application, it is an absolute requirement to eliminate the CsCl present in the preparation. This is commonly achieved by extensive dialysis against PBS-based solutions, but this process results in an incomplete purification and a loss of titer. It has been suggested that the dialysis step leads to an irreversible aggregation of AAV particles, making them unsuitable for efficient infection (Hermens *et al.*, 1999).

In the production protocol based on adenovirus infection, the CsCl gradient allows only incomplete separation of adenoviral contaminating particles, which have to be eliminated by a final step of heat inactivation at 56° C, but resulting in an additional loss of AAV particles. In general, this ultracentrifugation-based procedure is only applicable for limited volumes of crude lysate due to the limited volume capacity of rotors.

2.2.2 High Performance Liquid Chromatography

Alternative approaches for the purification and concentration of rAAV have been recently developed, based on high performance liquid chromatography (HPLC). Sulfonated cellulose column chromatography (cation-exchange) concentration of a crude lysate led to a nearly 100% recovery and a concentration factor of 50-fold (Tamayose, Hirai and Shimada, 1996). However, a small amount of adenovirus as well as serum and cellular proteins were co-eluted with AAV.

An immunoaffinity purification procedure was established based on a monoclonal antibody highly specific for AAV-2 capsids (A20) (Grimm *et al.*, 1998). In this procedure the monoclonal antibody A20 which only recognises assembled AAV-2 capsids is coupled to the chromatography matrix. This protocol led to 70% recovery of the loading vector and a four-fold concentration. Although the purity of the eluted virus was estimated to be 80%, this appealing procedure is unlikely to be envisaged for large-scale manufacturing, due to the cost and the availability of the A20 antibody.

Finally, a single step purification process based on heparin affinity chromatography produced satisfactory results (Clark *et al.*, 1999). Despite the fact that the production system was based on adenovirus superinfection, no contaminating adenovirus was detected in the peak rAAV HPLC fractions. Similarly, the rAAV fractions were essentially free of cellular and serum proteins, and the final yield was 70%. Final stock titers ranged around $10^{12\text{-}13}$ physical particles per ml (pp/ml) and *in vivo* biologic activity was retained.

2.2.3 Iodixanol Gradient

A recent method based on iodixanol gradient ultracentrifugation combined with subsequent heparin sepharose affinity column purification led to a 99% purity preparation (Zolotukhin *et al.*, 1999). Iodixanol is a physiological X-ray contrast solution, which does not require further purification steps and allows separation of the particles in a 1-hour centrifugation. With this method, the purity reached 99%, the particle-to-infectivity ratio was less than 100:1 and the yield per cell averaged 100 infectious units (iu), depending on the column used.

A similar approach was developed in which a column chromatography purification step was followed by a three hour-ultracentrifugation in a 60% iodixanol density gradient. In addition, a dilution step in PBS followed by re-concentration with a Centricon 100 concentrator appeared necessary because of the viscosity of the recovered fractions (60% iodixanol). The yield averaged 30 iu per cell and the particle-to-infectivity ratio was 500:1 (Hermens *et al.*, 1999).

In all the purification procedures described so far, both empty capsids and fully functional particles are produced. There is presently no technique allowing their separation, although the presence of the former may constitute a potential problem for *in vivo* applications, as capsid proteins are the major targets of immune response to AAV vectors.

2.3 rAAV STOCK TITRATION

Using the conventional rAAV preparation methods, the pp/iu ratio is 1000:1. Decreasing this ratio has been one of the goals of many groups. Several of them have reported, on the basis of various modifications, ratios as low as from 100:1 or even 20:1 (Vincent, Piraino and Wadsworth, 1997). Providing that the cell production remains unchanged in term of total pp, this represents a considerable effective scale-up (in terms of iu/cell produced or iu/ml of preparation obtained).

Most of the above protocols for production and purification lead to a yield of particles per cell that is much lower than for the wt, but it is often difficult to compare their relative efficiencies due to a lack of standardised methods of titration. The best parameter to estimate the efficiency of production is to estimate the amount of particles (physical and infectious) produced per producer cell. In addition, this allows the direct calculation of the particle-to-infectivity ratio. Therefore two titers must be determined: the physical particle titer (DNA-containing particles), based on the detection of viral genomes and the infectious titer, relying on a trans-complementing replication assay.

Because only a single genome copy is packaged into pre-formed capsids, each genome detected represents a complete virion. Physical particles are quantified by dot blot hybridisation of an aliquot of the preparation previously treated with DNAse and lysed to disrupt the capsids. This method, also referred as "DNAse-resistant particles" can be conveniently replaced and automated by real-time quantitative PCR, which allows rapid and reproducible quantification of rAAV (Clark *et al.*, 1999), although the cost of this method must be considered in view of large-scale manufacture.

As we discussed previously, the determination of the infectious titer is crucial, as it can be considered as the "biologically active" titer. To this purpose, a recombination centre assay (RCA) was developed (Snyder, 1996) and further modified (Salvetti *et al.*, 1998). In this assay, all the functions supporting rAAV replication are supplied in trans. A HeLa derived cell line expressing *rep* and *cap* genes is infected with serial dilutions of the rAAV and co-infected with a wtAd. After 24 hours, the cells are harvested, transferred, lysed and fixed onto nylon membranes that are subsequently hybridised against a probe recognising the transgene sequence. Each positive signal represents a successfully infected cell and is referred as one iu. However, this procedure is time consuming and cumbersome. Improvements in infectious particle titer determination were made, based either on high-throughput hybridisation methods (Atkinson *et al.*, 1998) or real-time PCR (Clark *et al.*, 1999).

Some authors find it convenient to determine the transducing titer of their preparation. This is commonly achieved by infecting cells with a rAAV fraction and revealing transgene product by the suitable method. The infection step is performed with and without adenovirus co-infection; two titers are obtained. The transducing titer in absence of adenovirus is usually 10 times lower than in its presence. The "+Ad" titer is questionable, as efforts are made to get rid of adenovirus contamination of the preparation in order to administrate pure rAAV. Moreover, the transducing titer may vary greatly with the sensitivity of the method of detection, from one laboratory to another, and also depends on the permissivity of the cell line used for titration.

The purity issue of rAAV stocks has become a major point of concern, since many conflicting data have been shown to be (at least partially) due to the wide variation in purity and titers of the preparations used. Artefactual positive transduction has been shown to be a common problem of insufficiently purified stock contaminated by the transgene product. This general phenomenon called "pseudotransduction" has been observed for at least two commonly used reporter systems: LacZ and alkaline phosphatase (Alexander, Russell and Miller, 1997). These observations have raised the absolute requirement of further purification (gradient purification and/or HPLC).

On the other hand, a recent study showed that a level of cellular contaminants, distinct from pseudotransduction, is necessary for efficient stimulation of rAAV mediated transduction. These contaminants, although still unidentified, are co-purified from the permissive 293 cells during the production/purification process and when added to cells along with the viral vector, led to enhancement of transduction (Tenenbaum *et al.*, 1999). Clark and colleagues, who successfully injected rAAV vector highly purified by chromatography in the mouse quadriceps (Clark *et al.*, 1999) drew contradictory conclusions, however.

3 AAV mediated gene delivery *in vivo*

In many disorders a long-term correction would be ideal. Because AAV based vectors have, in addition of a lack of toxicity, the potential of integration into the host genome, they represent the tool of choice to fulfil all these requirements. Many studies have emphasised the importance of the route of administration. Table 1 summarises some representative pre-clinical and clinical studies of rAAV-mediated gene therapy. In most of the studies presented below, some humoral immune reaction was observed and will be discussed later in this review.

3.1 MUSCLE

Many studies have shown the high transduction efficiency of muscle by AAV vectors. High and persistent expression was shown, based on reporter gene expression (Fisher *et al.*, 1997; Snyder *et al.*, 1997b; Xiao, Li and Samulski, 1996). Muscle injury-induced regeneration failed to enhance gene transfer efficiency (Snyder *et al.*, 1997b). Muscle can also constitute a "factory system" to deliver therapeutic proteins systemically. This was evaluated in a mouse model in which a long-term excretion of erythropoïetin (Epo) in the blood system was described following intra-muscular gene transfer of a murine Epo encoding AAV vector. Moreover, no destructive immune response was observed (Kessler *et al.*, 1996; Snyder *et al.*, 1997b).

Two pre-clinical studies in a large animal model recently demonstrated the therapeutic feasibility of gene therapy based on rAAV. Intramuscular injection of a rAAV encoding the canine coagulation factor IX in a canine model of haemophilia B allowed stable and dose dependent therapeutic, albeit low, circulating concentration of the factor IX over 17 months. Nevertheless, correction remained partial as measured by the reduction in whole blood clotting time (Chao *et al.*, 1999; Herzog *et al.*, 1999) and did not prevent bleeding episodes requiring blood transfusion. In both of the latter studies, the vector dose was divided and injected in several separate sites in the muscle. More importantly, these studies delineated that the curative doses in large animals and humans would need to be at or above 10^{13} particles per kg. Based on these data, a clinical trial was initiated and initial observations on the first three patients were recently published. Of the three patients with severe haemophilia B who received intramuscular injection of an AAV vector expressing the human factor IX, none showed signs of toxicity, gerlime transmission and anti-factor IX antibodies (Kay *et al.*, 2000).

Mice immunised by an intra-muscular injection of a rAAV expressing the HSV-2 glycoprotein B (gB) elicited gB-specific cytoxic T lymphocytes and antibodies (Manning *et al.*, 1997), demonstrating the possibility of using rAAV as a vehicle for delivery of vaccines.

The first demonstration of complete restoration of a physiological function was reported in a limb girdle muscular dystrophy (LGMD) 2F rodent model. In this 4-month study, a single intramuscular injection of a rAAV-δ-sarcoglycan led to almost complete recovery of muscle strength and prevented the muscle hypertrophy characteristic of the disease with normal muscle weight and size. Interestingly, over-expression of δ-sarcoglycan could not only completely restore the defect but did not appear to be detrimental for the animal (Xiao *et al.*, 2000).

3.2 CENTRAL AND PERIPHERAL NERVOUS SYSTEMS

Kaplitt *et al.* first demonstrated the efficiency and inocuity of rAAV in the central nervous system. A four-month expression of a therapeutic gene was observed without toxicity in a parkinsonian model in the rat (Kaplitt *et al.*, 1994). Since then, it has been shown that neurons are the main target of AAV infection (Bartlett, Samulski and McCown, 1998; Chamberlin *et al.*, 1998). Similar observations were made in a Parkinson primate model (During *et al.*, 1998a). Bone marrow (Peel *et al.*, 1997) and the neuro-sensory epithelium of the inner ear (Lalwani *et al.*, 1998) were efficiently targeted. Functional restoration of photoreceptor neurons was achieved in rodent models (Flannery *et al.*, 1997; Jomary *et al.*, 1997). Therapeutic levels were achieved in a mouse model of mucopolysaccharidosis VII (MPS VII) with a rAAV, leading to the elimination of the storage vesicles in the brain that are characteristic of the disease (Elliger *et al.*, 1999).

The threshold dose for detectable cell transduction with a reporter gene system was evaluated at 10^7 iu per site of injection. The transduced cells were in the vicinity of the site of inoculation and did not spread to distant sites in the brain. Expression was persistent (up to 15 months) but dropped considerably after 4 months. (Lo *et al.*, 1999).

3.3 LIVER

Because of its major role in a number of metabolic pathways, liver is a major potential target for gene transfer, Gene transduction by rAAV could be used either to restore a defective liver function or for being a source of delivery of a therapeutic protein to the blood system. Haemophilia B clotting factor IX was delivered successfully to the liver and led to stable and persistent expression in mice (Nakai *et al.*, 1998; Snyder *et al.*, 1997a). The latter study also indicated that transduction in neonatal hepatocytes was efficient.

Recent pre-clinical studies in murine and canine models of haemophilia B achieved curative levels of hFIX by intraportal infusion of rAAV-FIX. The bleeding disorder was corrected over a period of 17 months in mice, whereas only partial correction was achieved in dogs over an 8-month period. The main limitation in this study was that the titers were insufficient for scaling up the dose to fit the size of the dog model (Snyder *et al.*, 1999).

Dissociated heavy and light chain subunits of factor VIII have been cloned separately in two AAV vectors and co-injected in the portal vein of mice (Burton *et al.*, 1999). Biological activity was observed for almost a year, although only 5-10% of the total FVIII produced seem to account for this activity. A likely explanation is that the two separated chains might have to be associated inside the same cell, therefore requiring co-transduction events to occur.

3.4 LUNG

Cystic fibrosis is the most common monogenic inherited disease among Caucasians, affecting 1/ 2500 births. Because long-term expression is required in CF patients, AAV represents an attractive vehicle. Many attempts of rAAV mediated gene transfer into the lungs have already proven their relevancy. Initial work reported up to 6 months detection of the *cf* gene product (the cystic fibrosis transmembrane regulator or CFTR) in the airway epithelium after fiberoptic bronchoscope administration of a rAAV into the lung of adult rabbits (Flotte *et al.*, 1993a). Intratracheal instillation of a rAAV encoding the CFTR in neonatal rabbits was able to transduce bronchial epithelium through the alveolar phase of lung development, but expression did not persist (Rubenstein *et al.*, 1997). Primate lungs were efficiently targeted with an AAV-CFTR by instillation and messenger RNA could be detected up to 6 months post-administration. Unfortunately, there was no indication concerning the functionality of the CFTR (Conrad *et al.*, 1996).

These studies have highlighted the importance of the route of administration and also the necessity of high titer preparations of rAAV. Moreover, in CF patients, long term expression of the therapeutic gene is required. The lack of persistence of vector expression observed by several groups has emphasised the necessity of re-administration of the vector. Contradictory results were obtained concerning the efficiency of repeated administration into the lungs (Beck *et al.*, 1999; Halbert *et al.*, 1998). Finally, a phase I clinical trial supported the conclusion that rAAV-CFTR administration in the maxillary sinus of CF patients was safe and efficient. It demonstrated a dose dependent gene transfer and a functional effect which was limited to day 14 post-treatment (Wagner *et al.*, 1998a; Wagner *et al.*, 1998b).

TABLE 2 : Pre-clinical and clinical studies of gene therapy mediated by AAV.

Tissue of administration / species / disease	**Transgene product**	**References**
Muscle / dog / haemophilia B	*Coagulation factor IX*	(Chao *et al.*, 1999) (Herzog *et al.*, 1999)
Muscle / human / haemophilia B	*Coagulation factor IX*	(Kay *et al.*, 2000)
Muscle / hamster / LGMD2F	*δ-sarcoglycan*	(Xiao *et al.*, 2000)
Muscle / mouse / immunisation	*HSV-2 glycoprotein B*	(Manning *et al.*, 1997)
Liver / mouse / haemophilia B	*Coagulation factor IX*	(Snyder *et al.*, 1999) (Nakai *et al.*, 1998)
Liver / dog / haemophilia B	*Coagulation factor IX*	(Snyder *et al.*, 1999)
Liver / mouse / haemophilia A	*Coagulation factor VIII*	(Burton *et al.*, 1999)
Liver / mouse / diabetes	*insulin*	(Sugiyama *et al.*, 1997)
Cerebrospinal fluid / mouse / MPS VII	*β-glucuronidase*	(Elliger *et al.*, 1999)
Retina/rat/ autosomic dominant retinis pigmentosa	*Ribozyme*	(Lewin *et al.*, 1998)
Retina/mouse/retinal degeneration	*cGMP phosphodiesterase β*	(Jomary *et al.*, 1997)
Brain / monkey / Parkinson disease	*tyrosine hydroxylase*	(During *et al.*, 1998a)
Brain / rat / Parkinson disease	*tyrosine hydroxylase*	(Kaplitt *et al.*, 1994)
Brain / rat / tumour	*HSV TK*	(Mizuno *et al.*, 1998)
Pluripotent hematopoietic stem cells / monkey / ß-thalassemia	*β-globin*	(Schimmenti *et al.*, 1998)
Oral administration / rat / lactose intolerance	*β–galactosidase*	(During *et al.*, 1998b)
Eye / mouse / retinal degeneration	*Phosphodiesterase β*	(Jomary *et al.*, 1997)
Lung / human / cystic fibrosis	*CFTR*	(Wagner *et al.*, 1998a; Wagner *et al.*, 1998b)

3.5 OTHER TISSUES

Many other tissues have been subjected to gene transfer mediated by AAV vectors with various efficiencies. Peroral administration of a rAAV was evaluated in a rat model of lactose intolerance (During *et al.*, 1998b). Transduction of the gut was achieved for up to 6 months and led to therapeutic level of transgene expression over a period of 4 months. Cancer gene therapy has been the subject of many studies and rAAV-mediated transduction was shown to be efficient (Okada *et al.*, 1996; Su *et al.*, 1997). Systemic administration of rAAV appears to lead essentially to transduction in the liver (Koeberl *et al.*, 1999; Watson *et al.*, 1998). Finally, vaccination strategies based on AAV vectors have raised promising results (During *et al.*, 2000; Liu *et al.*, 2000).

4 The next steps with AAV vectors

4.1 THE IMMUNOLOGICAL ISSUE

Several studies have demonstrated that there is no cell-mediated immunity (cytotoxic T lymphocytes, CTL) following wt (Hernandez *et al.*, 1999) or recombinant AAV administration. Only transitory non-destructive inflammation was observed following rAAV injection in muscle (Jooss *et al.*, 1998; Xiao, Li and Samulski, 1996). No cellular or humoral immune response against the protein encoded by the transgene has been observed following rAAV mediated gene transfer. This was shown to be due to the inability of rAAV to transduce dendritic cells (Jooss *et al.*, 1998). If no cellular immune response has been observed, neutralising antibodies against AAV capsid have been detected in many cases (Halbert *et al.*, 1998; Xiao, Li and Samulski, 1996). Subsequent re-administration of the vector was unsuccessful in mouse muscle (Xiao, Li and Samulski, 1996) and lungs (Halbert *et al.*, 1998). Nevertheless, transient immunosuppression at the time of the initial administration allowed successful re-administration in mouse lung (Halbert *et al.*, 1998) and muscle (Manning *et al.*, 1998). Contradictory observations were obtained in rabbit airways, where re-administration of the vector was not hampered by the humoral immune response (Beck *et al.*, 1999) and in the mouse brain (Lo *et al.*, 1999). A similar unsuccessful gene transfer with rAAV was observed following intraportal rAAV administration in mice with pre-existing immunity to AAV. Moreover, transient immunosuppression at the time of the second administration did not lead to gene transfer (Moskalenko *et al.*, 2000). This study is more likely to mimic the situation that will be encountered in humans, since most of the candidates for gene therapy will turn out to be seropositive to AAV. It is likely, however, that the immune response and the feasibility of efficient re-administration will vary between tissues, and species comparisons and extrapolation might not be relevant.

A study in rhesus macaques showed that following rAAV delivery, only a humoral immune response was elicited toward capsid proteins, whereas co-infection with adenovirus led to a cell mediated immune response to the AAV capsid (Hernandez *et al.*, 1999). This raises the crucial question of the quality and purity of rAAV produced with adenovirus.

One possibility to minimising the immune response appears to be in the use of non-human AAV serotypes. Recombinant vectors based on AAV-4 (Chiorini *et al.*, 1997), AAV-3, AAV-6 (Rutledge, Halbert and Russell, 1998) and AAV-1 (Xiao *et al.*, 1999) were developed. One interesting point of the latter study reside in the observation that serotype 1 is more efficient in transducing muscle than AAV serotype 2, whereas the opposite is true in liver. Sequential treatment with one serotype followed by the other gave mixed results regarding the efficiency of the second administration. Cross neutralisation appeared to be dependent on the tissue and the sequence of serotype administration. Finally, pseudotypes of rAAV-2 with other serotype capsid proteins were engineered and use for *in vivo* mouse lung transduction (Halbert *et al.*, 2000). The authors found that AAV-6 pseudotype was as efficient as AAV-2 vectors but less immunogenic. In addition, prior exposure to one pseudotype did not affect efficiency of a secondary exposure to the other.

4.2 AAV VECTOR TARGETING AND REGULATION

Specific targeting is of primary importance in gene therapy. It has been achieved to some extent by the use of specific promoters and regulatory sequences, but this might not be satisfactory enough or simply not possible in some applications. Targeting specific receptors of the target cell population can represent a good alternative. In addition, it could overcome the poor infectability of cell types representing therapeutic targets.

In other situations, the lack of targeting specificity can lead to undesirable and harmful side effects for the patient (e.g. suicide gene transfer in cancer therapy). Re-targeting has been explored through different approaches. A CD34-targeted AAV was developed using a chimeric single chain antibody against the CD34-AAV capsid, which allowed hematopoïetic stem cells to be targeted, but the resulting titers were low and further improvements are needed (Yang *et al.*, 1998). A strategy based on a bispecific $F(ab'\gamma)_2$ antibody led to efficient retargeting of rAAV toward otherwise non-permissive cells, but vector mediated expression was low (Bartlett *et al.*, 1999). Retargeting AAV to other cellular receptors, such as the integrin receptor, by insertional mutagenesis of the capsid gene (RGD motif of the laminin fragment P1) was successfully achieved (Girod *et al.*, 1999).

In some situations, long term persistence of expression might be undesirable. Expression of the therapeutic protein can be required at certain stages of a disease, or at precise windows of time during the development, regulated expression of the transgene becoming therefore highly necessary. In this context, pharmacologically regulated transgene expression has been demonstrated in muscle (Rendahl *et al.*, 1998; Rivera *et al.*, 1999; Ye *et al.*, 1999) and brain (Haberman, McCown and Samulski, 1998).

5 Conclusion

To date, a considerable amount of data undoubtedly show that AAV is extremely powerful for efficient delivery of therapeutic genes *in vivo*. There is substantial evidence that neither adverse effects nor medical risk are associated with rAAV administration, making this vector safe for clinical applications in humans. No cellular immunity exists, thus avoiding the problems encountered with adenoviral vectors. Nevertheless, the question of the pre-existing immunity and re-administration in humans will need to be further explored. Recombination and spread out risks are weakly documented and remain to be more systematically explored. These issues are critical for human application and can only be elucidated in clinical trials.

6 Acknowledgement

I thank Susan Cure for critical reading of this manuscript.

7 References

Alexander, I. E., Russell, D. W., and Miller, A. D. (1997). Transfer of contaminants in adeno-associated virus vector stocks can mimic transduction and lead to artifactual results. *Hum Gene Ther* **8**(16), 1911-20.

Allen, J. M., Debelak, D. J., Reynolds, T. C., and Miller, A. D. (1997). Identification and elimination of replication-competent adeno- associated virus (AAV) that can arise by nonhomologous recombination during AAV vector production. *J Virol* **71**(9), 6816-22.

Atkinson, E. M., Debelak, D. J., Hart, L. A., and Reynolds, T. C. (1998). A high-throughput hybridization method for titer determination of viruses and gene therapy vectors. *Nucleic Acids Res* **26**(11), 2821-3.

Balague, C., Kalla, M., and Zhang, W. W. (1997). Adeno-associated virus Rep78 protein and terminal repeats enhance integration of DNA sequences into the cellular genome. *J Virol* **71**(4), 3299-306.

Bartlett, J. S., Kleinschmidt, J., Boucher, R. C., and Samulski, R. J. (1999). Targeted adeno-associated virus vector transduction of nonpermissive cells mediated by a bispecific F(ab'gamma)2 antibody. *Nat Biotechnol* **17**(2), 181-6.

Bartlett, J. S., Samulski, R. J., and McCown, T. J. (1998). Selective and rapid uptake of adeno-associated virus type 2 in brain. *Hum Gene Ther* **9**(8), 1181-6.

Beck, S. E., Jones, L. A., Chesnut, K., Walsh, S. M., Reynolds, T. C., Carter, B. J., Askin, F. B., Flotte, T. R., and Guggino, W. B. (1999). Repeated delivery of adeno-associated virus vectors to the rabbit airway. *J Virol* **73**(11), 9446-55.

Berns, K. I. (1990). Parvovirus replication. *Microbiol Rev* **54**(3), 316-29.

Berns, K. I. (1996). Parvoviridae: The virus and their replication. 3rd ed. *In* "Fields in Virology" (B. N. Fields , D. M. Knipe, and P. M. Howley, Eds.), Vol. 2, pp. 2173-2197. Lippincott - Raven, Philadelphia.

Blacklow, N. R., Hoggan, M. D., and Rowe, W. P. (1968). Serologic evidence for human infection with adenovirus-associated viruses. *J Natl Cancer Inst* **40**(2), 319-27.

Bohenzky, R. A., LeFebvre, R. B., and Berns, K. I. (1988). Sequence and symmetry requirements within the internal palindromic sequences of the adeno-associated virus terminal repeat. *Virology* **166**(2), 316-27.

Burton, M., Nakai, H., Colosi, P., Cunningham, J., Mitchell, R., and Couto, L. (1999). Coexpression of factor VIII heavy and light chain adeno-associated viral vectors produces biologically active protein. *Proc Natl Acad Sci U S A* **96**(22), 12725-12730.

Chamberlin, N. L., Du, B., de, L. S., and Saper, C. B. (1998). Recombinant adeno-associated virus vector: use for transgene expression and anterograde tract tracing in the CNS. *Brain Res* **793**(1-2), 169-75.

Chao, H., Samulski, R., Bellinger, D., Monahan, P., Nichols, T., and Walsh, C. (1999). Persistent expression of canine factor IX in hemophilia B canines. *Gene Ther* **6**(10), 1695-1704.

Chiorini, J. A., Yang, L., Liu, Y., Safer, B., and Kotin, R. M. (1997). Cloning of adeno-associated virus type 4 (AAV4) and generation of recombinant AAV4 particles. *J Virol* **71**(9), 6823-33.

Clark, K. R., Liu, X., McGrath, J. P., and Johnson, P. R. (1999). Highly purified recombinant adeno-associated virus vectors are biologically active and free of detectable helper and wild-type viruses. *Hum Gene Ther* **10**(6), 1031-9.

Clark, K. R., Voulgaropoulou, F., Fraley, D. M., and Johnson, P. R. (1995). Cell lines for the production of recombinant adeno-associated virus. *Hum Gene Ther* **6**(10), 1329-41.

Clark, K. R., Voulgaropoulou, F., and Johnson, P. R. (1996). A stable cell line carrying adenovirus-inducible rep and cap genes allows for infectivity titration of adeno-associated virus vectors. *Gene Ther* **3**(12), 1124-32.

Conrad, C. K., Allen, S. S., Afione, S. A., Reynolds, T. C., Beck, S. E., Fee, M. M., Barrazza, O. X., Adams, R., Askin, F. B., Carter, B. J., Guggino, W. B., and Flotte, T. R. (1996). Safety of single-dose administration of an adeno-associated virus (AAV)- CFTR vector in the primate lung. *Gene Ther* **3**(8), 658-68.

Conway, J., Rhys, C., Zolotukhin, I., Zolotukhin, S., Muzyczka, N., Hayward, G., and Byrne, B. (1999). High-titer recombinant adeno-associated virus production utilizing a recombinant herpes simplex virus type I vector expressing AAV-2 Rep and Cap. *Gene Ther* **6**(6), 986-993.

Duan, D., Li, Q., Kao, A. W., Yue, Y., Pessin, J. E., and Engelhardt, J. F. (1999). Dynamin is required for recombinant adeno-associated virus type 2 infection. *J Virol* **73**(12), 10371-6.

Duan, D., Sharma, P., Yang, J., Yue, Y., Dudus, L., Zhang, Y., Fisher, K. J., and Engelhardt, J. F. (1998). Circular intermediates of recombinant adeno-associated virus have defined structural characteristics responsible for long-term episomal persistence in muscle tissue. *J Virol* **72**(11), 8568-77.

During, M. J., Samulski, R. J., Elsworth, J. D., Kaplitt, M. G., Leone, P., Xiao, X., Li, J., Freese, A., Taylor, J. R., Roth, R. H., Sladek Jr, J. R., O'Malley, K. L., and Redmond Jr, D. E. (1998a). In vivo expression of therapeutic human genes for dopamine production in the caudates of MPTP-treated monkeys using an AAV vector. *Gene Ther* **5**, 820-827.

During, M. J., Xu, R., Young, D., Kaplitt, M. G., Sherwin, R. S., and Leone, P. (1998b). Peroral gene therapy of lactose intolerance using an adeno-associated virus vector. *Nat Med* **4**(10), 1131-5.

Dyall, J., Szabo, P., and Berns, K. I. (1999). Adeno-associated virus (AAV) site-specific integration: Formation of AAV-AAVS1 junctions in an in vitro system. *Proc Natl Acad Sci U S A* **96**(22), 12849-12854.

Elliger, S., Elliger, C., Aguilar, C., Raju, N., and Watson, G. (1999). Elimination of lysosomal storage in brains of MPS VII mice treated by intrathecal administration of an adeno-associated virus vector. *Gene Ther* **6**(6), 1175-1178.

Erles, K., Sebokova, P., and Schlehofer, J. R. (1999). Update on the prevalence of serum antibodies (IgG and IgM) to adeno-associated virus (AAV). *J Med Virol* **59**(3), 406-11.

Fan, P. D., and Dong, J. Y. (1997). Replication of rep-cap genes is essential for the high-efficiency production of recombinant AAV. *Hum Gene Ther* **8**(1), 87-98.

Ferrari, F. K., Samulski, T., Shenk, T., and Samulski, R. J. (1996). Second-strand synthesis is a rate-limiting step for efficient transduction by recombinant adeno-associated virus vectors. *J Virol* **70**(5), 3227-3234.

Fisher, K. J., Gao, G. P., Weitzman, M. D., DeMatteo, R., Burda, J. F., and Wilson, J. M. (1996). Transduction with recombinant adeno-associated virus for gene therapy is limited by leading-strand synthesis. *J Virol* **70**(1), 520-32.

Fisher, K. J., Jooss, K., Alston, J., Yang, Y., Haecker, S. E., High, K., Pathak, R., Raper, S. E., and Wilson, J. M. (1997). Recombinant adeno-associated virus for muscle directed gene therapy. *Nat Med* **3**(3), 306-12.

Flannery, J. G., Zolotukhin, S., Vaquero, M. I., LaVail, M. M., Muzyczka, N., and Hauswirth, W. W. (1997). Efficient photoreceptor-targeted gene expression in vivo by recombinant adeno-associated virus. *Proc Natl Acad Sci U S A* **94**(13), 6916-21.

Flotte, T. R., Afione, S. A., Conrad, C., McGrath, S. A., Solow, R., Oka, H., Zeitlin, P. L., Guggino, W. B., and Carter, B. J. (1993a). Stable in vivo expression of the cystic fibrosis transmembrane conductance regulator with an adeno-associated virus vector. *Proc Natl Acad Sci U S A* **90**(22), 10613-7.

Flotte, T. R., Afione, S. A., Solow, R., Drumm, M. L., Markakis, D., Guggino, W. B., Zeitlin, P. L., and Carter, B. J. (1993b). Expression of the cystic fibrosis transmembrane conductance regulator from a novel adeno-associated virus promoter. *J Biol Chem* **268**(5), 3781-90.

Gao, G. P., Qu, G., Faust, L. Z., Engdahl, R. K., Xiao, W., Hughes, J. V., Zoltick, P. W., and Wilson, J. M. (1998). High-titer adeno-associated viral vectors from a Rep/Cap cell line and hybrid shuttle virus. *Hum Gene Ther* **9**(16), 2353-62.

Girod, A., Ried, M., Wobus, C., Lahm, H., Leike, K., Kleinschmidt, J., Deleage, G., and Hallek, M. (1999). Genetic capsid modifications allow efficient re-targeting of adeno-associated virus type 2. *Nat Med* **5**(9), 1052-6.

Grimm, D., Kern, A., Pawlita, M., Ferrari, F., Samulski, R., and Kleinschmidt, J. (1999). Titration of AAV-2 particles via a novel capsid ELISA: packaging of genomes can limit production of recombinant AAV-2. *Gene Ther* **6**(7), 1322-1330.

Grimm, D., Kern, A., Rittner, K., and Kleinschmidt, J. A. (1998). Novel tools for production and purification of recombinant adenoassociated virus vectors. *Hum Gene Ther* **9**(18), 2745-60.

Haberman, R. P., McCown, T. J., and Samulski, R. J. (1998). Inducible long-term gene expression in brain with adeno-associated virus gene transfer. *Gene Ther* **5**(12), 1604-11.

Halbert, C. L., Rutledge, E. A., Allen, J. M., Russell, D. W., and Miller, A. D. (2000). Repeat transduction in the mouse lung by using adeno-associated virus vectors with different serotypes. *J Virol* **74**(3), 1524-32.
Halbert, C. L., Standaert, T. A., Wilson, C. B., and Miller, A. D. (1998). Successful readministration of adeno-associated virus vectors to the mouse lung requires transient immunosuppression during the initial exposure. *J Virol* **72**(12), 9795-805.
Hermens, W. T., ter Brake, O., Dijkhuizen, P. A., Sonnemans, M. A., Grimm, D., Kleinschmidt, J. A., and Verhaagen, J. (1999). Purification of recombinant adeno-associated virus by iodixanol gradient ultracentrifugation allows rapid and reproducible preparation of vector stocks for gene transfer in the nervous system. *Hum Gene Ther* **10**(11), 1885-91.
Hermonat, P. L., Quirk, J. G., Bishop, B. M., and Han, L. (1997). The packaging capacity of adeno-associated virus (AAV) and the potential for wild-type-plus AAV gene therapy vectors. *Febs Lett* **407**(1), 78-84.
Hernandez, Y. J., Wang, J., Kearns, W. G., Loiler, S., Poirier, A., and Flotte, T. R. (1999). Latent adeno-associated virus infection elicits humoral but not cell-mediated immune responses in a nonhuman primate model. *J Virol* **73**(10), 8549-58.
Herzog, R. W., Yang, E. Y., Couto, L. B., Hagstrom, J. N., Elwell, D., Fields, P. A., Burton, M., Bellinger, D. A., Read, M. S., Brinkhous, K. M., Podsakoff, G. M., Nichols, T. C., Kurtzman, G. J., and High, K. A. (1999). Long-term correction of canine hemophilia B by gene transfer of blood coagulation factor IX mediated by adeno-associated viral vector. *Nat Med* **5**(1), 56-63.
Hoggan, M. D., Blacklow, N. R., and Rowe, W. P. (1966). Studies of small DNA viruses found in various adenovirus preparations: physical, biological, and immunological characteristics. *Proc Natl Acad Sci U S A* **55**(6), 1467-74.
Holscher, C., Horer, M., Kleinschmidt, J. A., Zentgraf, H., Burkle, A., and Heilbronn, R. (1994). Cell lines inducibly expressing the adeno-associated virus (AAV) rep gene: requirements for productive replication of rep-negative AAV mutants. *J Virol* **68**(11), 7169-77.
Hong, G., Ward, P., and Berns, K. I. (1992). In vitro replication of adeno-associated virus DNA. *Proc Natl Acad Sci U S A* **89**(10), 4673-7.
Horer, M., Weger, S., Butz, K., Hoppe, S. F., Geisen, C., and Kleinschmidt, J. A. (1995). Mutational analysis of adeno-associated virus Rep protein-mediated inhibition of heterologous and homologous promoters. *J Virol* **69**(9), 5485-96.
Inoue, N., and Russell, D. W. (1998). Packaging cells based on inducible gene amplification for the production of adeno-associated virus vectors. *J Virol* **72**(9), 7024-31.
Jomary, C., Vincent, K. A., Grist, J., Neal, M. J., and Jones, S. E. (1997). Rescue of photoreceptor function by AAV-mediated gene transfer in a mouse model of inherited retinal degeneration. *Gene Ther* **4**(7), 683-90.
Jooss, K., Yang, Y., Fisher, K. J., and Wilson, J. M. (1998). Transduction of dendritic cells by DNA viral vectors directs the immune response to transgene products in muscle fibers. *J Virol* **72**(5), 4212-23.
Kaplitt, M. G., Leone, P., Samulski, R. J., Xiao, X., Pfaff, D. W., O'Malley, K. L., and During, M. J. (1994). Long-term gene expression and phenotypic correction using adeno-associated virus vectors in the mammalian brain. *Nat Genet* **8**(2), 148-54.
Kearns, W. G., Afione, S. A., Fulmer, S. B., Pang, M. C., Erikson, D., Egan, M., Landrum, M. J., Flotte, T. R., and Cutting, G. R. (1996). Recombinant adeno-associated virus (AAV-CFTR) vectors do not integrate in a site-specific fashion in an immortalized epithelial cell line. *Gene Ther* **3**(9), 748-55.
Kessler, P. D., Podsakoff, G. M., Chen, X., McQuiston, S. A., Colosi, P. C., Matelis, L. A., Kurtzman, G. J., and Byrne, B. J. (1996). Gene delivery to skeletal muscle results in sustained expression and systemic delivery of a therapeutic protein. *Proc Natl Acad Sci U S A* **93**(24), 14082-7.
Koeberl, D. D., Bonham, L., Halbert, C. L., Allen, J. M., Birkebak, T., and Miller, A. D. (1999). Persistent, therapeutically relevant levels of human granulocyte colony-stimulating factor in mice after systemic delivery of adeno-associated virus vectors. *Hum Gene Ther* **10**(13), 2133-40.
Kotin, R. M., Linden, R. M., and Berns, K. I. (1992). Characterization of a preferred site on human chromosome 19q for integration of adeno-associated virus DNA by non-homologous recombination. *Embo J* **11**(13), 5071-8.
Kotin, R. M., Menninger, J. C., Ward, D. C., and Berns, K. I. (1991). Mapping and direct visualization of a region-specific viral DNA integration site on chromosome 19q13-qter. *Genomics* **10**(3), 831-4.
Lalwani, A., Walsh, B., Reilly, P., Carvalho, G., Zolotukhin, S., Muzyczka, N., and Mhatre, A. (1998). Long-term in vivo cochlear transgene expression mediated by recombinant adeno-associated virus. *Gene Ther* **5**(2), 277-81.
Lamartina, S., Roscilli, G., Rinaudo, D., Delmastro, P., and Toniatti, C. (1998). Lipofection of Purified Adeno-Associated Virus Rep68 Protein: toward a Chromosome-Targeting Nonviral Particle. *J Virol* **72**(9), 7653-8.

Lewin, A. S., Drenser, K. A., Hauswirth, W. W., Nishikawa, S., Yasumura, D., Flannery, J. G., and LaVail, M. M. (1998). Ribozyme rescue of photoreceptor cells in a transgenic rat model of autosomal dominant retinitis pigmentosa. *Nat Med* **4**(8), 967-71.

Li, J., Samulski, R. J., and Xiao, X. (1997). Role for highly regulated rep gene expression in adeno-associated virus vector production. *J Virol* **71**(7), 5236-43.

Linden, R. M., Ward, P., Giraud, C., Winocour, E., and Berns, K. I. (1996). Site-specific integration by adeno-associated virus. *Proc Natl Acad Sci U S A* **93**(21), 11288-94.

Linden, R. M., Winocour, E., and Berns, K. I. (1996). The recombination signals for adeno-associated virus site-specific integration. *Proc Natl Acad Sci U S A* **93**(15), 7966-72.

Liu, X. L., Clark, K. R., and Johnson, P. R. (1999). Production of recombinant adeno-associated virus vectors using a packaging cell line and a hybrid recombinant adenovirus. *Gene Ther* **6**(2), 293-9.

Lo, W. D., Qu, G., Sferra, T. J., Clark, R., Chen, R., and Johnson, P. R. (1999). Adeno-associated virus-mediated gene transfer to the brain: duration and modulation of expression. *Hum Gene Ther* **10**(2), 201-13.

Mah, C., Qing, K., Khuntirat, B., Ponnazhagan, S., Wang, X. S., Kube, D. M., Yoder, M. C., and Srivastava, A. (1998). Adeno-associated virus type 2-mediated gene transfer: role of epidermal growth factor receptor protein tyrosine kinase in transgene expression. *J Virol* **72**(12), 9835-43.

Manning, W. C., Paliard, X., Zhou, S., Pat, B. M., Lee, A. Y., Hong, K., Walker, C. M., Escobedo, J. A., and Dwarki, V. (1997). Genetic immunization with adeno-associated virus vectors expressing herpes simplex virus type 2 glycoproteins B and D. *J Virol* **71**(10), 7960-2.

Manning, W. C., Zhou, S., Bland, M. P., Escobedo, J. A., and Dwarki, V. (1998). Transient immunosuppression allows transgene expression following readministration of adeno-associated viral vectors. *Hum Gene Ther* **9**(4), 477-85.

Matsushita, T., Elliger, S., Elliger, C., Podsakoff, G., Villarreal, L., Kurtzman, G. J., Iwaki, Y., and Colosi, P. (1998). Adeno-associated virus vectors can be efficiently produced without helper virus. *Gene Ther* **5**(7), 938-45.

McPherson, R. A., Rosenthal, L. J., and Rose, J. A. (1985). Human cytomegalovirus completely helps adeno-associated virus replication. *Virology* **147**(1), 217-22.

Miao, C. H., Snyder, R. O., Schowalter, D. B., Patijn, G. A., Donahue, B., Winther, B., and Kay, M. A. (1998). The kinetics of rAAV integration in the liver. *Nat Genet* **19**(1), 13-5.

Mizuno, M., Yoshida, J., Colosi, P., and Kurtzman, G. (1998). Adeno-associated virus vector containing the herpes simplex virus thymidine kinase gene causes complete regression of intracerebrally implanted human gliomas in mice, in conjunction with ganciclovir administration. *Jpn J Cancer Res* **89**(1), 76-80.

Moskalenko, M., Chen, L., van Roey, M., Donahue, B. A., Snyder, R. O., McArthur, J. G., and Patel, S. D. (2000). Epitope Mapping of Human Anti-Adeno-Associated Virus Type 2 Neutralizing Antibodies: Implications for Gene therapy and Virus Structure. *J Virol* **74**(4), 1761-1766.

Nakai, H., Herzog, R. W., Hagstrom, J. N., Walter, J., Kung, S. H., Yang, E. Y., Tai, S. J., Iwaki, Y., Kurtzman, G. J., Fisher, K. J., Colosi, P., Couto, L. B., and High, K. A. (1998). Adeno-associated viral vector-mediated gene transfer of human blood coagulation factor IX into mouse liver. *Blood* **91**(12), 4600-7.

Nakai, H., Iwaki, Y., Kay, M. A., and Couto, L. B. (1999). Isolation of recombinant adeno-associated virus vector-cellular DNA junctions from mouse liver. *J Virol* **73**(7), 5438-47.

Ni, T. H., McDonald, W. F., Zolotukhin, I., Melendy, T., Waga, S., Stillman, B., and Muzyczka, N. (1998). Cellular proteins required for adeno-associated virus DNA replication in the absence of adenovirus coinfection. *J Virol* **72**(4), 2777-87.

Ni, T. H., Zhou, X., McCarty, D. M., Zolotukhin, I., and Muzyczka, N. (1994). In vitro replication of adeno-associated virus DNA. *J Virol* **68**(2), 1128-38.

Okada, H., Miyamura, K., Itoh, T., Hagiwara, M., Wakabayashi, T., Mizuno, M., Colosi, P., Kurtzman, G., and Yoshida, J. (1996). Gene therapy against an experimental glioma using adeno-associated virus vectors. *Gene Ther* **3**(11), 957-64.

Peel, A. L., Zolotukhin, S., Schrimsher, G. W., Muzyczka, N., and Reier, P. J. (1997). Efficient transduction of green fluorescent protein in spinal cord neurons using adeno-associated virus vectors containing cell type-specific promoters. *Gene Ther* **4**(1), 16-24.

Pereira, D. J., McCarty, D. M., and Muzyczka, N. (1997). The adeno-associated virus (AAV) Rep protein acts as both a repressor and an activator to regulate AAV transcription during a productive infection. *J Virol* **71**(2), 1079-88.

Pieroni, L., Fipaldini, C., Monciotti, A., Cimini, D., Sgura, A., Fattori, E., Epifano, O., Cortese, R., Palombo, F., and La Monica, N. (1998). Targeted integration of adeno-associated virus-derived plasmids in transfected human cells. *Virology* **249**(2), 249-59.

Ponnazhagan, S., Erikson, D., Kearns, W. G., Zhou, S. Z., Nahreini, P., Wang, X. S., and Srivastava, A. (1997). Lack of site-specific integration of the recombinant adeno-associated virus 2 genomes in human cells. *Hum Gene Ther* **8**(3), 275-84.

Qing, K., Khuntirat, B., Mah, C., Kube, D. M., Wang, X. S., Ponnazhagan, S., Zhou, S., Dwarki, V. J., Yoder, M. C., and Srivastava, A. (1998). Adeno-associated virus type 2-mediated gene transfer: correlation of tyrosine phosphorylation of the cellular single-stranded D sequence- binding protein with transgene expression in human cells in vitro and murine tissues in vivo. *J Virol* **72**(2), 1593-9.

Qing, K., Mah, C., Hansen, J., Zhou, S., Dwarki, V., and Srivastava, A. (1999). Human fibroblast growth factor receptor 1 is a co-receptor for infection by adeno-associated virus 2. *Nat Med* **5**(1), 71-7.

Rendahl, K. G., Leff, S. E., Otten, G. R., Spratt, S. K., Bohl, D., Roey, M. V., Donahue, B. A., Cohen, L. K., Mandel, R. J., Danos, O., and Snyder, R. O. (1998). Regulation of gene expression in vivo following transduction by two separate rAAV vectors. *Nat Biotechnol* **16**(8), 757-61.

Richardson, W. D., and Westphal, H. (1984). Requirement for either early region 1a or early region 1b adenovirus gene products in the helper effect for adeno-associated virus. *J Virol* **51**(2), 404-10.

Rivera, V. M., Ye, X., Courage, N. L., Sachar, J., Cerasoli, F., Jr., Wilson, J. M., and Gilman, M. (1999). Long-term regulated expression of growth hormone in mice after intramuscular gene transfer. *Proc Natl Acad Sci U S A* **96**(15), 8657-62.

Rizzuto, G., Gorgoni, B., Cappelletti, M., Lazzaro, D., Gloaguen, I., Poli, V., Sgura, A., Cimini, D., Ciliberto, G., Cortese, R., Fattori, E., and La Monica, N. (1999). Development of animal models for adeno-associated virus site-specific integration. *J Virol* **73**(3), 2517-26.

Rubenstein, R. C., McVeigh, U., Flotte, T. R., Guggino, W. B., and Zeitlin, P. L. (1997). CFTR gene transduction in neonatal rabbits using an adeno-associated virus (AAV) vector. *Gene Ther* **4**(5), 384-92.

Russell, D. W., Alexander, I. E., and Miller, A. D. (1995). DNA synthesis and topoisomerase inhibitors increase transduction by adeno-associated virus vectors. *Proc Natl Acad Sci U S A* **92**(12), 5719-23.

Russell, D. W., and Hirata, R. K. (1998). Human gene targeting by viral vectors. *Nat Genet* **18**(4), 325-30.

Russell, D. W., Miller, A. D., and Alexander, I. E. (1994). Adeno-associated virus vectors preferentially transduce cells in S phase. *Proc Natl Acad Sci U S A* **91**(19), 8915-9.

Rutledge, E. A., Halbert, C. L., and Russell, D. W. (1998). Infectious clones and vectors derived from adeno-associated virus (AAV) serotypes other than AAV type 2. *J Virol* **72**(1), 309-19.

Salvetti, A., Oreve, S., Chadeuf, G., Favre, D., Cherel, Y., Champion, A. P., David, A. J., and Moullier, P. (1998). Factors influencing recombinant adeno-associated virus production. *Hum Gene Ther* **9**(5), 695-706.

Samulski, R. J., Berns, K. I., Tan, M., and Muzyczka, N. (1982). Cloning of adeno-associated virus into pBR322: rescue of intact virus from the recombinant plasmid in human cells. *Proc Natl Acad Sci U S A* **79**(6), 2077-81.

Samulski, R. J., Chang, L. S., and Shenk, T. (1987). A recombinant plasmid from which an infectious adeno-associated virus genome can be excised in vitro and its use to study viral replication. *J Virol* **61**(10), 3096-101.

Samulski, R. J., Zhu, X., Xiao, X., Brook, J. D., Housman, D. E., Epstein, N., and Hunter, L. A. (1991). Targeted integration of adeno-associated virus (AAV) into human chromosome 19. *Embo J* **10**(12), 3941-50.

Sanlioglu, S., Duan, D., and Engelhardt, J. F. (1999). Two independent molecular pathways for recombinant adeno-associated virus genome conversion occur after UV-C and E4orf6 augmentation of transduction. *Hum Gene Ther* **10**(4), 591-602.

Schimmenti, S., Boesen, J., Claassen, E. A., Valerio, D., and Einerhand, M. P. (1998). Long-term genetic modification of rhesus monkey hematopoietic cells following transplantation of adenoassociated virus vector-transduced CD34+ cells. *Hum Gene Ther* **9**(18), 2727-34.

Schlehofer, J. R. (1994). The tumor suppressive properties of adeno-associated viruses. *Mutat Res* **305**(2), 303-13.

Schlehofer, J. R., Ehrbar, M., and zur Hausen, H. (1986). Vaccinia virus, herpes simplex virus, and carcinogens induce DNA amplification in a human cell line and support replication of a helpervirus dependent parvovirus. *Virology* **152**(1), 110-7.

Shieh, M. T., WuDunn, D., Montgomery, R. I., Esko, J. D., and Spear, P. G. (1992). Cell surface receptors for herpes simplex virus are heparan sulfate proteoglycans. *J Cell Biol* **116**(5), 1273-81.

Snyder, R. O., Xiao X., and Samulski, R. J. (1996). Production of Recombinant Adeno-Associated Viral Vectors. *In* "Current Protocols in Human Genetics" (J. H. N. Dracopoli, B. Krof, D. Moir, C. Morton, C. Seidman, J. Seidman, and D. Smith, Ed.), pp. 12.1.1-24. John Wiley and Sons Publisher, New York.

Snyder, R. O., Miao, C., Meuse, L., Tubb, J., Donahue, B. A., Lin, H. F., Stafford, D. W., Patel, S., Thompson, A. R., Nichols, T., Read, M. S., Bellinger, D. A., Brinkhous, K. M., and Kay, M. A. (1999).

Correction of hemophilia B in canine and murine models using recombinant adeno-associated viral vectors. *Nat Med* **5**(1), 64-70.
Snyder, R. O., Miao, C. H., Patijn, G. A., Spratt, S. K., Danos, O., Nagy, D., Gown, A. M., Winther, B., Meuse, L., Cohen, L. K., Thompson, A. R., and Kay, M. A. (1997a). Persistent and therapeutic concentrations of human factor IX in mice after hepatic gene transfer of recombinant AAV vectors. *Nat Genet* **16**(3), 270-6.
Snyder, R. O., Spratt, S. K., Lagarde, C., Bohl, D., Kaspar, B., Sloan, B., Cohen, L. K., and Danos, O. (1997b). Efficient and stable adeno-associated virus-mediated transduction in the skeletal muscle of adult immunocompetent mice. *Hum Gene Ther* **8**(16), 1891-900.
Srivastava, A., Lusby, E. W., and Berns, K. I. (1983). Nucleotide sequence and organization of the adeno-associated virus 2 genome. *J Virol* **45**(2), 555-64.
Su, H., Lu, R., Chang, J. C., and Kan, Y. W. (1997). Tissue-specific expression of herpes simplex virus thymidine kinase gene delivered by adeno-associated virus inhibits the growth of human hepatocellular carcinoma in athymic mice. *Proc Natl Acad Sci U S A* **94**(25), 13891-6.
Sugiyama, A., Hattori, S., Tanaka, S., Isoda, F., Kleopoulos, S., Rosenfeld, M., Kaplitt, M., Sekihara, H., and Mobbs, C. (1997). Defective adenoassociated viral-mediated transfection of insulin gene by direct injection into liver parenchyma decreases blood glucose of diabetic mice. *Horm Metab Res* **29**(12), 599-603.
Summerford, C., Bartlett, J. S., and Samulski, R. J. (1999). AlphaVbeta5 integrin: a co-receptor for adeno-associated virus type 2 infection. *Nat Med* **5**(1), 78-82.
Summerford, C., and Samulski, R. J. (1998). Membrane-associated heparan sulfate proteoglycan is a receptor for adeno-associated virus type 2 virions. *J Virol* **72**(2), 1438-45.
Surosky, R. T., Urabe, M., Godwin, S. G., McQuiston, S. A., Kurtzman, G. J., Ozawa, K., and Natsoulis, G. (1997). Adeno-associated virus Rep proteins target DNA sequences to a unique locus in the human genome. *J Virol* **71**(10), 7951-9.
Tamayose, K., Hirai, Y., and Shimada, T. (1996). A new strategy for large-scale preparation of high-titer recombinant adeno-associated virus vectors by using packaging cell lines and sulfonated cellulose column chromatography. *Hum Gene Ther* **7**(4), 507-13.
Tenenbaum, L., Hamdane, M., Pouzet, M., Avalosse, B., Stathopoulos, A., Jurysta, F., Rosenbaum, C., Hanemann, C., Levivier, M., and Velu, T. (1999). Cellular contaminants of adeno-associated virus vector stocks can enhance transduction. *Gene Ther* **6**(6), 1045-1053.
Tobiasch, E., Burguete, T., Klein, B. P., Heilbronn, R., and Schlehofer, J. R. (1998). Discrimination between different types of human adeno-associated viruses in clinical samples by PCR. *J Virol Methods* **71**(1), 17-25.
Tratschin, J. D., Miller, I. L., and Carter, B. J. (1984). Genetic analysis of adeno-associated virus: properties of deletion mutants constructed in vitro and evidence for an adeno-associated virus replication function. *J Virol* **51**(3), 611-9.
Tratschin, J. D., Miller, I. L., Smith, M. G., and Carter, B. J. (1985). Adeno-associated virus vector for high-frequency integration, expression, and rescue of genes in mammalian cells. *Mol Cell Biol* **5**(11), 3251-60.
Vincent, K. A., Piraino, S. T., and Wadsworth, S. C. (1997). Analysis of recombinant adeno-associated virus packaging and requirements for rep and cap gene products. *J Virol* **71**(3), 1897-905.
Wagner, J. A., Moran, M. L., Messner, A. H., Daifuku, R., Conrad, C. K., Reynolds, T., Guggino, W. B., Moss, R. B., Carter, B. J., Wine, J. J., Flotte, T. R., and Gardner, P. (1998a). A phase I/II study of tgAAV-CF for the treatment of chronic sinusitis in patients with cystic fibrosis. *Hum Gene Ther* **9**(6), 889-909.
Wagner, J. A., Reynolds, T., Moran, M. L., Moss, R. B., Wine, J. J., Flotte, T. R., and Gardner, P. (1998b). Efficient and persistent gene transfer of AAV-CFTR in maxillary sinus. *Lancet* **351**(9117), 1702-3.
Walker, S. L., Wonderling, R. S., and Owens, R. A. (1997). Mutational analysis of the adeno-associated virus type 2 Rep68 protein helicase motifs. *J Virol* **71**(9), 6996-7004.
Wang, D., Fischer, H., Zhang, L., Fan, P., Ding, R. X., and Dong, J. (1999). Efficient CFTR expression from AAV vectors packaged with promoters--the second generation. *Gene Ther* **6**(4), 667-75.
Watson, G. L., Sayles, J. N., Chen, C., Elliger, S. S., Elliger, C. A., Raju, N. R., Kurtzman, G. J., and Podsakoff, G. M. (1998). Treatment of lysosomal storage disease in MPS VII mice using a recombinant adeno-associated virus. *Gene Ther* **5**(12), 1642-9.
Weindler, F. W., and Heilbronn, R. (1991). A subset of herpes simplex virus replication genes provides helper functions for productive adeno-associated virus replication. *J Virol* **65**(5), 2476-83.
Weitzman, M. D., Kyostio, S. R., Kotin, R. M., and Owens, R. A. (1994). Adeno-associated virus (AAV) Rep proteins mediate complex formation between AAV DNA and its integration site in human DNA. *Proc Natl Acad Sci U S A* **91**(13), 5808-12.
West, M. H., Trempe, J. P., Tratschin, J. D., and Carter, B. J. (1987). Gene expression in adeno-associated virus vectors: the effects of chimeric mRNA structure, helper virus, and adenovirus VA1 RNA. *Virology* **160**(1), 38-47.

Wickham, T. J., Filardo, E. J., Cheresh, D. A., and Nemerow, G. R. (1994). Integrin alpha v beta 5 selectively promotes adenovirus mediated cell membrane permeabilization. *J Cell Biol* **127**(1), 257-64.
Wickham, T. J., Mathias, P., Cheresh, D. A., and Nemerow, G. R. (1993). Integrins alpha v beta 3 and alpha v beta 5 promote adenovirus internalization but not virus attachment. *Cell* **73**(2), 309-19.
Wistuba, A., Kern, A., Weger, S., Grimm, D., and Kleinschmidt, J. A. (1997). Subcellular compartmentalization of adeno-associated virus type 2 assembly. *J Virol* **71**(2), 1341-1352.
Wu, P., de, F. C., Millard, W. J., King, M. A., Wang, S., Bryant, S. O., Gao, Y. P., Martin, E. J., and Meyer, E. M. (1996). An AAV promoter-driven neuropeptide Y gene delivery system using Sendai virosomes for neurons and rat brain. *Gene Ther* **3**(3), 246-53.
Xiao, W., Chirmule, N., Berta, S. C., McCullough, B., Gao, G., and Wilson, J. M. (1999). Gene therapy vectors based on adeno-associated virus type 1. *J Virol* **73**(5), 3994-4003.
Xiao, X., Li, J., and Samulski, R. J. (1996). Efficient long-term gene transfer into muscle tissue of immunocompetent mice by adeno-associated virus vector. *J Virol* **70**(11), 8098-108.
Xiao, X., Li, J., and Samulski, R. J. (1998). Production of high-titer recombinant adeno-associated virus vectors in the absence of helper adenovirus. *J Virol* **72**(3), 2224-32.
Xiao, X., Li, J., Tsao, Y. P., Dressman, D., Hoffman, E. P., and Watchko, J. F. (2000). Full functional rescue of a complete muscle (TA) in dystrophic hamsters by adeno-associated virus vector-directed gene therapy. *J Virol* **74**(3), 1436-42.
Yang, C. C., Xiao, X., Zhu, X., Ansardi, D. C., Epstein, N. D., Frey, M. R., Matera, A. G., and Samulski, R. J. (1997). Cellular recombination pathways and viral terminal repeat hairpin structures are sufficient for adeno-associated virus integration in vivo and in vitro. *J Virol* **71**(12), 9231-47.
Yang, Q., Mamounas, M., Yu, G., Kennedy, S., Leaker, B., Merson, J., Wong-Staal, F., Yu, M., and Barber, J. R. (1998). Development of novel cell surface CD34-targeted recombinant adenoassociated virus vectors for gene therapy. *Hum Gene Ther* **9**(13), 1929-37.
Ye, X., Rivera, V. M., Zoltick, P., Cerasoli, F., Jr., Schnell, M. A., Gao, G., Hughes, J. V., Gilman, M., and Wilson, J. M. (1999). Regulated delivery of therapeutic proteins after in vivo somatic cell gene transfer. *Science* **283**(5398), 88-91.
Zhang, L., Wang, D., Fischer, H., Fan, P. D., Widdicombe, J. H., Kan, Y. W., and Dong, J. Y. (1998). Efficient expression of CFTR function with adeno-associated virus vectors that carry shortened CFTR genes. *Proc Natl Acad Sci U S A* **95**(17), 10158-63.
Zolotukhin, S., Byrne, B., Mason, E., Zolotukhin, I., Potter, M., Chesnut, K., Summerford, C., Samulski, R., and Muzyczka, N. (1999). Recombinant adeno-associated virus purification using novel methods improves infectious titer and yield. *Gene Ther* **6**(6), 973-985.

ALPHAVIRUS VECTORS: HIGHLY EFFICIENT SYSTEMS FOR TRANSIENT GENE EXPRESSION

C. SMERDOU [1] and P. LILJESTRÖM [1,2]
[1] *Microbiology and Tumor Biology Center, Karolinska Institutet, Box 280, S-17177 Stockholm, Sweden.*
[2] *Department of Vaccine Research, Swedish Institute for Infectious Disease Control, S-17182 Solna, Sweden.*

Abstract

Alphaviruses are enveloped viruses containing a single positive strand RNA molecule as genome. Expression vectors based on alphaviruses have been developed from Sindbis virus (SIN), Semliki Forest virus (SFV), and Venezuelan Equine Encephalitis virus (VEE). Alphavirus self-replicating RNA containing heterologous genes can be synthesized *in vitro* from plasmids in which the recombinant alphavirus replicon cDNA sequences are under the control of a prokaryotic promoter. When this self-amplifying RNA is transfected into cells a high level expression of the heterologous proteins/RNA is obtained. The alphavirus vectors can be delivered directly as naked RNA, as naked DNA and as viral particles. In order to encapsidate the alphaviral recombinant RNA into particles, packaging systems have been developed which supply the structural proteins of the virus in *trans*. Delivery of alphavirus as naked DNA is based on DNA/RNA layered vectors in which the recombinant replicon is transcribed from an RNA polymerase-II promoter in the nucleus of the transfected cell. Alphavirus vectors have been used to express proteins with many different purposes, which include large-scale production, characterization, functional studies and virus assembly. The ability of alphavirus to infect neurons has opened many possibilities for the study of protein function in these cells. In addition, alphavirus vectors can also be used for gene delivery *in vivo* with potential applications in vaccination and gene therapy, although some improvements in the present technology are necessary for this latter purpose. Alphavirus sequences have also been used to engineer cell lines for diagnostic virus detection.

1. Introduction

Alphaviruses are enveloped positive strand RNA viruses with a world-wide distribution. They have been classified in three main groups according to their sequence homology (for a review see Strauss and Strauss, 1994) where Sindbis virus (SIN), Semliki Forest

M. Al-Rubeai (ed.), Cell Engineering, 182–210.

virus (SFV) and Venezuelan Equine Encephalitis virus (VEE) are the main representative members of each group. Alphaviruses can infect many animal hosts such as birds and mammals, including humans. They are arthropod borne, being usually transmitted by mosquitoes. SIN and SFV are considered avirulent to humans, while VEE is pathogenic. Alphaviruses present several features that make them good candidates for the development of expression systems: i) a very broad host range; ii) high expression levels of the structural proteins in infected cells, a mechanism which can been used for expression of heterologous proteins at similar levels; iii) a relatively small genome (around 11-12 kb) from which a cDNA can be synthesized and easily manipulated *in vitro*, and iv) the fact that the naked viral genomic RNA is infectious and can give rise to a productive infection when transfected into cells. Expression vectors have been developed for SIN, SFV and VEE (Liljeström and Garoff, 1991; Pushko et al., 1997; Xiong et al., 1989). These vectors can be used for delivery of genes into cells as recombinant viral particles, naked RNA or DNA. In the latter case the replicon vector is transcribed in the nucleus of the transfected cell from an RNA polymerase II promoter. These delivery systems and their applications will be discussed in this chapter.

2. The biology of alphaviruses

Alphavirus particles contain a single positive strand RNA molecule as genome. The genomic RNA associates with 240 subunits of the capsid protein to form the nucleocapsid, which is surrounded by a lipid bilayer containing 240 heterodimers of the spike proteins E1 and E2 (Strauss and Strauss, 1994). Upon entry of the virus into the cell the genomic RNA is released into the cytoplasm where the viral replication takes place. The genomic RNA, which is capped and polyadenylated, can function as a mRNA. In the first place the 5' two-thirds of the genomic RNA is translated to produce the non-structural proteins (nsp) 1-4 which constitute the viral replicase (Figure 1). One of the functions of this enzyme is to amplify the genomic RNA by copying it into negative strands, which can serve as templates for the production of new positive strand RNA genomes. The replicase has also a second function that drives the synthesis of a subgenomic RNA from an internal promoter on the minus strand RNA (subgenomic promoter). This subgenomic RNA has the same sequence as the last third of the virus genome and encodes the structural proteins of the virus. These are synthesized as a polyprotein precursor (NH_2-C-p62-6K-E1-COOH) which is cotranslationally cleaved by the capsid proteinase and signal peptidase to produce the capsid protein (C) and the envelope proteins p62, 6K and E1. In SFV and SIN, a sequence at the 5' end of the capsid gene has the property of functioning as a translational enhancer, significantly increasing the expression level of the structural proteins (Frolov and Schlesinger, 1994a; Frolov and Schlesinger, 1996; Sjöberg and Garoff, 1996; Sjöberg et al., 1994). In the infected cell nucleocapsids are generated in the cytoplasm by the interaction of the C protein with the amplified viral genomes. The spike proteins are translocated to the endoplasmic reticulum (ER), where dimerization of p62 and E1 takes place. The p62/E1 heterodimers are routed to the plasma membrane and budding occurs at this location through the interaction between nucleocapsids and spikes (Lee et al., 1996; Lopez et al.,

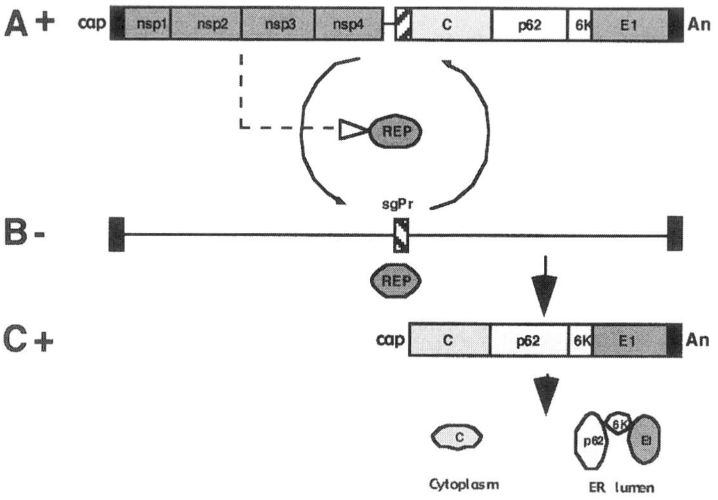

Figure 1. Alphavirus replication. The alphavirus genome is a positive strand RNA molecule of approximately 11-12 kb (A). The viral replicase (REP) is encoded by the nonstructural protein genes (nsp 1-4) located at the 5'-end of the genome. In the infected/transfected cell REP is translated in the cytoplasm from the genomic RNA. This enzyme recognizes specific sequences at the 3' end of the genomic RNA and synthesizes a complementary strand of negative polarity (B). This RNA will also be copied by REP to produce more positive genomic RNA that can be encapsidated into viral particles. The replicase also recognizes a subgenomic promoter (sg Pr) sequence in the negative strand RNA from which a smaller positive subgenomic RNA is synthesized (C). This subgenomic RNA will be translated into a polypeptide that is processed into the different structural proteins of the virus (capsid, p62, 6K and E1). The capsid (c) is synthesized in the cytoplam, while p62, E1, and 6K are translocated to the endoplasmic reticulum (ER). The black squares at the 5' and 3' ends of the RNAs are sequences necessary for RNA replication; An, polyA.

1994; Skoging et al., 1996; Suomalainen et al., 1992; Zhao et al., 1994). During transport to the cell surface p62 is cleaved to the mature form E2 by a host cell protease. This cleavage is necessary for the infectivity of the particles (Jain et al., 1991; Lobigs and Garoff, 1990; Salminen et al., 1992). Another important feature of the alphaviruses is that a few hours after infection a strong shut-off of host cell protein synthesis is induced with the result that only the structural proteins of the virus are synthesized. These can constitute up to 50% of the total protein content in an infected cell. Alphavirus infection also induces a cytopathic effect leading to cell death by apoptosis in 48-72 hours.

3. Expression systems based on alphaviruses

In order to develop an alphavirus expression system the viral genomic RNA was first cloned as a cDNA in a plasmid downstream of a prokaryotic promoter, such as SP6 or T7. From this kind of plasmid the alphaviral genomic RNA can be transcribed *in vitro* using SP6/T7 polymerase. The RNA obtained can be transfected into cells by different methods that most commonly include the use of cationic lipids or electroporation. With the latter method transfection efficiencies of 100% are routinely achieved (Liljeström et al., 1991). Alphavirus genomic cDNA clones have been synthesized up to date for SFV (Liljeström et al., 1991), SIN (Rice et al., 1987), VEE (Davis et al., 1989), and Ross River virus (RRV) (Kuhn et al., 1991). *In vitro* manipulation of these cDNA clones has allowed the study a number of different viral functions, which include the localization of the minimal replication and packaging sequences. Two different types of vectors have been developed to express foreign proteins in the context of an alphavirus genome. In the first one, the so-called replication-packaging proficient vectors, the expression of the foreign gene is achieved by the inclusion of a second subgenomic promoter followed by the heterologous sequence in a nonessential region of the genome (Figure 2A and 2B). The extra-sequences can be inserted 3' of the structural genes or at the end of the replicase genes (before the viral subgenomic promoter). Transfection of cells with this type of vector results in the expression of both the heterologous protein and the viral structural proteins. The latter are able to package the recombinant genomic RNA into viral particles that can be further propagated. This approach has been used to construct SIN (Hahn et al., 1992; Raju and Huang, 1991) and VEE vectors (Caley et al., 1999; Davis et al., 1996). These vectors are very efficient to express foreign proteins and easy to produce, but show in general two types of problems. The first one is related to their stability, which is reduced after several passages, probably due to restrictions in the packaging capacity of the viral particles. The second problem is associated with their infectious nature, which for safety reasons will limit their possible use in vaccination. Replication-packaging proficient vectors have also been used to express short epitopes on the surface of the viral proteins E2 or E3, using permissive sites in these proteins where short sequences can be inserted without disrupting their folding and functionality (London et al., 1992).

The second type alphavirus expression vectors are the so-called suicidal vectors. In these vectors the viral structural genes are replaced by the foreign gene, which is placed downstream of the subgenomic promoter (Figure 2C-2E). When RNA transcribed from this type of vector is transfected into cells only the replicase and the heterologous protein or RNA will be synthesized (Figure 3A). In this system no viral particles can be produced due to the lack of the structural proteins. These vectors have been developed for SIN (Xiong et al., 1989), SFV (Liljeström and Garoff, 1991), and VEE (Pushko et al., 1997). The expression level that can be obtained with alphavirus vectors has been reported as 60-100 pg protein/cell when *LacZ* was used as a reporter gene (Berglund et al., 1998; Bredenbeek et al., 1993; Liljeström and Garoff, 1991). However, by making an in-frame fusion of the gene of interest at the 3' end of the viral capsid gene the level of expression can be increased around ten-fold (Figure 2F and 2G) (Berglund et al., 1998; Frolov and Schlesinger, 1994a; Sjöberg et al., 1994). It has been shown that the 5'end of the capsid gene is enough to provide this translation enhancing effect. This region includes the first 102 nucleotides of the SFV capsid gene (Sjöberg et al., 1994) or the first 226 nt of the SIN capsid gene (Frolov and Schlesinger, 1994a), respectively. No similar translation enhancer has yet been identified in VEE. A model to explain the mechanism of action of the translation enhancer has been proposed and it is based on the formation of a stable hairpin structure in this area. This secondary structure could slow

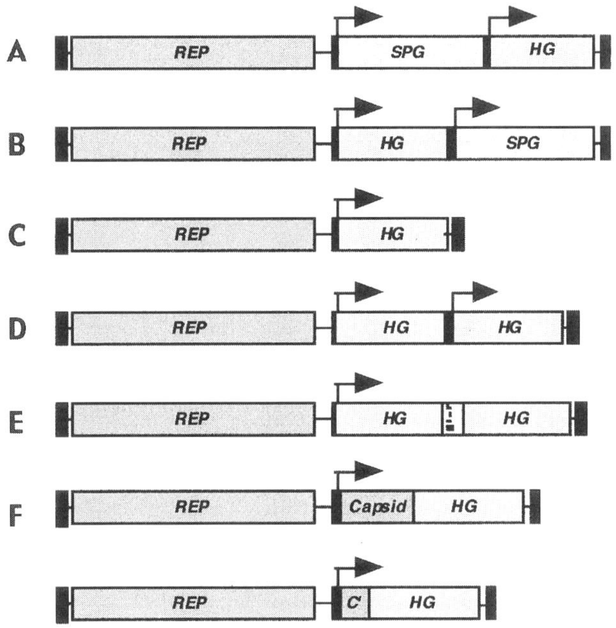

Figure 2. Alphavirus vectors. A and B, alphavirus replication–packaging proficient vectors with a second subgenomic promoter located 3'or 5' of the structural proteins genes, respectively; C-E, alphavirus suicidal vectors containing one or two heterologous genes; F and G, enhanced alphavirus suicidal vectors containing a fusion of the heterologous gene with the whole viral capsid gene or the 5' translation enhancer region of this gene, respectively. REP, genes coding for the viral replicase; SPG, genes coding for the viral structural proteins; HG, heterologous gene; IRES, internal ribosome binding site; C', 5' end of the capsid gene containing the translation enhancer; black squares with horizontal arrows, subgenomic promoters; black squares at the ends of the vectors, 5' and 3' sequences necessary for vector RNA replication.

down the movement of the ribosomes, allowing a more efficient recruitment of low concentration translation factors (Frolov and Schlesinger, 1994a; Frolov and Schlesinger, 1996). This would have an optimal effect if the RNA containing the hairpin is very abundant in the cell, as it is the case with the subgenomic RNA in the transfected/infected cells. The fusion of the whole capsid gene with the foreign sequence can provide the translation enhancer effect and at the same time the self-cleaving activity of the carboxy-terminal domain of the capsid, usually resulting in the release of the foreign protein (Sjöberg et al., 1994). However, if only the minimal translation enhancer from the capsid is used, the foreign protein will be expressed as a fusion polypeptide with the amino-terminal end of the capsid. An alternative way to remove the capsid enhancer from the fusion protein is to use the sequence of the 2A autoprotease from foot-and-mouth disease virus (FMDV) as a linker. The 2A protease can cleave in *cis* at its carboxy-terminal end resulting in the release of the foreign protein containing only an additional proline residue at the amino-terminal end (Ryan and Drew, 1994; Ryan et al., 1991; Smerdou and Liljeström, 1999)

4. Production of suicidal alphavirus particles

In order to package recombinant alphavirus suicidal RNA into viral particles several strategies have been developed to provide the viral structural proteins in *trans*. The production of recombinant alphavirus particles is of great importance since they constitute a very efficient way to deliver the alphavirus vector RNA into the target cells. Packaging systems currently in use are based on the expression of the viral structural proteins from helper RNAs which can be cotransfected into cells with the vector RNA or intrinsically produced in stable cell lines.

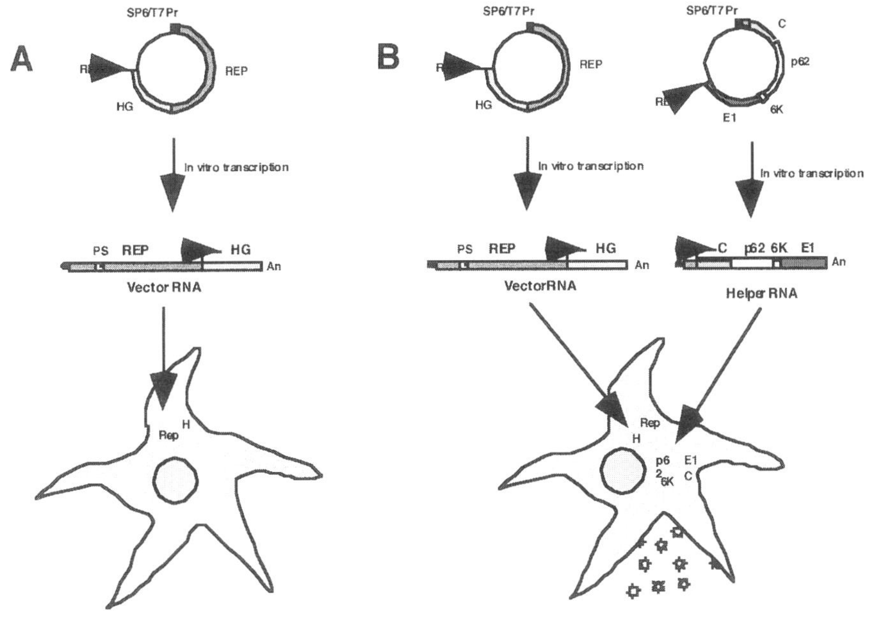

Figure 3. A. Transfection of cells with recombinant alphavirus RNA. The alphavirus suicidal vector RNA (vector RNA) can be synthesized by *in vitro* transcription from a plasmid in which the alphavirus sequences are under the control of a prokaryotic promoter (SP6 or T7). The vector RNA can be introduced into eukaryotic cells by lipofection or electroporation. Once inside the cell the replicase (Rep) will be made. This enzyme will amplify the vector RNA and also synthesize a small subgenomic RNA from which the heterologous gene product (H) will be translated at high levels. No viral particles are produced with this system, since the viral structural genes are absent. B. Packaging of alphavirus recombinant RNA into viral particles. The structural proteins of the virus can be provided in *trans* from a second RNA (helper RNA) that can also be synthesized *in vitro* from a plasmid containing the SP6 or T7 promoter. When cells are cotransfected with the vector RNA and the helper RNA, the replicase will be synthesized from the first one and amplify both RNA species. The replicase can also synthesize subgenomic RNAs from both vector and helper RNAs. The helper subgenomic RNA will be translated to produce the viral structural proteins: capsid (C), p62 (E2 precursor), 6K and E1. The specific interaction of the capsid protein with the packaging signal (PS) present in the vector RNA will result in the production of nucleocapsids that will bud at the cell surface by interacting with the viral spike proteins (E2, E1 and 6K). The viral particles produced in this way are able to infect cells but cannot spread since the genes coding for the structural proteins are not present in the encapsidated RNA. HG, heterologous gene; RE, restriction enzyme site used to linearize the plasmids previous to transcription; black circle at the left of the RNAs, CAP; An, polyA.

4.1 TRANSFECTION OF CELLS WITH ONE OR MORE HELPER RNAs

Alphavirus packaging systems were initially based on the cotransfection of cells with the alphaviral suicidal vector RNA and a single helper RNA containing the viral structural genes downstream of the subgenomic promoter (Figure 3B). This helper RNA also contains the 5'and 3' sequences necessary for replication, to allow its amplification by the viral replicase. However, most of the sequence coding for the replicase was deleted, including the packaging signals (Frolova et al., 1997; White et al., 1998). By cotransfecting cells with the vector RNA carrying the foreign gene and the helper RNA, the replicase encoded by the former will amplify both RNA species. In addition, the replicase will also make two subgenomic RNAs that will be translated to produce the foreign and structural proteins. At this stage the interaction of the capsid proteins with the packaging signal present only in the genomic RNA will give rise to the formation of nucleocapsids. These will bud at the cell surface by interacting with the spike proteins. The particles generated in this way are suicidal since they contain a genome that, while able to replicate, cannot result in a productive infection. This type of helper systems have been developed for SFV (Liljeström and Garoff, 1991), SIN (Bredenbeek et al., 1993), and VEE (Pushko et al., 1997) and allow production of high titer stocks of alphavirus recombinant particles (in the range of 10^9 particles per 10^6 transfected cells). In these packaging systems, however, recombination between the helper and vector RNAs can take place in the cotransfected cells, giving rise to the formation of wild type genomes that can be incorporated into viral particles and propagated. Recombination takes place at a relative high frequency in alphaviruses (Berglund et al., 1993; Haijou et al., 1996; Hill et al., 1997; Raju et al., 1995; Schlesinger and Weiss, 1994; Weiss and Schlesinger, 1991) and replication proficient virus (RPV) has been detected with the three alphavirus helper systems described here. The frequency of recombination between the vector and helper RNA has been estimated to be 10^{-6} or less (Berglund et al., 1993). Because VEE can cause severe human illness additional attenuating mutations were introduced in the E2 and E1 genes present in the VEE helper RNA. In this case any potential recombinant virus arising during the packaging reaction would have a high probability of being attenuated (Davis et al., 1996; Davis et al., 1991).

In order to reduce the amount of RPVs in the SFV packaging system a new helper RNA was developed by the introduction of a mutation abolishing intracellular cleavage of p62, rendering the particles conditionally infectious (helper-2) (Berglund et al., 1993). Particles generated with helper-2 are non-infectious but can be artificially "activated" by incubation with chymotrypsin. This packaging system increased considerably the biosafety of the SFV vectors, since no RPVs have been detected to date. However, recombination between the vector and the helper RNAs can still take place, although generation of RPVs would also require reversions or suppressor mutations able to regenerate a functional p62 cleavage site. Such mutations have been found, which either regenerated the p62 cleavage site or restored infectivity without cleavage of p62, probably through conformational changes of the protein (Tubulekas and Liljeström, 1998). New helpers were constructed containing additional nucleotide changes to reduce the frequency of reversion, but in all cases new suppressors were found in a different domain of the protein which led to extensive attenuation (Tubulekas and Liljeström,

1998). Although the probability of simultaneously having recombination and suppressor mutations in this system is very low it cannot be completely excluded.

A more recent strategy to improve the biosafety of the alphaviruses packaging systems is based on splitting the helper RNA into two independent molecules, increasing the number of recombination events necessary to generate wild type genomes. This strategy has been developed for SIN (Frolov et al., 1997), SFV (Smerdou and Liljeström, 1999) and VEE (Pushko et al., 1997), and is based on expressing the capsid and the spike proteins from two different helpers. In the split helper systems cells have to be transfected with three independent RNAs (the vector replicon and the two helper RNAs) and the generation of wild type genomes requires at least two independent recombination events between the three RNAs. In this way the probability of RPV generation is considerably reduced. The expression of the capsid protein from an independent helper RNA does not have any effect on its expression level. However, when the spikes are expressed without the capsid the translation enhancer effect provided by the capsid gene is lost, and the levels of expression considerably reduced (Smerdou and Liljeström, 1999). In order to solve this problem different strategies have been employed for SIN and SFV. In the SIN split helper system the spikes were fused with a heterologous capsid protein from RRV, in which the RNA binding domain had been deleted. The 5'end of the RRV capsid gene provided the translation enhancer effect, while the carboxy-terminal domain of the RRV capsid was able to cleave itself off the spike polypeptide (Frolov et al., 1997). With this system high titer stocks of SIN recombinant particles (1-2x10^9 particles/ml) were obtained without detection of RPVs. A minor drawback of this system is that the mutated RRV capsid can interact with the SIN spikes to form chimeric viral particles that are not completely devoid of RNA. The SFV split helper is based on the fusion of the spikes with the minimal enhancer sequence of the homologous capsid (the first 102 nucleotides of the SFV capsid gene). The cleavage between the translation enhancer and the spikes in this system is achieved by using the foot-and-mouth disease virus (FMDV) 2A protease as a linker (Smerdou and Liljeström, 1999). This protease has only 17 amino acids and cleaves in *cis* at its carboxy-terminal end (Ryan and Drew, 1994; Ryan et al., 1991). The 2A-protease cleavage activity is highly efficient in releasing the spikes proteins from the translation enhancer. High stock titers of recombinant SFV particles (10^9particles/ml) can be obtained with this system without RPV generation. In order to further increase the safety of the SFV split helper a mutation was introduced in the active site of the capsid protease gene in the helper coding for this protein (Smerdou and Liljeström, 1999). This mutation had no effect in the context of the split helper, but abolishes the self-cleaving activity of the capsid, necessary if a wild type genome is generated through recombination. A split helper system has also been developed for VEE, but in this case no translation enhancer was used to express the spike polyprotein at high level, since it does not seem to be necessary in the context of this virus. Also high titers of recombinant particles (10^8 particles/ml) were generated with the VEE split helper packaging system without detection of RPVs (Pushko et al., 1997).

4.2 PACKAGING CELL LINES

As an alternative to the packaging systems based on the cotransfection of cells with vector and helper RNAs, the latter can be constitutively produced in stable cell lines engineered for that purpose. These cell lines can express the viral structural proteins upon infection or transfection with an alphavirus replicon, being able to package it in recombinant particles. Such cell lines have been recently developed for SIN (Polo et al., 1999). They were obtained by stabely transforming cells with a plasmid that contains the SIN helper sequences downstream of an RNA polymerase-II promoter. In these cells the helper RNA is transcribed in the nucleus and transported to the cytoplasm in a constitutive way. However, synthesis of the structural proteins does not take place, since they must be translated from the subgenomic RNAs produced by the replicase. When the packaging cells are transfected with a SIN replicon the vector-encoded replicase can amplify both the vector and helper RNAs present in the cytoplasm and also synthesize the corresponding subgenomic RNAs. The structural proteins can be translated and package the recombinant vector RNA. The particles generated in this way can propagate in this cell line since all the cells can provide the structural proteins. Several packaging cell lines have been generated for SIN, which include the use of one or two helpers (Polo et al., 1999). In the latter case the split helper approach described above for SIN was used. Upon infection of this cell line at a low multiplicity of infection (moi 0.2) recombinant particles were produced with titers of $0.5\text{-}1 \times 10^7$ particles/ml 48 hours post infection, and RPVs could not be found. Higher particle titers were obtained in packaging cell lines containing a single helper expressing all the structural proteins, but this was accompanied by the presence of RPVs. These packaging cell lines were also able to package SFV replicons with similar efficiency.

5. Alphavirus vectors can also be delivered as naked DNA

An alternative way to transfection of cells with alphavirus self-replicating RNA is the use of alphavirus DNA based vectors. Such vectors are based on plasmids in which the alphavirus replicon sequence is placed downstream of a eukaryotic promoter. When this type of plasmid is transfected into a cell, the vector RNA will be transcribed in the nucleus by the host cell RNA polymerase II and then transported to the cytoplasm, where the replicase is produced by translation of the first part of the RNA. This replicase subsequently initiates the amplification of the vector RNA, and the synthesis of the subgenomic RNA from which the foreign gene will be expressed (Figure 4). Several SIN and SFV DNA vectors have been described. One SIN DNA vector was constructed by placing the SIN replicon sequence downstream of the Rous sarcoma virus (RSV) LTR promoter (Herweijer et al., 1995). When the luciferase gene was used as a reporter gene in this vector a level of expression 10- to 240-fold higher than with a conventional plasmid was detected in rat myotubes transfected *in vitro*. The vector could also be used *in vivo*, as shown by expression of luciferase in mouse quadriceps. In contrast to conventional plasmids the expression level in the injected tissue decreased slowly and became undetectable at day 16 post injection, which underlies the transient nature of the

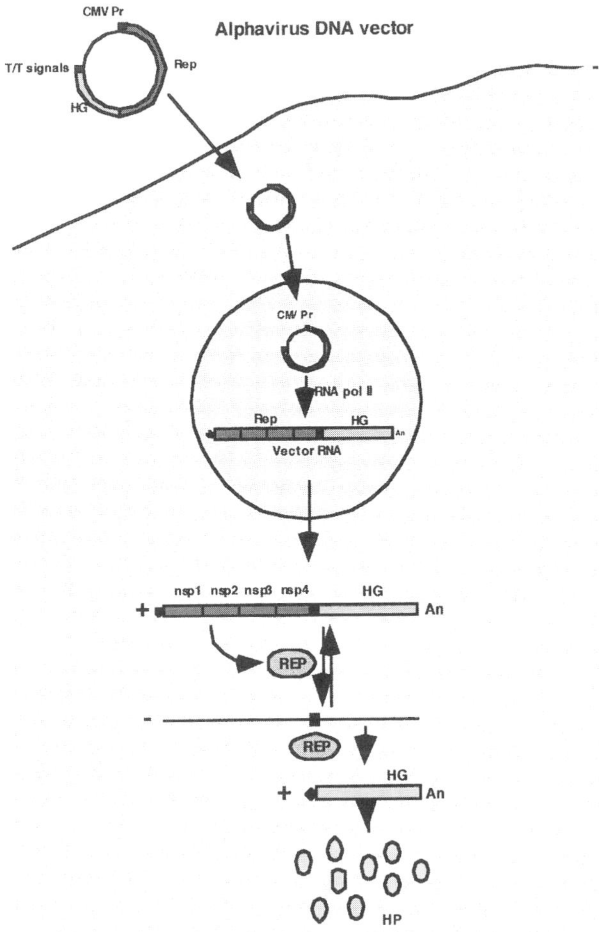

Figure 4. Alphavirus DNA/RNA layered vectors. Alphavirus DNA vectors carry the sequence of the alphavirus replicon downstream of an eukaryotic promoter such as the CMV promoter (CMV Pr). Upon transfection into cells these plasmids are transported into the nucleus where the cellular RNA polymerase II (RNA pol II) will transcribe the vector RNA from the CMV Pr. This RNA is then transported to the cytoplasm where the nsp1-4 genes are translated to form the replicase (REP). This enzyme will amplify the vector RNA and synthesize a subgenomic RNA from which expression of the heterologous gene (HG) will take place at high levels. T/T, transcription termination signals; nsp, nonstructural proteins; An, polyA; HP, heterologous protein.

alphavirus DNA vector. Other eukaryotic promoters have been tested to drive the expression of the SIN replicon from DNA plasmids, including the cytomegalovirus immediate early promoter (CMV IE) and the Moloney murine leukemia virus LTR promoter (Dubensky et al., 1996). The CMV IE promoter has shown to be the more efficient. The expression levels obtained with SIN DNA vectors were 2 to 10-fold higher than with conventional expression plasmids, depending on the promoter used.

SFV DNA vectors have also been recently developed (Berglund et al., 1998; DiCiommo and Bremner, 1998; Kohno et al., 1998). In these vectors the CMV IE promoter was used to drive the transcription of the SFV replicon sequence. They also contained an SV40 transcription termination cassette downstream of the poly (dA-dT) at the end of the replicon. The expression of foreign protein from one of these SFV DNA vectors has been extensively characterized by using *LacZ* as a reporter gene (Berglund et al., 1998). One important conclusion was that there are no significant differences in the level of β-galactosidase expression in cells transfected with the SFV-LacZ DNA plasmid as compared to cells transfected with SFV-LacZ RNA or infected with SFV-LacZ recombinant particles. This indicates that the three alternative systems that can be used to deliver SFV replicons into cells have similar efficiency to express foreign proteins. Inclusion of the SFV translation enhancer sequences (first 144 nucleotides of the capsid gene) upstream and in frame with the reporter gene provided a ten-fold increase in the expression level (Berglund et al., 1998). The transient nature of the SFV-derived plasmids was indicated by a strong cytopathic effect in the transfected cells 72 h after transfection (Berglund et al., 1998; Kohno et al., 1998).

6. Multiple applications of heterologous gene expression

Alphavirus expression systems have proven to be a very efficient tool to express heterelogous proteins in many different types of animal cells. Expression in mammalian cells is usually transient, although new vectors have been recently developed which can maintain a more prolonged expression with a reduced cytopathogenicity for the cells. Alphavirus vectors have been used to express proteins *in vitro* with many different applications, such as large-scale production and purification, functional characterization, intracellular localization, and virus assembly. The possibility of expressing foreign proteins or RNAs *in vivo* has also led to the use of alphavirus vectors in vaccination and gene therapy studies. The alphavirus genomes have recently been used to engineer cell lines for different applications such as production of alphavirus suicide particles (described above in this chapter), and virus detection. An *in vivo* approach has also been used to engineer mosquitoes resistant to certain viral diseases through a persistent infection with a recombinant alphavirus vector. Each one of these applications will be discussed bellow.

6.1 LARGE SCALE PRODUCTION OF PROTEINS

The high expression level of foreign protein obtained with alphavirus vectors in mammalian cells has led to the optimization of these systems for large scale production of proteins. Although alphavirus vectors have a very broad host range, the infectivity and the expression level of the recombinant proteins can vary depending on the cell line. Several studies have shown that BHK cells express higher levels of foreign protein upon infection with recombinant SFV vectors as compared to other cell lines such as HeLa, NSO, COS M6, TK6, NIE 115, or CHO (Blasey et al., 1998; Scheer et al., 1999). In addition, BHK cells seem to express equal levels of foreign protein both in adherent and suspension cultures, which is important for scale-up purposes (Blasey et al., 1997). Optimal conditions for infection of BHK cells in suspension with SFV recombinant particles have been determined, and include 10% FCS in a medium at pH of 6.9, a high multiplicity of infection (40) and a high concentration of viral particles. Using these conditions the enzyme human cyclooxygenase-2 has been produced in suspension spinners at 1 liter scale, yielding substantial amounts of recombinant protein (16 mg protein/10^9 cells) (Blasey et al., 1997). Large scale production of the mouse serotonin 5-HT3 receptor expressed from an SFV vector has also been achieved by infection of BHK cells in an 11.5 liter reactor containing 10^9 cells/L. In this case a yield of 1-2 mg of receptor/L was obtained (Blasey et al., 1998; Hovius et al., 1998; Lundström et al., 1997). Further optimization of the system showed that a pH of 7.3 during infection and production could result in a two fold increase in the expression levels of the 5-HT3 receptor (Blasey et al., 1998). Recently it has been shown that production of recombinant protein from mammalian cell lines infected with SFV vectors can be considerably increased by lowering the incubation temperature from 37°C to 33°C (Schlaeger and Lundström, 1998). This was achieved in serum-free suspension cultures and allowed a prolongation of the postinfection time up to 50 h, considerably improving the capacity of the system for *in vitro* production of proteins. This effect has been observed only in BHK and CHO cells, but not in other cell lines such as the human cell line HEK293 or in the subclone 293 (EBNA).

6.2 FUNCTIONAL STUDIES OF PROTEINS EXPRESSED FROM ALPHAVIRUS VECTORS

Functional studies have been mainly focused on the expression and characterization of different receptors in mammalian cell lines (for a review see Lundström, 1999). Although alphavirus vectors induce cytopathogenicity in the infected/transfected cells there is a time window of at least 24 hours postinfection in which functional assays can be performed. Many G protein coupled receptors have been expressed using SFV recombinant particles. Among these are the human neurokinin-1 and -2 receptors, which were expressed at high density on the surface of different cell lines (Lundström et al., 1995a; Lundström et al., 1994; Lundström et al., 1995b). The functional activity of these proteins was shown by the stimulation of Ca^{2+} mobilization in the infected cells, indicating functional coupling to G proteins. The human dopamine D3 receptor and several mutants of this protein were also expressed using the SFV system (Lundström et al., 1995a; Lundström and Turpin, 1996; Lundström et al., 1998). A mutant of the D3 receptor containing the Ser9Gly mutation showed a significantly higher dopamine affinity as compared to the wild type receptor (Lundström and Turpin, 1996). This mutation has been associated to a higher risk of schizofrenia. Several other mutants of the D3 receptor were also expressed and their affinity for different ligands was assayed. These studies provided some information about the receptor binding site, which could be useful in the design of specific drugs able to discriminate between the dopamine D3 and D2 receptors (Lundström et al., 1998). The possibility of coinfecting cells with alphavirus particles expressing different proteins has allowed the study the interaction between G protein coupled receptors and different protein G subunits. The α_{1b}-

adrenergic receptor was expressed alone or together with the G protein α_q, β or γ_2 subunits by infecting BHK cells with one or more SFV vectors. The stimulation of these cells with epinefrine led to an increase in inositol phosphate accumulation, confirming the functional coupling of the receptor to G protein. This functional response was enhanced when the α_{1b}-adrenergic receptor was co-expressed with the G protein α_q subunit indicating an interaction between the two proteins (Scheer et al., 1999). Other G protein-coupled receptors expressed with the SFV system include mutants of the human metabotropic glutamate receptor 8 (Malherbe et al., 1999), the human gonadotropin releasing hormone receptor (Marheineke et al., 1998) and several odorant receptors (Monastyrskaia et al., 1999). Ligand-gated ion channels such as the mouse serotonine 5-HT3 receptor or the rat and human purinoreceptor P2X subtypes have also been expressed and characterized using the SFV system (Evans et al., 1996; Hovius et al., 1998; Lundström et al., 1997; Michel et al., 1996).

Alphavirus vectors have been used to express and characterize a great number of viral proteins. Examples include the HIV-1 envelope proteins (Altmeyer et al., 1999; Paul et al., 1993; Verrier et al., 1997), the hepatitis C virus envelope glycoproteins (Dubuisson et al., 1994), the Japanese encephalitis virus structural proteins (Pugachev et al., 1995), the rubella virus structural proteins (Chen et al., 1995), the capsid protein of hepatitis E virus (Torresi et al., 1999; Torresi et al., 1997), the replication proteins from human cytomegalovirus (McCue and Anders, 1998), the LaCrosse virus envelope proteins (Kamrud et al., 1998), the NS3-NS4A protease from hepatitis C virus (Filocamo et al., 1999), and the nonstructural protein (NSs) from Rift Valley fever virus (Yadani et al., 1999). Many other viral proteins have also been expressed in alphavirus for vaccination studies, as it will be discussed with more detail bellow. Assembly studies of viral proteins expressed from alphavirus vectors have included human papilloma virus type 16 capsid protein (Heino et al., 1995), capsid proteins from rotaviruses (Nilsson et al., 1998), Moloney leukemia virus envelope proteins (Opstelten et al., 1998), and mutant capsid proteins from SFV (Skoging and Liljeström, 1998; Skoging et al., 1996; Suomalainen et al., 1992).

In a different type of study SIN based vectors were also used to produce a functional monoclonal antibody specific for SIN E2 protein through cotransfection of BHK cells with two SIN vectors expressing the light and heavy chain of the antibody, respectively (Liang et al., 1997).

6.3 EXPRESSION OF PROTEINS IN NEURONS

Several studies have shown that alphavirus vectors can infect primary neurons with high efficiency (Gwag et al., 1998; Olkkonen et al., 1993). SFV recombinant particles have been used to express cell membrane associated proteins such as rab8 and VIP21 from MDCK cells in cultured rat hippocampal neurons (de Hoop et al., 1994; Olkkonen et al., 1993). In this case 50-60% of the neurons expressed the reporter proteins when infected with a multiplicity of one. Some interesting studies have included the expression in neurons of proteins associated with the onset of familial Alzheimer disease. These include presenilin 1 (Cook et al., 1996) and the amyloid precursor protein (APP) (Simons et al., 1995), as well as mutant forms of these proteins associated with the disease. These experiments have been useful to determine the intracellular localization of these proteins and their mutant forms. The localization of the enzyme catechol O-methyltransferase was also analyzed in primary neurons infected with recombinant SFV particles (Ulmanen et al., 1997). In a functional study an SFV vector expressing the synaptic SNARE protein SNAP-25 in hippocampal neurons was able to perturb the synaptic release machinery in the infected cells (Owe-Larsson et al., 1999). This indicates that alphavirus vectors can be used as an effective tool to study the function of proteins involved in neurotransmitter release. Recombinant SIN viruses are also equally efficient for selective gene transfer into neurons as shown by the expression of the marker proteins β-galactosidase and green fluorescent protein in neuron-glia cocultures (Gwag et al., 1998). The selectivity for neurons was confirmed by the negligible infection observed in the glial cells present in the culture. Other studies have shown that alphavirus recombinant vectors can infect neurons in rat brain (Altman-Hamamdzic et al., 1997; Lundström et al., 1999) and more recently in hippocampal slice cultures (Ehrengruber et al., 1999; Maletic-Savatic et al., 1999). In these studies expression of the green fluorescent protein from SFV or SIN particles allowed an excellent visualization of neurons in living slices. Interestingly, the infected neurons appear morphologically normal and viable up to 5 days postinfection, although cytopathic effects start after this time and eventually lead to cell death (Ehrengruber et al., 1999). The extensive time window in which the infected neurons are viable can be very useful to characterize the physiological changes that take place upon expression or suppression of specific gene products.

6.4.1 VACCINATION WITH ALPHAVIRUS VECTORS

Recombinant alphavirus vectors have been used in numerous vaccination studies, being able to elicit strong immune responses against a variety of antigens. These vectors can be delivered as recombinant viral particles, naked RNA or DNA. The efficiency of each of the delivery systems, as well as their advantages and disadvantages will be discussed below.

6.4.1. Vaccination with alphavirus viral particles

Immunization with recombinant alphavirus particles has shown that these type of vectors are very effective in inducing long lasting humoral and cellular immune responses against heterologous antigens. Some studies have involved vaccination with recombinant replication-packaging proficient vectors which have the potential to propagate in the vaccinated animals (Caley et al., 1999; Caley et al., 1997; Davis et al., 1996; Hahn et al., 1992; London et al., 1992; Pugachev et al., 1995; Tsuji et al., 1998). These vectors are efficient in inducing immune responses against heterologous antigens, but their use as vaccines cannot be considered due to their propagating nature. In addition, these vectors have stability problems and can induce strong immune responses against the vector proteins, preventing their use in repetitive immunizations.

On the other hand, suicide recombinant alphavirus particles can induce strong immune responses against heterologous antigens with very low immune responses against the vector itself (Berglund et al., 1999; Pushko et al., 1997). SFV particles expressing the nucleoprotein of influenza virus (SFV-NP) or β-galactosidase (SFV-LacZ) have been used to characterize the immune responses generated with this type of vectors in vaccinated mice (Berglund et al., 1999; Zhou et al., 1994; Zhou et al., 1995). The vectors elicited long lasting humoral and cellular immune responses against the recombinant antigens. It was particularly interesting that as few as 100 SFV-NP infectious units inoculated intravenously could generate a strong $CD8^+$ cytotoxic T-cell (CTL) response. In addition, immunization of mice with a combination of SFV–NP and SFV-HA particles (the latter expressing hemagglutinin from influenza virus) was able to confer total protection when the animals were challenged with the homologous influenza virus. A similar result was observed in mice immunized with VEE particles expressing influenza HA (Davis et al., 1996; Pushko et al., 1997). Other examples in which suicide alphavirus recombinant particles have been able to induce immune responses in mice include SFV particles expressing proteins prM and E of the flavivirus Murray Valley encephalitis virus (MVE) (Colombage et al., 1998), HIV-1 envelope glycoproteins (Brand et al., 1998), prME and NS1 proteins from louping ill virus (LIV) (Fleeton et al., 1999b; Fleeton et al., 1999c), respiratory syncytial virus glycoproteins (Chen et al., 1999), SIN particles expressing the glycoprotein B of herpes simplex virus (Polo et al., 1999), and VEE particles expressing Lassa virus nucleocapsid (Pushko et al., 1997).

Suicide alphavirus particles have been proposed as good candidates for human clinical applications due to their high immunogenicity and high level of safety. Although no human trials have been done to date, some vaccination studies with SFV or VEE recombinant particles have been performed in nonhuman primates. In one case SFV recombinant particles expressing the envelope protein of simian immunodeficiency virus PBj14 (SIV-PBj14) were able to induce antibodies and protection from lethal disease in pigtail macaques (Mossman et al., 1996). Several other studies with SFV expressing different HIV-1 and SIV proteins were able to prime specific immune responses in primates (Berglund et al., 1997; Notka et al., 1999; Osterhaus et al., 1999). A complete protection against Marburg virus (MBGV) was achieved in cynomolgus macaques with VEE particles expressing the glycoprotein of MBGV (Hevey et al., 1998).

6.4.2. Vaccination with alphavirus naked RNA

In vivo administration of RNA for vaccination has not been widely used, due to the instability of this type of molecules, as well as to the high cost involved in the synthesis of RNA *in vitro*. When compared to DNA vectors, naked RNA has the advantage of a higher safety level, due to the elimination of chromosomal integration risk, potentially present when DNA vectors are used *in vivo*. In contrast to delivery of conventional mRNA, the use of self-amplifying RNA derived from alphavirus vectors could provide a high level of expression in the transfected cells with a low input of RNA. In addition, delivery of alphavirus as naked RNA would have a higher degree of safety as compared to viral particles, since the risk of wild type generation is completely eliminated. Delivery of alphavirus RNA *in vivo* has been tested in mice by intramuscular injection of recombinant SIN RNA carrying the luciferase gene (SIN-luc) (Johanning et al., 1995). The expression level obtained with the SIN-luc RNA was 24 -fold higher than that observed with a conventional mRNA and lasted at least 5-fold longer. At 28 days postinjection, however, the levels of luciferase expression were undetectable, probably due to the cytopathic effects produced by the self-replicative vector in transfected cells. The induction of apoptosis in cells transfected with alphavirus replicons has been described for SIN and SFV, being responsible for the transient nature of these vectors (Frolov and Schlesinger, 1994b; Glasgow et al., 1998; Ying et al., 1999).

Naked recombinant RNA has also been used for vaccination studies. When mice were immunized with 10 µg of recombinant SFV-NP RNA a long lasting specific antibody response was elicited (Zhou et al., 1994). This response was lower than that induced by SFV-NP viral particles, probably due to their higher efficiency of entry into the cells. Similar results have been reported in mice immunized with recombinant SFV RNA expressing HIV protein gp160 (Brand et al., 1998). Immunization with SFV RNA expressing the HIV-1 envelope protein has also been used to generate specific monoclonal antibodies in mice (Giraud et al., 1999). In a protection study mice immunized with recombinant SFV RNA expressing influenza virus hemagglutinin (HA) and subsequently challenged with the homologous influenza strain were able to completely clear the virus from the lungs and partially reduce the amount of virus in the trachea, which indicates that immunization with SFV RNA can control the spread of disease (Dalemans et al., 1995). SFV recombinant RNAs expressing the flavivirus LIV prME or NS1 proteins or the human respiratorial syncitial virus fusion protein have also induced protective immune responses in mice (Fleeton et al., 1999a).

6.4.3. Vaccination with alphavirus naked DNA

Vaccination with naked DNA has been used to induce long-lasting immune responses and protective immunity against a variety of pathogens in several animal models. A potential problem associated with plasmid vaccination is the risk of chromosomal integration in transfected cells *in vivo*, which could lead to the generation of tumors. However, evidence of integration has not been found when PCR-based studies were done in mice immunized with high doses of plasmid DNA (Nichols et al., 1995), suggesting that DNA vaccines have a high degree of safety. A second problem related to DNA vaccines is the possible induction of tolerance due to expression of antigen at low levels in the organism during prolonged periods of time (Mor et al., 1996; Wolff et al., 1992). Vaccination with alphavirus DNA vectors could theoretically avoid both of these problems since: (i) they express high level of foreign protein; and (ii) cell death by apoptosis will be induced in transfected cells after a few days.

Several studies have been performed in order to test the induction of immune responses with alphavirus DNA vectors. In one of these studies a SIN plasmid expressing hepatitis B virus *c* or *e* genes was able to induce an specific antibody response against the hepatitis antigens when inoculated intramuscularly in mice (Dubensky et al., 1996). A quite extensive characterization of the humoral and cellular immune responses induced by alphavirus recombinant DNA vectors has been carried out by immunizing mice with SIN DNA vectors expressing the glycoprotein B (gB) of herpes simplex virus (HSV-1) (Hariharan et al., 1998), or with SFV-derived plasmids expressing β-galactosidase or the influenza virus antigens NP or HA (Berglund et al., 1998). In these plasmids the transcription of SIN and SFV recombinant replicons was driven by the CMV IE promoter. When the alphavirus DNA vectors were compared to conventional plasmids similar levels of protein expression were observed *in vitro*.

However, in immunization studies, the alphavirus vectors showed a clear advantage over the conventional plasmids, reflected by the fact that up to 1000-fold more of the conventional DNA was needed to induce the same level of responsiveness in mice. In addition, the alphavirus based DNA vectors were able to induce immune responses in a larger number of animals as compared to conventional plasmids. By including the capsid translation enhancer in the SFV DNA vector the level of the immune response was increased when doses lower then 1 μg of DNA were used. This indicates that an increase in the level of antigen expression has an additive effect in the immune response in the context of the SFV vector. In general, the SFV-LacZ DNA vector induced a similar antibody response than SFV-LacZ RNA, but lower than that obtained with recombinant SFV-LacZ particles (Berglund et al., 1998). The CTL response was however very similar when SFV-NP DNA and recombinant particles were compared. The higher antibody response elicited with particles is probably associated with a higher efficiency of infection as compared to transfection, resulting in a greater production of antigen. Induction of a cellular response is, however, dependent on the presentation of antigen by a few APCs and small amounts of naked alphavirus DNA can probably be enough to induce a strong T-cell response. Protection against viral challenge was also induced in both systems.

There are a couple of examples in which conventional plasmids have been able to induce higher immune antibody responses than alphavirus DNA vectors expressing the same antigen. These include a SIN DNA vector expressing the envelope proteins of Seoul hantavirus (SEOV) (Kamrud et al., 1999), and an SFV DNA vector expressing LIV prME proteins (Fleeton et al., 1999c). However, in the these studies alphavirus vectors expressing different antigens from the same viruses (SEOV or LIV, respectively) induced better humoral immune responses than the conventional vectors, suggesting that the nature of the antibody response may vary depending on the expressed antigen. In spite of these results and of one single report in which an alphavirus DNA vector based on SIN expressing encephalomyocarditis virus *VP1* gene was unable to induce a measurable immune response in mice (Lee et al., 1997), it can be concluded that alphavirus DNA-based vectors generally induce stronger humoral and cellular immune responses than conventional DNA vaccines. An important question to be addressed is why the alphavirus DNA vectors are more efficient than conventional plasmids in inducing immune responses when they do not differ significantly in the level of heterologous protein expression. A possible explanation is that the self-replicating vectors could be involved in the induction of cytokines. It is known that α and β interferons are induced following infection with RNA viruses, including alphaviruses, (Brandt et al., 1994; Vilcek and Sen, 1996) and this could also be the case in cells transfected with the alphavirus-based plasmids. The self-amplifying process of the replicon transcribed in the cell mimics that of the wild-type virus, and this would result in the induction of interferon (IFN). This molecule can activate natural killer cells and plasma IFN would be induced (Gates et al., 1984; Kaluza et al., 1987), including IFN-γ (Donnelly et al., 1997), which can activate macrophages and stimulate differentiation of Th0 cells to become inflammatory CD4 T-cells (Th1). Additionally, it may be that the induction of cell death by apoptosis could enhance antigenic crosspriming.

6.5 ENGINEERING MOSQUITOES WITH ALPHAVIRUS SEQUENCES FOR CONTROL OF DISEASE

Alphaviruses are naturally transmitted by mosquitoes, which become chronically infected by these viruses, and this feature has been used to design recombinant replicons able to control viral infection in these insects with the aim of preventing disease transmission to humans. It has been shown that SIN vectors can express heterologous proteins or RNAs in mosquito cells and transduced mosquitoes (Higgs et al., 1995; Kamrud et al., 1995; Olson et al., 1992; Olson et al., 1994; Seabaugh et al., 1998). Antisense RNAs to different genes of viruses transmitted by mosquitoes have been expressed from SIN vectors in transduced mosquitoes (Higgs et al., 1998; Olson et al., 1996; Powers et al., 1994; Rayms-Keller et al., 1995). A SIN vector expressing an antisense RNA targeted to the premembrane coding region of DEN-2 was able to inhibit replication of this virus in the salivary glands of the transduced mosquitoes (Olson et al., 1996). This resulted in the inability of these insects to transmit the virus. Similar results were obtained with SIN expressing antisense RNAs to the premembrane (prM) or polymerase (NS5) coding regions of a vaccine strain of yellow fever virus (Higgs et al., 1998).

6.6 ALPHAVIRUS VECTORS FOR GENE THERAPY

A few alphavirus recombinant vectors with potential use in gene therapy have been described. They include vectors with possible antiviral and antitumoral applications. Antiviral approaches included the expression of single chain antibodies or ribozymes from alphavirus replicons (Jiang et al., 1995; Smith et al., 1997). Vectors designed for antitumoral therapy have been based on the expression of specific tumor antigens or cytokines. In one case mice inoculated with SFV suicide particles expressing a weak tumor rejection antigen from mouse mastocytoma (P815), could be protected from tumor challenge (Colmenero et al., 1999), indicating that an immune response against the tumor had been generated. The antitumoral effect can also be achieved by RNA immunization (Ying et al., 1999). In this case SFV-LacZ RNA was able to protect mice from challenge with colon carcinoma cells expressing β-galactosidase, as well as to prolong the survival of animals with pre-established tumors. Finally, the use of alphavirus vectors expressing cytokines has also been applied to control tumor growth. In this case an SFV vector expressing mouse IL-12 was able to inhibit the growth and neovascularization of B16 murine melanomas when injected directly into the tumor (Asselin-Paturel et al., 1998; Zhang et al., 1997).

The use of alphavirus vectors for certain gene therapy applications, however, requires some improvements in the present technology. One is the need to target the vectors to specific cells or tissues, and another is the development of noncytopathic vectors able to maintain a prolonged expression of the therapeutic product in the transfected/infected cells. Both of these problems have been recently addressed.

6.6.1. Targeting alphavirus vectors

The construction of alphavirus targeted vectors requires both the elimination of the endogenous broad tropism as well as the introduction of a novel tropism. A possible strategy to redirect the alphavirus tropism is the generation of "minimal" viruses through encapsidation of the recombinant alphavirus genome inside an envelope containing a nonrelated viral protein involved in receptor binding and entry. One example of such "minimal" viruses was generated by transfecting cells with an SFV recombinant RNA vector expressing the vesicular stomatitis virus glycoprotein (VSV-G) (Rolls et al., 1994). Minimal viruses displaying VSV-G on the surface were infectious and could be neutralized with specific anti-VSV serum, indicating that an interaction with the specific VSV-G receptor was probably mediating the infection. By including a second subgenomic promoter followed by a heterologous gene in the SFV-VSV-G vector, these "minimal" viruses were able to express other foreign proteins at high levels (Rolls et al., 1996). A similar strategy was used to encapsidate SFV recombinant RNA in an envelope containing the murine leukemia virus (MLV) envelope protein with a truncated cytoplasmic tail (Lebedeva et al., 1997).

The problem of targeting alphavirus vectors has also been addressed through modification of the alphavirus spike protein responsible for binding to receptors in the target cells. Random insertion mutagenesis in p62 gave rise to several SIN mutants with impaired binding to vertebrate cells (Dubuisson and Rice, 1993). Some of these mutants exhibited normal particle assembly, release and penetration. The possibility of inserting foreign sequences in permissive sites in the p62 glycoprotein has led to the construction of SIN virus vectors able to target specific cells. Using that strategy, human choriocarcinoma cells could specifically be targeted by SIN vectors containing the alpha- and beta- hCG genes inserted in the envelope protein sequence (Sawai and Meruelo, 1998). The same principle has been used by Ohno and coworkers to create alphavirus vectors with a tropism that can be modulated. In this case two IgG binding domains of protein A from *Staphylococcus aureus* were inserted in the E2 region of SIN (Ohno et al., 1997). Chimeric particles obtained in this way had a minimal infectivity in BHK and several human cell lines, but could be targeted to specific cells lines by combining them with monoclonal antibodies specific for cell surface antigens. These chimeric particles could be targeted virtually to any cell type, provided that a specific antibody is available. One problem that has been observed with this system is that the infectivity mediated by the antibody is not as efficient as that mediated by the normal spike-receptor interaction, limiting the possible applications of these vectors. Using these chimeric particles a SIN recombinant vector expressing herpes simplex virus thymidine kinase was specifically introduced in several human tumor cell lines, which after infection showed an increased sensitivity to the prodrug ganciclovir (Ijima et al., 1999).

Targeting of alphavirus naked RNA or DNA to specific cells or tissues offers additional difficulties. Some possible strategies include complexing the alphavirus nucleic acid with specific monoclonal antibodies (mAbs) or using virosomes containing specific proteins in the surface. Using the first approach Sawai and coworkers were able to deliver a SIN recombinant RNA to specific cancer cells *in vitro* (Sawai et al., 1998). In this case the RNA was previously biotinylated and then coupled to the mAb through a streptavidin-protein A fusion protein. Other possible approaches include the use of specific promoters that would be activated only in certain cells or at a certain time, including also externally inducible promoters.

6.6.2. Non-cytopathic alphavirus vectors

Alphavirus vectors are cytopathic for most vertebrate cells. This cytopathogenicity is mediated by apoptosis and leads to cell death usually within 2-3 days after infection/transfection (Berglund et al., 1998; Frolov and Schlesinger, 1994b; Glasgow et al., 1997; Glasgow et al., 1998; Sammin et al., 1999; Ying et al., 1999). The transient nature of the alphaviral vectors does not represent a problem for some applications such as vaccination, where it could be advantageous. However, for gene therapy applications vectors able to sustain a more prolonged expression of therapeutic products would be desired. In order to render the alphavirus vectors noncytopathic two different approaches have been used. The first one is based on the expression of anti-apoptotic genes in the infected/transfected cells (for a review see Griffin and Hardwick, 1997). One of such genes is the proto-oncogene *bcl-2* which encodes a 26 kD inner mitochondrial membrane protein able to block apoptosis in neurons. Cells transfected with *bcl-2* were able to tolerate a persistent infection when infected with wild type SIN or SFV (Levine et al., 1993; Liao et al., 1997; Scallan et al., 1997). bcl-2 has also been expressed from a replication-packaging proficient SIN vector, resulting in a reduced mortality rate in inoculated neonatal mice (Levine et al., 1996), or a reduced cytotoxicity in BHK cells infected *in vitro* (Sawai et al., 1999). Similar results were obtained in animals inoculated with SIN recombinant particles expressing beclin, a protein that interacts with bcl-2 in mammalian cells (Liang et al., 1998). Cleavage of bcl-2 by caspases can inhibit the anti-apoptotic activity of this protein (Cheng et al., 1997). Because in certain cases alphaviruses could trigger a bcl-2 caspase mediated inactivation (Grandgirard et al., 1998) the expression of bcl-2 cleavage resistant mutants could provide increased protection to alphavirus induced apoptosis (Cheng et al., 1997). Another anti-apoptotic factor that has been expressed in alphavirus includes Bax, a member of the bcl-2 family of proteins that can inhibit neuronal cell death in mice infected with SIN (Lewis et al., 1999). The anti-apoptotic function of Bax seems to be determined by cell specific factors, being able to function as a pro-apoptotic factor only in some types of neurons.

A second strategy to render the alphavirus vectors noncytopathic is based on the isolation of mutant viruses able to establish a persistent infection in mammalian cells. In cells infected with SIN heavily enriched with defective-interfering (DI) particles it was possible to isolate one mutant virus able to establish persistent infection (Sin-1) (Weiss et al., 1980). A single mutation in the nsp2 gene (Pro726 → Ser) was responsible for that capacity (Dryga et al., 1997). More recently, noncytopathic SIN replicons have been isolated by expressing the puromycin N-acetyltranferase gene (*pac*) from the subgenomic promoter in the SIN expression vector. When cells transfected with recombinant SIN-*pac* were incubated in the presence of the antibiotic, mutant replicons could be rescued from the surviving cells (Agapov et al., 1998; Frolov et al., 1999). These mutants contained single amino acid changes in nsp2, either in the same position as in Sin-1 (Pro726 → Leu) or in a different position (Asn779 → Lys). Additional mutant SIN replicons were engineered with different amino acids at position 726 in the nsp2 gene which allowed to establish a direct correlation between the level of viral RNA replication and the cytopathogenicity (Frolov et al., 1999). Several proteins have been expressed from SIN noncytopathic vectors which include β-galactosidase, green fluorescent protein, T7 RNA polymerase, CD46 (measles virus receptor), yellow fever virus prM, E, and NS1 proteins, etc (Agapov et al., 1998; Lindenbach and Rice, 1997; Olivo et al., 1998). Some of the drawbacks associated to these noncytopathic vectors are the low level of foreign protein expression (around 4% of that obtained with their cytopathic counterparts) and the lack of stability. This latter problem can lead to a progressive loss of foreign gene expression after several passages, which could limit the duration of expression *in vivo*. A DNA version of a SIN attenuated vector has also been engineered by cloning the replicon sequence downstream of the RSV LTR promoter (Agapov et al., 1998).

6.6.3. Using SFV to package retroviral particles

Production of retrovirus particles usually requires the transfection of a retrovirus recombinant proviral DNA into a packaging cell line that provides the retroviral structural proteins in *trans*. Li and Garoff have developed a new strategy to package retroviral particles based on the SFV expression vector (Li and Garoff, 1996). This system is based on the cotransfection of cells with three independent recombinant SFV RNAs which carry the *gag-pol*, *env* and a recombinant retrovirus genome from Moloney leukemia virus (MLV), respectively. The MLV structural proteins and the retroviral recombinant genome will be produced in the cytoplasm of the transfected cells from the SFV vectors. This results in the production of retroviral particles containing the MLV genome with titers of 10^7 particles/10^6 transfected cells and without detection of RPVs. Both particles carrying the ecotropic or the amphotropic envelope could be generated by using different SFV-*env* vectors, which makes the system especially attractive for the purpose of human gene therapy. An advantage of this system is that it can be used to package retroviral genomes carrying heterologous genes containing introns, which can increase the expression levels of the foreign gene (Li and Garoff, 1998). This is not possible by using conventional retrovirus packaging cell lines, since in these cells retrovirus genomes are generated in the nucleus and subjected to splicing. SFV vectors can also be used to produce retroviral particles by transfecting retroviral packaging cell lines with only one single SFV replicon carrying a retroviral recombinant genome (Wahlfors and Morgan, 1999; Wahlfors et al., 1997). In this case the retroviral structural proteins are provided in *trans* by the host cell.

6.7 ENGINEERING CELL LINES FOR VIRUS DETECTION

Alphavirus sequences have been used to engineer several cell lines for detection of homologous or heterologous viruses. A cell line for titration of SIN has been developed by Olivo and coworkers (Olivo et al., 1994). In this cell line the RSV-LTR promoter controls the transcription of a SIN defective RNA carrying the luciferase gene downstream of the subgenomic promoter (SIN-*luc*). SIN-*luc* RNA is produced constitutively in these cells, but expression of luciferase takes place only upon infection with SIN particles carrying a functional replicase. Cell lines for detection of herpesviruses have been developed more recently (Ivanova et al., 1999). In this case the cells were stabely transformed with a plasmid in which the SIN-LacZ sequence was placed downstream of a herpes simplex virus (HSV) early promoter. The SIN-LacZ RNA will not be normally produced in these cells, since transcription from the HSV early promoter requires the expression of HSV immediate early genes. However, upon infection of these cells with HSV, SIN-LacZ RNA will be transcribed, producing the alphavirus replicase that will amplify the replicon and synthesize a subgenomic RNA, allowing the synthesis of β-galactosidase. These cell lines were improved by transformation with a SIN helper cDNA that provides the SIN structural proteins upon HSV infection. These proteins can package the SIN-LacZ RNA into particles that by infecting the surrounding cells will amplify the signal. A similar cell line that allowed detection of human cytomegalovirus (HCMV) was also developed (Ivanova et al., 1999).

7. Acknowledgements

We thank the Swedish Medical Research Council, the Swedish Council for Engineering Sciences, the Swedish Cancer Research Foundation and the European Union Biotechnology Program for support.

8. References

Agapov, E. V., Frolov, I., Lindebach, B. D., Prágal, B. M., Schlesinger, S., and Rice, C. M. (1998). NoncytopathicSindbis virus RNA vectors for heterologous gene expression. Proc. Natl. Acad. Sci. USA. **95:**12989-12994.

Altman-Hamamdzic, S., Groseclose, C., Ma, J. X., Hamamdzic, D., Vrindavanam, N. S., Middaugh, L. D., Parratto, N. P., and Sallee, F. R. (1997). Expression of β -galactosidase in mouse brain: utilization of a novel nonreplicative Sindbis virus vector as a neuronal gene delivery system. Gene Ther. **4:**815-822.

Altmeyer, R., Mordelet, E., Girard, M., and Vidal, C. (1999). Expression and detection of macrophage-tropic HIV-1 gp120 in the brain using conformation-dependent antibodies. Virology. **259:**314-323.

Asselin-Paturel, C., Lassau, N., Guinebretiere, J. M., Zhang, J., Gay, F., Bex, F., Hallez, S., Leclere, J., Peronneau, P., Mami-Chouaib, F., and Chouaib, S. (1998). Transfer of the murine interleukin-12 gene in vivo by a Semliki Forest virus vector induces B16 tumor regression through inhibition of tumor blood vessel formation monitored by Doppler ultrasonography. Gene Ther. **5:**606-615.

Berglund, P., Fleeton, M. N., Smerdou, C., and Liljeström, P. (1999). Immunization with recombinant Semliki Forest virus induces protection against influenza challenge in mice. Vaccine. **17:**497-507.

Berglund, P., Quesada-Rholander, M., Putkonen, P., Biberfeld, G., Thorstensson, R., and Liljeström, P. (1997). Outcome of immunization of cynomolgus monkeys with recombinant Semliki Forest virus encoding human immunodeficiency virus type 1 envelope protein and challenge with a high dose of SHIV-4 virus. AIDS Res. Hum. Retrovir. **13:**1487-1495.

Berglund, P., Sjöberg, M., Garoff, H., Atkins, G. J., Sheahan, B. J., and Liljeström, P. (1993). Semliki Forest virus expression system: production of conditionally infectious recombinant particles. Bio/Technology. **11:**916-920.

Berglund, P., Smerdou, C., Fleeton, M. N., Tubulekas, I., and Liljeström, P. (1998). Enhancing immune responses using suicidal DNA vaccines. Nat. Biotech. **16:**562-565.

Blasey, H. D., Lundström, K., Tate, S., and Bernard, A. R. (1997). Recombinant protein production using the Semliki Forest virus expression system. Cytotechnol. **24:**65-72.

Brand, D., Lemiale, F., Turbica, I., Buzelay, L., Brunet, S., and Barin, F. (1998). Comparative analysis of humoral immune responses to HIV type 1 envelope glycoproteins in mice immunized with a DNA vaccine, recombinant Semliki Forest virus RNA, or recombinant Semliki Forest virus particles. AIDS Res. Hum. Retrovir. **14:**1369-1377.

Brandt, E. R., Linnane, A. W., and Devenish, R. J. (1994). Expression of IFN A genes in subpopulations of peripheral blood cells. Brit. J. Haematol. **86:**717-725.

Bredenbeek, P. J., Frolov, I., Rice, C. M., and Schlesinger, S. (1993). Sindbis virus expression vectors: packaging of RNA replicons by using defective helper RNAs. J. Virol. **67:**6439-6446.

Caley, I. J., Betts, M. R., Davis, N. L., Swanstrom, R., Frelinger, J. A., and Johnston, R. E. (1999). Venezuelan equine encephalitis virus vectors expressing HIV-1 proteins: vector design strategies for improved vaccine efficacy. Vaccine. **17:**3124-3135.

Caley, I. J., Betts, M. R., Irlbeck, D. M., Davis, N. L., Swanstrom, R., Frelinger, J. A., and Johnston, R. E. (1997). Humoral, mucosal, and cellular immunity in response to a human immunodeficiency virus type 1 immunogen expressed by a Venezuelan equine encephalitis virus vaccine vector. J. Virol. **71:**3031-3038.

Chen, J. P., Miller, D., Katow, S., and Frey, T. K. (1995). Expression of the rubella virus structural proteins by an infectious Sindbis virus vector. Archiv. Virol. **140:**2075-2084.

Chen, M., Hu, K. F., Örvell, C., Morein, B., and Liljeström, P. (1999). Intranasal vaccination with recombinant Semliki Forest virus protects against respiratory syncitial virus infection: Effects of IL-12 co-delivery. Submitted.

Cheng, E. H., Kirsch, D. G., Clem, R. J., Ravi, R., Kastan, M. B., Bedi, A., Ueno, K., and Hardwick, J. M. (1997). Conversion of Bcl-2 to a Bax-like death effector by caspases. Science. **278:**1966-1968.

Colmenero, P., Liljeström, P., and Jondal, M. (1999). Induction of P815 tumor immunity by recombinant Semliki Forest virus expressing the P1A gene. Gene Ther. **6:**1728-1733.

Colombage, G., Hall, R., Pavy, M., and Lobigs, M. (1998). DNA-based and alphavirus-vectored immunisation with prM and E proteins elicits long-lived and protective immunity against the flavivirus Murray Valley encephalitis virus. Virology. **250:**151-163.

Cook, D. G., Sung, J. C., Golde, T. E., Felsenstein, K. M., Wojczyk, B. S., Tanzi, R. E., Trojanowski, J. Q., Lee, V. M., and Doms, R. W. (1996). Expression and analysis of presenilin 1 in a human neuronal system: localization in cell bodies and dendrites. Proc. Natl. Acad. Sci. USA. **93:**9223-9228.

Dalemans, W., Delers, A., Delmelle, C., Denamur, F., Meykens, R., Thiriart, C., Veenstra, S., Francotte, M., Bruck, C., and Cohen, J. (1995). Protection against homologous influenza challenge by genetic immunization with SFV-RNA encoding Flu-HA. Ann. N. Y. Acad. Sci. **772:**255-256.

Davis, N. L., Brown, K. W., and Johnston, R. E. (1996). A viral vaccine vector that expresses foreign genes in lymph nodes and protects against mucosal challenge. J. Virol. **70:**3781-3787.

Davis, N. L., Powell, N., Greenwald, G. F., Willis, L. V., Johnson, B. J. B., Smith, J. F., and Jonhston, R. E. (1991). Attenuating mutations in the E2 glycoprotein gene of Venezuelan equine encephalitis virus: Construction of single and multiple mutants in a full-length cDNA clone. Virology. **183:**20-31.

Davis, N. L., Willis, L. V., Smith, J. F., and Johnston, R. E. (1989). In vitro synthesis of infectious Venezuelan equine encephalitis virus RNA from a cDNA clone: Analysis of a viable deletion mutant. Virology. **171:**189-204.

de Hoop, M. J., Olkkonen, V. M., Ikonen, E., Williamson, E., von Poser, C., Meyn, L., and Dotti, C. G. (1994). Semliki Forest virus as a tool for protein expression in cultured rat hippocampal neurons. Gene Ther. **1:**S28-31.

DiCiommo, D. P. and Bremner, R. (1998). Rapid, high level protein production using DNA-based Semliki Forest virus vectors. J. Biol. Chem. **273:**18060-18066.

Donnelly, S. M., Sheahan, B. J., and Atkins, G. J. (1997). Long-term effects of Semliki Forest virus infection in the mouse central nervous system. Neuropathol. Appl. Neurobiol. **23:**235-241.

Dryga, S. A., Dryga, O. A., and Schlesinger, S. (1997). Identification of mutations in a Sindbis virus variant able to establish persistent infection in BHK cells: the importance of a mutation in the nsP2 gene. Virology. **228:**74-83.

Dubensky, T. W. Jr., Driver, D. A., Polo, J. M., Belli, B. A., Latham, E. M., Ibanez, C. E., Chada, S., Brumm, D., Banks, T. A., Mento, S. J., Jolly, D. J., and Chang, S. M. (1996). Sindbis virus DNA-based expression vectors: utility for in vitro and in vivo gene transfer. J. Virol. **70:**508-519.

Dubuisson, J., Hsu, H. H., Cheung, R. C., Greenberg, H. B., Russell, D. G., and Rice, C. M. (1994). Formation and intracellular localization of hepatitis C virus envelope glycoprotein complexes expressed by recombinant vaccinia and Sindbis viruses. J. Virol. **68:**6147-6160.

Dubuisson, J. and Rice, C. M. (1993). Sindbis virus attachment: isolation and characterization of mutants with impaired binding to vertebrate cells. J. Virol. **67:**3363-3374.

Ehrengruber, M. U., Lundström, K., Schweitzer, C., Heuss, C., Schlesinger, S., and Gähwiler, B. H. (1999). Recombinant Semliki Forest virus and Sindbis virus efficiently infect neurons in hippocampal slice cultures. Proc. Natl. Acad. Sci. USA. **96:**7041-7046.

Evans, R. J., Lewis, C., Virginio, C., Lundström, K., Buell, G., Surprenant, A., and North, R. A. (1996). Ionic permeability of, and divalent cation effects on, two ATP-gated cation channels (P2X receptors) expressed in mammalian cells. J. Physiology. **497:**413-422.

Filocamo, G., Pacini, L., Nardi, C., Bartholomew, L., Scaturro, M., Delmastro, P., Tramontano, A., De Francesco, R., and Migliaccio, G. (1999). Selection of functional variants of the NS3-NS4A protease of hepatitis C virus by using chimeric Sindbis viruses. J. Virol. **73:**561-575.

Fleeton, M. N., Chen, M., Klingström, J., Berglund, P., Murphy, A. M., Rhodes, G., Parker, S. E., Atkins, G. J., and Liljeström, P. (1999a). Protection from viral infection upon vaccination with naked self-replicating RNA. Submitted.

Fleeton, M. N., Liljeström, P., Sheahan, B. J., and Atkins, G. J. (1999b). Protective responses induced in mice by louping ill virus prME and NS1 antigens expressed from Semliki Forest virus vectors. J. Gen. Virol. **81**:000-000.

Fleeton, M. N., Sheahan, B. J., Gould, E. A., Atkins, G. J., and Liljeström, P. (1999c). Recombinant Semliki Forest virus particles encoding the prME or NS1 proteins of louping ill virus protect mice from lethal challenge. J. Gen. Virol. **80:**1189-1198.

Frolov, I., Agapov, E., Hoffman, T. A. J., Prágal, B. M., Lippa, M., Schlesinger, S., and Rice, C. M. (1999). Selection of RNA replicons capable of persistent noncytopathic replication in mammalian cells. J. Virol. **73:**3854-3865.

Frolov, I., Frolova, E., and Schlesinger, S. (1997). Sindbis virus replicons and Sindbis virus: assembly of chimeras and of particles deficient in virus RNA. J. Virol. **71:**2819-2829.

Frolov, I. and Schlesinger, S. (1994a). Translation of Sindbis virus mRNA: effects of sequences downstream of the initiating codon. J. Virol. **68:**8111-8117.

Frolov, I. and Schlesinger, S. (1994b). Comparison of the effects of Sindbis virus and Sindbis virus replicons on host cell protein synthesis and cytopathogenicity in BHK cells. J. Virol. **68:**1721-1727.

Frolov, I. and Schlesinger, S. (1996). Translation of Sindbis virus mRNA: analysis of sequences downstream of the initiating AUG codon that enhance translation. J. Virol. **70:**1182-1190.

Frolova, E., Frolov, I., and Schlesinger, S. (1997). Packaging signals in alphaviruses. J. Virol. **71:**248-258.

Gates, M. C., Sheahan, B. J., and Atkins, G. J. (1984). The pathogenicity of the M9 mutant of Semliki Forest virus in immune-compromised mice. J. Gen. Virol. **65:**73-80.

Giraud, A., Ataman-Onal, Y., Battail, N., Piga, N., Brand, D., Mandrand, B., and Verrier, B. (1999). Generation of monoclonal antibodies to native human immunodeficiency virus type 1 envelope glycoprotein by immunization of mice with naked RNA. J. Virol. Methods. **79:**75-84.

Glasgow, G. M., McGee, M. M., Sheahan, B. J., and Atkins, G. J. (1997). Death mechanisms in cultured cells infected by Semliki Forest virus. J. Gen. Virol. **78:**1559-1563.

Glasgow, G. M., McGee, M. M., Tarbatt, C. J., Mooney, D. A., Sheahan, B. J., and Atkins, G. J. (1998). The Semliki Forest virus vector induces p53-independent apoptosis. J. Gen. Virol. **79:**2405-2410.

Grandgirard, D., Studer, E., Monney, L., Belser, T., Fellay, I., Borner, C., and Michel, M. R. (1998). Alphaviruses induce apoptosis in Bcl-2-overexpressing Cells: evidence for a caspase-mediated, proteolytic inactivation of Bcl-2. EMBO J. **17:**1268-1278.

Griffin, D. E. and Hardwick, J. M. (1997). Regulators of apoptosis on the road to persistent alphavirus infection. Ann. Rev. Microbiol. **51:**565-592.

Gwag, B. J., Kim, E. Y., Ryu, B. R., Won, S. J., Ko, H. W., Oh, Y. J., Cho, Y. G., Ha, S. J., and Sung, Y. C. (1998). A neuron-specific gene transfer by a recombinant defective Sindbis virus. Molec. Brain Res. **63:**53-61.

Hahn, C. S., Hahn, Y. S., Braciale, T. J., and Rice, C. M. (1992). Infectious Sindbis virus transient expression vectors for studying antigen processing and presentation. Proc. Natl. Acad. Sci. USA. **89:**2679-83.

Haijou, M., Hill, K. R., Subramaniam, S. V., Hu, J. Y., and Raju, R. (1996). Nonhomologous RNA-RNA recombination events at the 3' nontranslated region of the Sindbis virus genome: hot spots and utilization of nonviral sequences. J. Virol. **70:**5153-5164.

Hariharan, M. J., Driver, D. A., Townsend, K., Brumm, D., Polo, J. M., Belli, B. A., Catton, D. J., Hsu, D., Mittelstaedt, D., McCormack, J. E., Karavodin, L., Dubensky, T. W. Jr., Chang, S. M., and Banks, T. A. (1998). DNA immunization against herpes simplex virus: enhanced efficacy using a Sindbis virus-based vector. J. Virol. **72:**950-958.

Heino, P., Dillner, J., and Schwartz, S. (1995). Human papillomavirus type 16 capsid proteins produced from recombinant Semliki Forest virus assemble into virus-like particles. Virology. **214:**349-359.

Herweijer, H., Latendresse, J. S., Williams, P., Zhang, G. F., Danko, I., Schlesinger, S., and Wolff, J. A. (1995). A plasmid-based self-amplifying Sindbis virus vector. Hum. Gen. Ther. **6:**1161-1167.

Hevey, M., Negley, D., Pushko, P., Smith, J., and Schmaljohn, A. (1998). Marburg virus vaccines based upon alphavirus replicons protect guinea pigs and nonhuman primates. Virology. **251:**28-37.

Higgs, S., Olson, K. E., Klimowski, L., Powers, A. M., Carlson, J. O., Possee, R. D., and Beaty, B. J. (1995). Mosquito sensitivity to a scorpion neurotoxin expressed using an infectious Sindbis virus vector. Insect Mol. Biol. **4:**97-103.

Higgs, S., Rayner, J. O., Olson, K. E., Davis, B. S., Beaty, B. J., and Blair, C. D. (1998). Engineered resistance in Aedes aegypti to a West African and a South American strain of yellow fever virus. Am. J. Trop. Med. Hyg. **58:**663-670.

Hill, K. R., Hajjou, M., Hu, J. Y., and Raju, R. (1997). RNA-RNA recombination in Sindbis virus: roles of the 3' conserved motif, poly(A) tail, and nonviral sequences of template RNAs in polymerase recognition and template switching. J. Virol. **71:**2693-2704.

Hovius, R., Tairi, A. P., Blasey, H., Bernard, A., Lundström, K., and Vogel, H. (1998). Characterization of a mouse serotonin 5-HT3 receptor purified from mammalian cells. J. Neurochem. **70:**824-834.

Ijima, Y., Ohno, K., Ikeda, H., Sawai, K., Levin, B., and Meruelo, D. (1999). Cell-specific targeting of a thymidine kinase/ganciclovir gene therapy system using a recombinant Sindbis virus vector. Int. J. Cancer. **80:**110-118.

Ivanova, L., Schlesinger, S., and Olivo, P. D. (1999). Regulated expression of a Sindbis virus replicon by herpesvirus promoters. J. Virol. **73:**1998-2005.

Jain, S. K., DeCandido, S., and Kielian, M. (1991). Processing of the p62 envelope precursor protein of Semliki Forest virus. J. Biol. Chem. **266:**5756-5761.

Jiang, W., Venugopal, K., and Gould, E. A. (1995). Intracellular interference of tick-borne flavivirus infection by using a single-chain antibody fragment delivered by recombinant Sindbis virus. J. Virol. **69:**1044-1049.

Johanning, F. W., Conry, R. M., Lo Buglio, A. F., Wright, M., Sumerel, L. A., Pike, M. J., and Curiel, D. T. (1995). A Sindbis virus mRNA polynucleotide vector achieves prolonged and high level heterologous gene expression in vivo. Nucl. Acids Res. **23:**1495-1501.

Kaluza, G., Lell, G., Reinacher, M., Stitz, L., and Willems, W. R. (1987). Neurogenic spread of Semliki Forest virus in mice. Archiv. Virol. **93:**97-110.

Kamrud, K. I., Hooper, J. W., Elgh, F., and Schmaljohn, C. S. (1999). Comparison of the protective efficacy of naked DNA, DNA-based Sindbis replicon, and packaged Sindbis replicon vectors expressing hantavirus structural genes in hamsters. Virology. **263:**209-219.

Kamrud, K. I., Olson, K. E., Higgs, S., Carlson, J. O., and Beaty, B. J. (1998). Use of the Sindbis replicon system for expression of LaCrosse virus envelope proteins in mosquito cells. Archiv. Virol. **143:**1365-1377.

Kamrud, K. I., Powers, A. M., Higgs, S., Olson, K. E., Blair, C. D., Carlson, J. O., and Beaty, B. J. (1995). The expression of chloramphenicol acetyltransferase in mosquitoes and mosquito cells using a packaged Sindbis replicon system. Exp. Parasitol. **81:**394-403.

Kohno, A., Emi, N., Kasai, M., Tanimoto, M., and Saito, H. (1998). Semliki Forest virus-based DNA expression vector transient protein production followed by cell death. Gene Ther. **5:**415-418.

Kuhn, R. J., Niesters, H. G., Hong, Z., and Strauss, J. H. (1991). Infectious RNA transcripts from Ross River virus cDNA clones and the construction and characterization of defined chimeras with Sindbis virus. Virology. **182:**430-441.

Lebedeva, I., Fujita, K., Nihrane, A., and Silver, J. (1997). Infectious particles derived from Semliki Forest virus vectors encoding murine leukemia virus envelopes. J. Virol. **71:**7061-7067.

Lee, A. H., Suh, Y. S., Sung, J. H., Yang, S. H., and Sung, Y. C. (1997). Comparison of various expression plasmids for the induction of immune response by DNA immunization. Mol. Cells. **7:**495-501.

Lee, S., Owen, K. E., Choi, H. K., Lee, H., Lu, G. G., Wengler, G., Brown, D. T., Rossmann, M. G., and Kuhn, R. J. (1996). Identification of a protein binding site on the surface of the alphavirus nucleocapsid and its implication in virus assembly. Structure. **4:**531-541.

Levine, B., Goldman, J. E., Jiang, H. H., Griffin, D. E., and Hardwick, J. M. (1996). Bcl-2 protects mice against fatal alphavirus encephalitis. Proc. Nat. Acad. Sci. USA. **93:**4810-4815.

Levine, B., Huang, Q., Isaacs, J. T., Reed, J. C., Griffin, D. E., and Hardwick, J. M. (1993). Conversion of lytic to persistent alphavirus infection by the bcl-2 cellular oncogene. Nature. **361:**739-742.

Lewis, J., Oyler, G. A., Ueno, K., Fannjiang, Y. R., Chau, B. N., Vornov, J., Korsmeyer, S. J., Zou, S. F., and Hardwick, J. M. (1999). Inhibition of virus-induced neuronal apoptosis by Bax. Nat. Med. **5:**832-835.

Li, K. J. and Garoff, H. (1996). Production of infectious recombinant Moloney murine leukemia virus particles in BHK cells using Semliki Forest virus-derived RNA expression vectors. Proc. Natl. Acad. Sci. USA. **93:**11658-11663.

Li, K. J. and Garoff, H. (1998). Packaging of intron containing genes into retrovirus vectors by alphavirus vectors. Proc. Natl. Acad. Sci. USA. **95:**3650-3654.

Liang, X. H., Jiang, H. H., and Levine, B. (1997). Expression of a biologically active antiviral antibody using a Sindbis virus vector system. Mol. Immunol. **34:**907-917.

Liang, X. H., Kleeman, L. K., Jiang, H. H., Gordon, G., Goldman, J. E., Berry, G., Herman, B., and Levine, B. (1998). Protection against fatal Sindbis encephalitis by beclin, a novel Bcl-2-interacting protein. J. Virol. **72:**8586-8596.

Liao, C. L., Lin, Y. L., Wang, J. J., Huang, Y. L., Yeh, C. T., Ma, S. H., and Chen, L. K. (1997). Effect of enforced expression of human bcl-2 on Japanese encephalitis virus-induced apoptosis in cultured cells. J. Virol. **71:**5963-5971.

Liljeström, P. and Garoff, H. (1991). A new generation of animal cell expression vectors based on the Semliki Forest virus replicon. Bio/Technology. **9:**1356-1361.

Liljeström, P., Lusa, S., Huylebroeck, D., and Garoff, H. (1991). In vitro mutagenesis of a full-length cDNA clone of Semliki Forest virus: the 6,000-molecular-weight membrane protein modulates virus release. J. Virol. **65:**4107-4113.

Lindenbach, B. D. and Rice, C. M. (1997). Trans-complementation of yellow fever virus NS1 reveals a role in early RNA replication. J. Virol. **71:**9608-9617.

Lobigs, M. and Garoff, H. (1990). Fusion function of the Semliki Forest virus spike is activated by proteolytic cleavage of the envelope glycoprotein precursor p62. J. Virol. **64:**1233-1240.

London, S. D., Schmaljohn, A. L., Dalrymple, J. M., and Rice, C. M. (1992). Infectious enveloped RNA virus antigenic chimeras. Proc. Natl. Acad. Sci. USA. **89:**207-211.

Lopez, S., Yao, J. S., Kuhn, R. J., Strauss, E. G., and Strauss, J. H. (1994). Nucleocapsid-glycoprotein interactions required for assembly of alphaviruses. J. Virol. **68:**1316-1323.

Lundström, K. (1999). Alphavirus as tools in neurobiology and gene therapy. J. Rec. Signal Transd. Res. **19:**673-686.

Lundström, K., Michel, A., Blasey, H., Bernard, A. R., Hovius, R., Vogel, H., and Surprenant, A. (1997). Expression of ligand-gated ion channels with the Semliki Forest virus expression system. J. Rec. Signal Transd. Res. **17:**115-126.

Lundström, K., Mills, A., Allet, E., Ceszkowski, K., Agudo, G., Chollet, A., and Liljeström, P. (1995a). High-level expression of G protein-coupled receptors with the aid of the Semliki Forest virus expression system. J. Rec. Signal Transd. Res. **15:**23-32.

Lundström, K., Mills, A., Buell, G., Allet, E., Adami, N., and Liljeström, P. (1994). High-level expression of the human neurokinin-1 receptor in mammalian cell lines using the Semliki Forest virus expression system. Eur. J. Biochem. **224:**917-921.

Lundström, K., Richards, J. G., Pink, J. R., and Jenck, F. (1999). Efficient *in vivo* expression of a reporter gene in rat brain after injection of recombinant replicon-deficient Semliki Forest virus. Gene Ther. Mol. Biol. **3:**15-23.

Lundström, K. and Turpin, M. P. (1996). Proposed shizophrenia-related gene polymorphism: expression of the Ser9Gly mutant human dopamine D3 receptor with the Semliki Forest virus system. Biochem. Biophys. Res. Commun. **225:**1068-1072.

Lundström, K., Turpin, M. P., Large, C., Robertson, G., Thomas, P., and Lewell, X. Q. (1998). Mapping of dopamine D3 receptor binding site by pharmacological characterization of mutants expressed in CHO cells with the Semliki Forest virus system. J. Rec. Signal Transd. Res. **18:**133-150.

Lundström, K., Vargas, A., and Allet, B. (1995b). Functional activity of a biotinylated human neurokinin 1 receptor fusion expressed in the Semliki Forest virus system. Biochem. Biophys. Res. Comm. **208:**260-266.

Maletic-Savatic, M., Malinow, R., and Svoboda, K. (1999). Rapid dendritic morphogenesis in CA1 hippocampal dendrites induced by synaptic activity. Science. **283:**1923-1927.

Malherbe, P., Kratzeisen, C., Lundström, K., Richards, J. G., Faull, R. L. M., and Mutel, V. (1999). Cloning and functional expression of alternative spliced variants of the human metabotropic glutamate receptor 8. Mol. Brain Res. **67:**201-210.

Marheineke, K., Lenhard, T., Haase, W., Beckers, T., Michel, H., and Reilander, H. (1998). Characterization of the human gonadotropin-releasing hormone receptor heterologously produced using the baculovirus/insect cell and the Semliki Forest virus systems. Cell. Mol. Neurobiol. **18:**509-524.

McCue, L. A. and Anders, D. G. (1998). Soluble expression and complex formation of proteins required for HCMV DNA replication using the SFV expression system. Protein Expression & Purification. **13:**301-312.

Michel, A. D., Lundström, K., Buell, G., N., Surprenant, A., Valera, S., and Humphrey, P. P. (1996). The binding characteristics of a human bladder recombinant P2X purinoreceptor, labelled with [^{3}H]-αβmeATP, [^{35}S]-ATPγS or [^{33}P]-ATP. British J. Pharmacol. **117:**1254-1260.

Monastyrskaia, K., Goepfert, F., Hochstrasser, R., Acuna, G., Leighton, J., Pink, J. R., and Lundström, K. (1999). Expression and intracellular localisation of odorant receptors in mammalian cell lines using Semliki Forest virus vectors. J. Rec. Signal Transd. Res. **19:**687-701.

Mor, G., Yamshchikov, G., Sedegah, M., Takeno, M., Wang, R., Houghten, R. A., Hoffman, S., and Klinman, D. M. (1996). Induction of neonatal tolerance by plasmid DNA vaccination of mice. J. Clin. Invest. **98:**2700-2705.

Mossman, S., Bex, F., Berglund, P., Arthos, J., O'Neil, S. P., Riley, D., Maul, D. H., Bruck, C., Momin, P., Burny, A., Fultz, P. N., Mullins, J. I., Liljeström, P., and Hoover, E. A. (1996). Protection against lethal SIVsmmPBj14 disease by a recombinant Semliki Forest virus gp160 vaccine and by gp120 subunit vaccine. J. Virol. **70:**1953-1960.

Nichols, W. W., Ledwith, B. J., Manam, S. V., and Troilo, P. J. (1995). Potential DNA vaccine integration into host cell genome. Ann. N. Y. Acad. Sci. **772:**30-39.

Nilsson, M., von Bonsdorff, C. H., Weclewicz, K., Cohen, J., and Svensson, L. (1998). Assembly of viroplasm and virus-like particles of rotavirus by a Semliki Forest virus replicon. Virology. **242:**255-265.

Notka, F., Stahl-Hennig, C., Dittmer, U., Wolf, H., and Wagner, R. (1999). Construction and characterization of recombinant VLPs and Semliki Forest virus live vectors for comparative evaluation in the SHIV monkey model. Biol. Chem. **380:**341-352.

Ohno, K., Sawai, K., Iijima, Y., Levin, B., and Meruelo, D. (1997). Cell-specific targeting of Sindbis virus vectors displaying IgG-binding domains of protein A. Nat. Biotechnol. **15:**763-767.

Olivo, P. D., Collins, P. L., Peeples, M. E., and Schlesinger, S. (1998). Detection and quantitation of human respiratory syncytial virus (RSV) using minigenome cDNA and a Sindbis virus replicon: A prototype assay for negative-strand RNA viruses. Virology. **251:**198-205.

Olivo, P. D., Frolov, I., and Schlesinger, S. (1994). A cell line that expresses a reporter gene in response to infection by Sindbis virus: a prototype for detection of positive strand RNA viruses. Virology. **198:**381-384.

Olkkonen, V. M., Liljeström, P., Garoff, H., Simons, K., and Dotti, C. G. (1993). Expression of heterologous proteins in cultured rat hippocampal neurons using the Semliki Forest virus vector. J. Neurosci. Res. **35:**445-451.

Olson, K. E., Carlson, J. O., and Beaty, B. J. (1992). Expression of the chloramphenicol acetyltransferase gene in Aedes albopictus (C6/36) cells using a non-infectious Sindbis virus expression vector. Insect Mol. Biol. **1:**49-52.

Olson, K. E., Higgs, S., Gaines, P. J., Powers, A. M., Davis, B. S., Kamrud, K. I., Carlson, J. O., Blair, C. D., and Beaty, B. J. (1996). Genetically engineered resistance to Dengue-2 virus transmission in mosquitoes. Science. **272:**884-886.

Olson, K. E., Higgs, S., Hahn, C. S., Rice, C. M., Carlson, J. O., and Beaty, B. J. (1994). The expression of chloramphenicol acetyltransferase in Aedes albopictus (C6/36) cells and Aedes triseriatus mosquitoes using a double subgenomic recombinant Sindbis virus. Insect Biochem. & Mol. Biol. **24:**39-48.

Opstelten, D. J., Wallin, M., and Garoff, H. (1998). Moloney murine leukemia virus envelope protein subunits, gp70 and Pr15E, form a stable disulfide-linked complex. J. Virol. **72:**6537-6545.

Osterhaus, A. D. M. E., van Baalen, C. A., Gruters, R. A., Schutten, M., Siebelink, C. H. J., Hulskotte, E. G. J., Tijhaar, E. J., Randall, R. E. R., van Amerongen, G., Fleuchaus, A., Erfle, V., and Sutter, G. (1999). Vaccination with Rev and Tat against AIDS. Vaccine. **17:**2713-2714.

Owe-Larsson, B., Berglund, M. M., Kristensson, K., Garoff, H., Larhammar, D., Brodin, L., and Low, P. (1999). Perturbation of the synaptic release machinery in hippocampal neurons by overexpression of SNAP-25 with the Semliki Forest virus vector. Eur. J. Neurosci. **11:**1981-1987.

Paul, N. L., Marsh, M., McKeating, J. A., Schulz, T. F., Liljeström, P., Garoff, H., and Weiss, R. A. (1993). Expression of HIV-1 envelope glycoproteins by Semliki Forest virus vectors. AIDS Res. Hum. Retroviruses. **9:**963-970.

Polo, J. M., Belli, B. A., Driver, D. A., Frolov, I., Sherrill, S., Hariharan, M. J., Townsend, K., Perri, S., Mento, S. J., Jolly, D. J., Chang, S. M. W., Schlesinger, S., and Dubensky, T. W. Jr. (1999). Stable alphavirus packaging cells lines for Sindbis virus and Semliki Forest virus-derived vectors. Proc. Natl. Acad. Sci. USA. **96:**4598-4603.

Powers, A. M., Olson, K. E., Higgs, S., Carlson, J. O., and Beaty, B. J. (1994). Intracellular immunization of mosquito cells to LaCrosse virus using a recombinant Sindbis virus vector. Virus Res. **32:**57-67.

Pugachev, K. V., Mason, P. W., and Frey, T. K. (1995). Sindbis vectors suppress secretion of subviral particles of Japanese encephalitis virus from mammalian cells infected with SIN-JEV recombinants. Virology. **209:**155-166.

Pugachev, K. V., Mason, P. W., Shope, R. E., and Frey, T. K. (1995). Double-subgenomic Sindbis virus recombinants expressing immunogenic proteins of Japanese encephalitis virus induce significant protection in mice against lethal JEV infection. Virology. **212:**587-594.

Pushko, P., Parker, M., Ludwig, G. V., Davis, N. L., Johnston, R. E., and Smith, J. F. (1997). Replicon-helper systems from attenuated Venezuelan equine encephalitis virus - Expression of heterologous genes in vitro and immunization against heterologous pathogens in vivo. Virology. **239:**389-401.

Raju, R. and Huang, H. V. (1991). Analysis of Sindbis virus promoter recognition in vivo, using novel vectors with two subgenomic mRNA promoters. J. Virol. **65:**2501-10.

Raju, R., Subramaniam, S. V., and Haijou, M. (1995). Genesis of Sindbis virus by in vivo recombination of nonreplicative RNA precursors. J. Virol. **69:**7391-7401.

Rayms-Keller, A., Powers, A. M., Higgs, S., Olson, K. E., Kamrud, K. I., Carlson, J. O., and Beaty, B. J. (1995). Replication and expression of a recombinant Sindbis virus in mosquitoes. Insect Mol. Biol. **4:**245-251.

Rice, C. M., Levis, R., Strauss, J. H., and Huang, H. V. (1987). Production of infectious RNA transcripts from Sindbis virus cDNA clones: mapping of lethal mutations, rescue of a temperature-sensitive marker, and in vitro mutagenesis to generate defined mutants. J. Virol. **61:**3809-3819.

Rolls, M. M., Haglund, K., and Rose, J. K. (1996). Expression of additional genes in a vector derived from a minimal RNA virus. Virology. **218:**406-411.

Rolls, M. M., Webster, P., Balba, N. H., and Rose, J. K. (1994). Novel infectious particles generated by expression of the vesicular stomatitis virus glycoprotein from a self-replicating RNA. Cell. **79:**497-506.

Ryan, M. D. and Drew, J. (1994). Foot-and-mouth disease virus 2A oligopeptide mediated cleavage of an artificial polyprotein. EMBO J. **13:**928-933.

Ryan, M. D., King, A. M. Q., and Thomas, G. P. (1991). Cleavage of foot-and-mouth disease virus polyprotein is mediated by residues located within a 19 amino acid sequence. J. Gen. Virol. **72:**2727-2732.

Salminen, A., Wahlberg, J. M., Lobigs, M., Liljeström, P., and Garoff, H. (1992). Membrane fusion process of Semliki Forest virus II: Cleavage dependent reorganization of the spike protein complex controls virus entry. J. Cell Biol. **116:**349-357.

Sammin, D. J., Butler, D., Atkins, G. J., and Sheahan, B. J. (1999). Cell death mechanisms in the olfactory bulb of rats infected intranasally with Semliki Forest virus. Neuropathology and Applied Neurobiology. **25:**236-243.

Sawai, K., Ikeda, H., Ishizu, A., and Meruelo, D. (1999). Reducing cytotoxicity induced by Sindbis viral vectors. Mol. Genet. Metabol. **67:**36-42.

Sawai, K. and Meruelo, D. (1998). Cell-specific transfection of choriocarcinoma cells by using Sindbis virus hCG expressing chimeric vector. Biochem. Biophys. Res. Commun. **248:**315-323.

Sawai, K., Ohno, K., Ijima, Y., Levin, B., and Meruelo, D. (1998). A novel method of cell-specific mRNA transfection. Mol. Genet. & Metabol. **64:**44-51.

Scallan, M. F., Allsopp, T. E., and Fazakerley, J. K. (1997). bcl-2 acts early to restrict Semliki Forest virus replication and delays virus-induced programmed cell death. J. Virol. **71:**1583-1590.

Scheer, A., Björklöf, K., Cotecchia, S., and Lundström, K. (1999). Expression of the alpha(1b)-adrenergic receptor and G protein subunits in mammalian cell lines using Semliki Forest virus expression system. J. Rec. Signal Transd. Res. **19:**369-378.

Schlaeger, E. J. and Lundström, K. (1998). Effect of temperature on recombinant protein expression in Semliki Forest virus infected mammalian cell lines growing in serum-free suspension cultures. Cytotechnol. **28:**205-211.

Schlesinger, S. and Weiss, B. G. (1994). Recombination between Sindbis virus RNAs. Archiv. Virol. **9:**213-220.

Seabaugh, R. C., Olson, K. E., Higgs, S., Carlson, J. O., and Beaty, B. J. (1998). Development of a chimeric Sindbis virus with enhanced per Os infection of Aedes aegypti. Virology. **243:**99-112.

Simons, M., Tienari, P. J., Dotti, C. G., and Beyreuther, K. (1995). Two-dimensional gel mapping of the processing of the human amyloid precursor protein in rat hippocampal neurons. FEBS Letters. **368:**363-366.

Sjöberg, E. M. and Garoff, H. (1996). The translation-enhancing region of the Semliki Forest virus subgenome is only functional in the virus-infected cell. J. Gen. Virol. **77:**1323-1327.

Sjöberg, E. M., Suomalainen, M., and Garoff, H. (1994). A significantly improved Semliki Forest virus expression system based on translation enhancer segments from the viral capsid gene. Bio/Technology. **12:**1127-1131.

Skoging, U. and Liljeström, P. (1998). Role of the C-terminal tryptophan residue for the structure-function of alphavirus capsid protein. J. Mol. Biol. **279:**865-872.

Skoging, U., Vihinen, M., Nilsson, L., and Liljeström, P. (1996). Aromatic interactions define the binding of the alphavirus spike to its nucleocapsid. Structure. **4:**519-529.

Smerdou, C. and Liljeström, P. (1999). Two-helper RNA system for production of recombinant Semliki forest virus particles. J. Virol. **73:**1092-1098.

Smith, S. M., Maldarelli, F., and Jeang, K. T. (1997). Efficient expression by an alphavirus replicon of a functional ribozyme targeted to human immunodeficiency virus type 1. J. Virol. **71:**9713-9721.

Strauss, J. H. and Strauss, E. G. (1994). The alphaviruses: Gene expression, replication, and evolution. Microbiol. Revs. **58:**491-562.

Suomalainen, M., Liljeström, P., and Garoff, H. (1992). Spike protein-nucleocapsid interactions drive the budding of alphaviruses. J. Virol. **66:**4737-4747.

Torresi, J., Li, F., Locarnini, S. A., and Anderson, D. A. (1999). Only the non-glycosylated fraction of hepatitis E virus capsid (open reading frame 2) protein is stable in mammalian cells. J. Gen. Virol. **80:**1185-1188.

Torresi, J., Meanger, J., Lambert, P., Li, F., Locarnini, S. A., and Anderson, D. A. (1997). High level expression of the capsid protein of hepatitis E virus in diverse eukaryotic cells using the Semliki Forest virus replicon. J. Virol. Meth. **69:**81-91.

Tsuji, M., Bergmann, C. C., Takita-Sonoda, Y., Murata, K., Rodrigues, E. G., Nussenzweig, R. S., and Zavala, F. (1998). Recombinant Sindbis viruses expressing a cytotoxic T-lymphocyte epitope of a malaria parasite or of influenza virus elicit protection against the corresponding pathogen in mice. J. Virol. **72:**6907-6910.

Tubulekas, I. and Liljeström, P. (1998). Suppressors of cleavage-site mutations in the p62 envelope protein of Semliki Forest virus reveal dynamics in spike structure function. J. Virol. **72:**2825-2831.

Ulmanen, I., Peränen, J., Tenhunen, J., Tilgmann, C., Karhunen, T., Panula, P., Bernasconi, L., Aubry, J. P., and Lundström, K. (1997). Expression and intracellular localization of catechol O-methyltransferase in transfected mammalian cells. Eur. J. Biochem. **243:**452-459.

Verrier, F. C., Charneau, P., Altmeyer, R., Laurent, S., Borman, A. M., and Girard, M. (1997). Antibodies to several conformation-dependent epitopes of gp120/gp41 inhibit CCR-5-dependent cell-to-cell fusion mediated by the native envelope glycoprotein of a primary macrophage-tropic HIV-1 isolate. Proc. Natl. Acad. Sci. USA. **94:**9326-9331.

Wahlfors, J. J. and Morgan, R. A. (1999). Production of minigen-containing retroviral vectors using an alphavirus/retrovirus hybrid vector system. Hum. Gene Ther. **10:**1197-1206.

Wahlfors, J. J., Xanthopoulos, K. G., and Morgan, R. A. (1997). Semliki Forest virus-mediated production of retroviral vector RNA in retroviral packaging cells. Hum. Gene Ther. **8:**2031-2041.

Weiss, B., Rosenthal, R., and Schlesinger, S. (1980). Establishment and maintenance of persistent infection by Sindbis virus in BHK cells. J. Virol. **33:**463-474.

Weiss, B. G. and Schlesinger, S. (1991). Recombination between Sindbis virus RNAs. J. Virol. **65:**4017-4025.

White, C. L., Thomson, M., and Dimmock, N. J. (1998). Deletion analysis of a defective interfering Semliki Forest virus RNA genome defines a region in the nsP2 sequence that is required for efficient packaging of the genome into virus particles. J. Virol. **72:**4320-4326.

Wolff, J. A., Ludtke, J. J., Acsadi, G., Williams, P., and Jani, A. (1992). Long-term persistence of plasmid DNA and foreign gene expression in mouse muscle. Hum. Mol. Genet. **1:**363-369.

Xiong, C., Levis, R., Shen, P., Schlesinger, S., Rice, C. M., and Huang, H. V. (1989). Sindbis virus: an efficient, broad host range vector for gene expression in animal cells. Science. **243:**1188-1191.

Yadani, F. Z., Kohl, A., Prehaud, C., Billecocq, A., and Bouloy, M. (1999). The carboxy-terminal acidic domain of Rift Valley fever virus NSs protein is essential for the formation of filamentous structures but not for the nuclear localization of the protein. J. Virol. **73:**5018-5025.

Ying, H., Zaks, T. Z., Wang, R. F., Irvine, K. R., Kammula, U. S., Marincola, F. M., Leitner, W. W., and Restifo, N. P. (1999). Cancer therapy using a self-replicating RNA vaccine. Nat. Med. **5:**823-827.

Zhang, J., Asselin-Paturel, C., Bex, F., Bernard, J., Chehimi, J., Willems, F., Caignard, A., Berglund, P., Liljeström, P., Burny, A., and Chouaib, S. (1997). Cloning of human Il-12 p40 and p35 DNA into the Semliki Forest virus vector: expression of Il-12 in human tumor cells. Gene Ther. **4:**367-374.

Zhao, H., Lindqvist, B., Garoff, H., von Bonsdorff, C. H., and Liljeström, P. (1994). A tyrosine-based motif in the cytoplasmic domain of the alphavirus envelope protein is essential for budding. EMBO J. **13:**4204-4211.

Zhou, X., Berglund, P., Rhodes, G., Parker, S. E., Jondal, M., and Liljeström, P. (1994). Self-replicating Semliki Forest virus RNA as recombinant vaccine. Vaccine. **12:**1510-1514.

Zhou, X., Berglund, P., Zhao, H., Liljeström, P., and Jondal, M. (1995). Generation of cytotoxic and humoral immune responses by nonreplicative recombinant Semliki Forest virus. Proc. Natl. Acad. Sci. USA. **92:**3009-3013.

TRANSIENT EXPRESSION IN MAMMALIAN CELLS : APPLICATIONS AND PERSPECTIVES

Alain R. BERNARD
Serono Pharmaceutical Research Institute
14 chemin des Aulx
CH - 1228 Plan-Les-Ouates, Switzerland
E-mail : alain.bernard@serono.com

Abstract

In this article, I review the various methods of transient gene expression in cultured cells currently available for the large scale (10-100 mg) production of recombinant proteins of pharmaceutical interest. The pros and cons of each technology are briefly outlined. I also summarize their current bottlenecks and attempt to propose directions for further progress and development. Finally a comparative study between methods is illustrated with a particular receptor protein

1. Introduction

The human genome is virtually decoded and the race is open to search for novel genes and assigning functions to thousands of novel DNA sequences (Drews, 2000). On the other hand, economic pressures force pharmaceutical companies to shorten the drug discovery cycle time and identify new candidate drugs faster and cheaper. This means having access to proteins of therapeutic interest ("targets") very rapidly and in sufficient amounts, i.e. 10-100 mg. So the biotechnologists are faced with two types of demands that have recently emerged : 1) Throughput : the need to produce a large number of different proteins at a scale sufficient for enabling functional studies and 2) Speed : the need to produce functional human proteins very rapidly. Transient expression technologies provide the basis for achieving these goals. The challenges are thus to expand their original domain of application (traditionally the identification of one unknown or undefined gene product at microscale) to the high throughput, high speed, large scale (milligram to gram quantities) production of a large number of genes.

In parallel, gene and cell therapy and DNA vaccination concepts have provoked the development of "gene delivery" technologies (Douar, 2000). These, at least in part, synergise with transient expression systems but are obviously submitted to very different constraints and aim at different goals (protein expression in a patient's body or in a patient's cells) and for these reasons I will not address the related technologies in this article.

Other articles in this book specifically describe each technology in detail and its applications (Jordan, 2000; Smerdou and Liljeström, 2000). Here, I will review their

M. Al-Rubeai (ed.), Cell Engineering, 211–218.

respective advantages and disadvantages and outline their current limitations (and, when possible, how they could be overcome) and the perspectives that each one offers. In addition, I will report on a comparison study between DNA and viral vectors used for the expression of the same recombinant protein.

Transient gene expression technologies were reviewed recently (Wurm and Bernard, 1999) and can be classified according to how the foreign gene is transferred inside the cells: DNA, RNA or viral vectors and common features to all of them are:
Extremely short time frame for the generation of product (days)
- Suitability to high throughput, i.e. expression of many genes or mutants in parallel at the same time.
- Suitability to a wide range of host cell systems
- Simplicity, in particular in the construction of expression vectors
- Intrinsic genetic stability and consistency due to extremely short time frame between generation of vector and product recovered

In terms of scalability, it appears that the viral based processes, in particular those using vaccinia have historically been first to reach large scale productions and this is clearly the result of the maturity of the vaccine production methodologies. DNA and RNA mediated transient expression systems are lagging behind, but are gradually catching up.

2. Plasmid DNA Vectors

DNA is normally produced by extracting it from *E.coli* strains harbouring the plasmid. Not much attention has been paid, up to now, to the quality of DNA derived from this process and to the reproducibility of the DNA preparation, when applied to cell transfection (Weber *et al.*, 1995). Lessons can be learnt from the gene therapy and DNA vaccine areas where several studies have been carried out (Schoor et al, 1995). What are the criteria that define accurately a DNA preparation : migration in agarose gels, UV absorbance, presence of contaminants (E.coli DNA, RNA, nucleases, proteins, endotoxins....), sterility ? Are there other features or conditions that determine transfection and expression efficiencies ? Are some features of the plasmid DNA more important and do they have different optima for specific transfection methods ? These questions need answers, some of them are addressed in depth in this book by M. Jordan (Jordan, 2000) but a systematic approach is required if we want to be able to reproducibly transfect cells in culture and then use them for reliable protein production on a large scale. Although it seems difficult to see presently how large scale RNA preparations could be used for that purpose as described in another chapter of this book (Smerdou and Liljeström, 2000), the same issues will have to be solved for this type of vectors.

Also, even if considerable progress has recently been made in our ability to produce large amounts of pure DNA (Schorr *et al.*, 1995) and in the development of elegant chemistries to optimize gene delivery (Felgner *et al.*, 1999), minimizing the mass of DNA required down to microgram quantities per liter of culture (equivalent to approximately femtogram quantities per cell) would certainly be a major breakthrough simplifying the large scale transfection process.

In fact, the actual yield of "operational" plasmid DNA (ie used as a template by the transfected cell for expressing the heterologous protein) vs plasmid DNA put into contact with the cell is probably extremely low considering the large molar excess of plasmid molecules DNA ($10^5 - 10^7$ per cell) used in the transfection mix.

Plasmid DNA can be introduced into mammalian cells by a wide variety of methods but three particular ones seem more amenable to large scale operations : Calcium-phosphate DNA co-precipitation (Graham and Van der Erb, 1973), polyethyleneimine (PEI) (Boussif *et al.*, 1995; Schlaeger and Christensen, 1999) or Ro-1539 DNA complexes (Haldankar and McClure, 1999; Schlaeger *et al*, 1999) and electroporation (Neumann *et al.*, 1982; Chu *et al.*, 1987; Parham *et al.*, 1998). Although many more methods are described in the literature, these three are most attractive either from a cost-effectiveness or from an efficiency point of view. Commercial transfection reagents are usually sold in small quantities and are not designed to be used in large scale bioreactors. More importantly, the three transfection techniques described above are currently developed for large scale transfection of both adherent and suspension cells. For cells grown in suspension, they have been used at scales above 10 litre (Etcheverry, 1999; Schlaeger *et al.*, 1999). Expression levels of 1-10 mg/L of product have been reported for secreted proteins, corresponding to specific productivities of about 0.1-1 pg/cell/day .

The molecular mechanisms of DNA entry inside cells are established : endocytosis for calcium phosphate precipitates and PEI complexes (Loyter *et al.*, 1982), naked DNA diffusion through pores for electroporation (Chu *et al.*, 1987). However, how and which specific features of the DNA or of the complexes affect the efficiency of the transfer inside the cell is only poorly understood in each of these methods. Once it has entered the cell, the fate of DNA is also significantly influenced by its format, whether complexed with calcium phosphate or anchored to another polymer like PEI or in free form.

Toxicity of reagents used for transfection also frequently limits their use. The generic approach is to dilute or to exchange the culture medium of transfected cultures with fresh medium soon after transfection (typically a few hours post addition of the transfection mix). Ideally, this step should be eliminated by employing completely non toxic reagents or by the ability to remove the toxicity by simple actions such as pH shifts (Jordan, 2000).

The most common cell types used in large scale DNA transfections are the HEK293 cell lines. The serum free culture of these cells in suspension is now well established (Graham, 1987; Ostrove *et al.*, 1998; Etcheverry; 1999; Geisse *et al.*, 1999; Schlaeger and Christensen, 1999). In these media, cell aggregation can be virtually negligible and clumping and aggregation have only to considered seriously when induced by the transfection reagents and / or transfection conditions. Solutions are usually found by manipulating media ingredients or physical conditions of the culture such as agitation speed (Tolbert *et al.*, 1980; Peshwa *et al.*, 1993). Several different cell lines have been derived for the HEK293 parent. The two most popular variants for transient expression are HEK293 T which express the SV-40 large T antigen (Kim and Lee, 1997) and HEK293-EBNA which express the nuclear antigen from Epstein-Barr Virus EBNA (Cachianes *et al.*, 1993).

Other cell types such as COS or CHO have been reported (Heffernan and Dennis, 1990; Blasey *et al.*, 1995a; Parham *et al.*, 1997; Jirka *et al.*, 1999) but they appear to either be less robust or yield less reproducible results than the HEK293 type.

These cell lines, as well as true anchorage dependant mammalian cell lines have also been used for conventional recombinant protein production or mass production of virus based vectors for gene therapy and they are usually cultured on microcarriers sytems (Ostrove *et al.*, 1998) or in stacked, multilayer systems (Blasey *et al.*, 1995b). Cells can be maintained in these production vessels for a limited period in static, batch-type culture or, for extended periods, in a perfused mode. The principal disadvantage of stacked multi-layer systems is cost.

Our laboratory and a few others have been working recently on a variation of the large scale transfection method (Blasey *et al.*, 1999; Geisse *et al.*, 1999; Schlaeger and Christensen, 1999). In this approach, HEK293-EBNA cells are transfected with plasmid DNA at conventional small scale, either by electroporation or by the CaPO4 method. The HEK293-EBNA cells are transfected with a plasmid such as pCEP4 (Invitrogen) which contains : the replication origin region (OriP) of EBV., a multiple cloning site, the CMV promotor, the bacterial AMP resistance marker and the mammalian HygB resistance gene. Most common expression vectors, like those used for constructing stable recombinant cell lines contain strong viral promoters, such as SV40 or CMV (Gorman *et al.*, 1990; Paborsky *et al.*, 1990; Cachianes *et al.*, 1993; Geisse *et al.*, 1997). Also, recently, the elongation factor 1 promotor, a non-viral promoter has been shown to also drive high levels of transcription (Kim *et al.*, 1990).

Hygromycin selection post transfection yields a cell population which can be scaled up in a production process. Thereafter, cultures can be continued like a stable cell line for at least 6 months without significant drop in expression level (Cachianes *et al.*, 1993). Both genomic and cDNA sequences can be expressed in this system. We and others have used this system extensively to express receptors and other proteins at levels between 0.2-10 µg/ml (Cachianes *et al.*, 1993; Boie *et al.*, 1995; Robbins and Horlick, 1998; Rominger *et a.l*, 1998; Etcheverry, 1999;).

3. Viral Vectors

Various viral vectors have been used for the large scale expression of recombinant proteins : Adenovirus, Vaccinia, Alphaviruses and others. Adenovirus, although now almost exclusively developed for gene therapy, was used to produce recombinant proteins in a 5 litre bioreactor (Massie *et al.*, 1995; Garnier *et al.*, 1996). A Vaccinia process was scaled up to continuous perfusion at the 40 litre working volume to produce HIV vaccine on large scale (Barrett *et al.*, 1989; Falkner *et al.*, 1992). These two viruses share the advantage of easy virus scale up to any volume required. However, viral replication and the related biological risks require the implementation of special safety measures when operated at industrial scale. Sindbis Virus, Venezuelan Equine Encephalitis and Semliki Forest Virus (SFV) are the most studied alphaviruses (Liljeström and Garoff, 1991) used for heterologous expression. Their broad host range make them particularly attractive for functional studies and investigation of the biological role of proteins (Polo *et al.*, 1999). Their use as vectors for transient gene expression is reviewed in another chapter (Smerdou and Liljeström, 2000). The particular advantages of these viruses are linked to their suitability for a wide range of purposes: large scale protein production for biochemical and in-vitro functional studies, in-vivo studies in animal models of diseases and, potentially, gene therapy and vaccination. They can be used in several different formats as packaged, recombinant viruses, as RNA vectors or as DNA vectors (Smerdou and Liljeström, 2000). Their particular disadvantages are linked mostly to their use in a packaged virus form for which biological safety should be stringent as these viruses can practically infect most mammalian cells and as their wild type equivalents are virulent pathogens.

To make this system even more safe and widely used would require to make packaging cell lines or to generate non lytic viruses. This has been achieved by Dubensky and colleagues for Sindbis (Polo *et al.*, 1999) and will, undoubtedly find many useful applications.

4. Comparative Study of DNA, RNA and Virus Vectors.

In the last part of this article, we report on our comparative experience with the HEK293-EBNA and the Semliki Forest Virus systems using a test case of the 5-HT$_3$ receptor. These results have been presented elsewhere in detail (Blasey *et al.*, 1999a) and only a summary is given here. The 5-HT$_3$ receptor is a ligand-gated cation channel of pharmacological interest (Tyers, 1991).

As described earlier, a recombinant non-infectious SFV expressing 5HT3 was packaged by BHK cells and used to infect cells after activation with chymotrypsin (Blasey *et al.*, 1999b).

As previously described (Blasey *et al.*, 1999a), HEK293-EBNA were transfected by electroporation with the plasmid pCEP4-5HT3, the culture was selected with 300 µg/ml hygromycin B from day 2 post transfection and a cell bank was frozen at passage 3 post transfection. The cells were then gradually scaled up in suspension culture up to a 12 l bioreactor.

With SFV, we measured an expression level of $3\text{-}4 \cdot 10^6$ receptors per cell and this level was increased to $8 \cdot 10^6$ receptors per cell by a fine tuning of pH post infection. This resulted in the production 10 to 25 mg receptor protein per reactor run.

With the HEK293-EBNA cells the expression level was above $9 \cdot 10^6$ receptors per cell which was much higher than what could be observed if the cells were cultured in anchorage dependant mode in a CellCube reactor (Blasey *et al.*, 1999a).

In a parallel experiment we also expressed the 5HT3 receptor by transfecting the pSFV-5HT$_3$ derived RNA vector into BHK cells (Blasey *et al.*, 2000). Expression levels reached up to $3 \; 10^6$ receptors per cell ; however, using the SFV to deliver the RNA was significantly more efficient compared to RNA transfection.

In summary, the two transient expression systems SFV and HEK293-EBNA allowed very high level 5-HT$_3$ receptor expression in a short period of time : days to weeks. The expressed receptor featured ligand-binding (Hovius *et al.*, 1998) and channel properties (Tairi *et al.*, 1998) similar to the native receptor.

5. Conclusions

A large variety of technologies are available for transient expression of recombinant proteins whether in cultured cell lines, in primary cells (cell therapy) or in-vivo (gene therapy). One novel technology was not described here because it is really novel and nascent : the cell free preparation of recombinant proteins (Fernholz, 1999; Michel-Reydellet *et al.*, 2000; Swartz *et al*, 2000) and it might well be extremely powerful once it is mature. With this exception, today, the issue is not to develop a new technology but rather more to choose the right system for each specific goal. Of course, the choice depends also on the type of target protein (intracellular, cell surface or secreted). Nevertheless, whatever system is chosen, it still remains to be optimized. I have outlined several areas which deserve more investigation to further enable a widespread use of these technologies. Given the important applications of these technologies, namely of assigning a function to every human gene and of discovering novel therapeutic targets for human diseases, significant progress in transient gene expression will be made in the near future.

6. Acknowledgements

I wish to thank Dr H.D. Blasey (Serono) for constructive discussions of the manuscript.

7. References

Barrett, N., Mitterer, A., Mundt, W., Eibl, J., Eibl, M., Gallo, R.C., Moss, B., and Dorner, F. (1989) Large scale production and purification of a vaccinia recombinant-derived HIV-1 gp160 and analysis of its immunogenicity, *AIDS Res.Hum.Retroviruses* **5**, 159-171.

Blasey, H.D., Aubry, J.P., Mazzei, G.J., and Bernard, A.R. (1995a) Large scale transient expression with COS cells, *Cytotech.* **18**, 183-192.

Blasey, H.D., Isch, C. and Bernard, A.R. (1995b) CellCube: a new system for large scale growth of adherent cells, *Biotech. Techniques* **9** (10), 725-728.

Blasey, H.D., Hovius, R., Vogel, H. and Bernard, A.R. (1999a) Transient expression technologies, their application and scale up: 5HT3 serotonin receptor case study; *Biochem. Soc. Trans.*, **27** (6), 956-960.

Blasey, H.D., Hovius, R., Rey, L., Vogel, H. and Bernard, A.R. (1999a) Mammalian transient gene expression - Potential, practical application and perspectives, In: *"Animal Cell Technology: products from cells, cells as products"* A. Bernard *et al.* (eds.), Kluwer Academic Publishers, pp. 331-337.

Blasey, H.D., Hovius, R, Vogel, H. and Bernard, A.R. (2000) High throughput recombinant protein production: transient gene expression strategies, *Abstract Paper, Cell Culture Engineering VII Conference*, Feb. 5-10, Santa Fe.

Boie, Y., Sawyer, N., Slipetz, D.M., Metters, K.M. and Abramovitz, M. (1995) Molecular cloning and characterization of the human prostanoid DP receptor, *J. Biol. Chem.* **270** (32), 18910-18916.

Boussif, O., Lezoualc'h, F., Zanta, M., Mergny, M., Scheman, D., Demeneix, B., and Behr, J.-P. (1995) A versatile vector for gene and oligonucleotide transfer into cells in culture and in vivo: Polyethyleneimmine, *Proc. Natl. Acad. Sci. USA* **92**, 7297-7301.

Cachianes, G., Ho, C., Weber, R.F., Williams, S.R., Goeddel, D.V. and Leung, D.W. (1993) Epstein-Barr virus-derived vectors for transient and stable expression of recombinant proteins, *BioTechniques* **15**, 255-259.

Chu, G., Hayakawa, H. and Berg, P. (1987) Electroporation for the efficient transfection of mammalian cells with DNA, *Nucleic Acids Res.* **15**, 1311-1326.

Douar, A.M. (2000) Adeno-associated virus: a promising tool for gene delivery, *This book*.

Drews, J. (2000) Drug Discovery today - and tomorrow, *Drug Discovery Today* **5** (1), 2-4.

Etcheverry, T. (1999) Recombinant proteins at your fingertips: transient transfections in 10 liter bioreactors, *Abstract Paper 4th gene delivery and cellular protein expression conference,* Oct 17-21, Lake Tahoe, California.

Falkner, F.G., Turecek, P.L., MacGillivray, R.T., Bodemer, W., Scheiflinger, F., Kandels, S., Mitterer, A., Kistner, O., Barrett, N., Eibl, J., and Dorner, F. (1992) High level expression of active human prothrombin in a vaccinia virus expression system, *Thromb Haemost.* **68,** 119-124.

Felgner, P.L., Zelphati, O. and Liang, X. (1999) Gene chemistry: breaking new ground, *Abstract Paper, 4th gene delivery and cellular protein expression conference,* Oct 17-21, Lake Tahoe, California.

Fernholz, E. (1999) About preparative protein expression in vitro, *Abstract Paper 4th gene delivery and cellular protein expression conference*. Oct 17-21, Lake Tahoe, California.

Garnier, A., Cote, J., Nadeau, I., Kamen, A. and Massie, B. (1996) Scale-up of the adenovirus expression system for the production of recombinant protein in human 293S cells *Biotechnol.Bioeng.* **51**, 613-623.

Geisse, S. Orakçi, O., Pabler, A., Patoux, A., Rinaldi, D., Watkins, J., Schmitz, R. and Gram H. (1999) Float! - and express more : converting the HEK.EBNA/oriP system to suspension culture, In: *"Animal Cell Technology : products from cells, cells as products"* A. Bernard *et al.* (eds.), Kluwer Academic Publishers pp. 339-345.

Geisse, S., Patoux, A., Rohr, F., Glueck, D., Hagen, S., Rinaldi, D. and Gram, H. (1997) EBNA-1/oriP driven expression in HEK293 cells, *Abstract Paper Transient gene expression in animal cells Conference,* May 11-14, Island of Jersey.

Gorman, C.M, Gies, D.R and McCray, G. (1990) Transient production of proteins using an adenovirus transformed cell line, *DNA and Protein Engineering Techniques* **2,** 1-28.

Graham, F.L., and Van der Eb, A.J. (1973) A new technique for the assay of infectivity of human adenovirus 5 DNA, *Virology* **52,** 456-467.

Graham, F.L. (1987) Growth of 293 cells in suspension culture, *J. Gen. Virol.* **68,** 937-940

Haldankar, R. and McClure, D.B. (1999) Transient transfection in mammalian cells: optimizing productivity and scale-up, *Abstract Paper 4th gene delivery and cellular protein expression conference,* Oct 17-21 Lake Tahoe, California.

Heffernan, M. and Dennis, J.W. (1990) Polyoma and hamster papovavirus large T antigen-mediated replication of expression shuttle vectors in Chinese hamster ovary cells, *Nucleic Acids Research* **19**, 85-92.

Hovius, R., Tairi, A.-P., Blasey, H., Bernard, A., Lundström, K. and Vogel, H. (1998) Characterization of a mouse serotonin 5-HT3 receptor purified from mammalian cells *J. Neurochem.* **70**, 824-834.

Jirka, J.F., Reedy, T.A., Cohen, L., Shao, L.M., Vaz, A. and Hawrylik, S.J. (1999) Automated scale-up of transiently expressed monoamine oxydase B in COS-7 cells on the Cellmate robotic cell culture system, *Abstract Paper4th gene delivery and cellular protein expression conference,* Oct 17-21, Lake Tahoe, California.

Jordan, M. (2000) Transient gene expression in mammalian cells based on the calcium phosphate transfection method, *This book.*

Kim, C.H., Oh, Y. and Lee, T.H. (1997) Codon optimization for high-level expression of human erythropoietin (EPO) in mammalian cells - human or yeast codon usage effect on over-expression in 293T cell culture, *Gene* **199,** 293-301.

Kim, D.W., Uetsuki, T., Kaziro, Y, Yamaguchi, N. and Sugano, S. (1990) Use of the human elongation factor 1 alpha as a versatile and efficient expression system, *Gene* **91,** 217-223.

Liljeström, P. and Garoff, H. (1991) A new generation of animal cell expression vectors based on the Semliki Forest virus system, *Bio/Technology* **9,** 1356-1361.

Loyter, A., Scangos, G.A. and Ruddle, F.H. (1982) Mechanism of DNA uptake by mammalian cells : Fate of exogenously added DNA monitored by the use of fluorescent dyes, *Proc. Natl. Acad. Sci.* **79,** 422-426.

Lundström, K. (1997) Alphaviruses as expression vectors, *Curr. Opin. Biotechnol.* **8,** 578-582.

Massie, B., Dionne, J., Lamarche, N., Fleurent, J. and Langelier, Y. (1995) Improved adenovirus vector provides herpes simplex virus ribonucleotide reductase R1 and R2 subunits very efficiently, *Bio/technology* **13,** 602-608.

Michel-Reydellet, N., Kim, D.-M. and Swartz, J.R. (2000) Improvement of cell-free protein synthesis by modification of E. coli's genotype, *Abstract Paper, 219th ACS national meeting*, San Francisco - March 26-30.

Neumann, E., Schefer-Ridder, M., Wang, Y. and Hofschneider, P.H. (1982) Gene transfer into mouse lyoma cells by electroporation in high electric fields, *EMBO* **1** (7), 841-845.

Ostrove, J.M., Iyer, P., Marshall, J. and Vacante, D. (1998) Comparison of manufacturing techniques for adenovirus production, In: *"New Developments and New Applications in Animal Cell Technology"* OW Merten *et al.* (eds.), Kluwer Academic Publishers, pp. 515-521.

Paborsky, L.R., Fendly, B.M., Fisher, K.L., Lawn, R.M., Marks, B.J., McCray, G., Tate, K.M., Vehar, G.A. and Gorman, C.M. (1990) Mammalian cell transient expression of tissue factor for the production of antigen, *Protein Engineering* **3,** 547-553.

Parham, J.H., Marie, A., Overton, L.K. and Hutchins, J.T. (1998) Optimization of transient gene expression in mammalian cells and potential for scale-up using flow electroporation, *Cytotech.* **28,** 147-155.

Parham, J., Iannone, M. and Hutchins, J. (1997) Enhancement of transient gene expression by cell cycle synchronization and butyrate treatment, *Abstract Paper Transient gene expression in animal cells Conference,* May 11-14, Island of Jersey.

Peshwa, M.V., Kyung, Y.-S., McClure. D.B. and Hu, W.S. (1993) Cultivation of mammalian cells as aggregates in bioreactors: Effect of calcium concentration on spatial distribution of viability, *Biotechnol. Bioeng.* **41**, 179-187.

Polo, J.M., Gardner, J., Ji, Y., Belli, B.A., Driver, D.A., Scherrill, S., Perri, S., Barnett, S., Otten, G., Liu, M.A. and Dubensky, T.W. (1999) Alphavirus DNA and particle replicon vectors for vaccine applications, *Abstract Paper, 4th gene delivery and cellular protein expression conference.* Oct 17-21 Lake Tahoe, California.

Robbins, A.K. and Horlick, R.A. (1998) Macrophage scavenger receptor confers an adherent phenotype to cells in culture, *Biotechniques* **25** (2), 240-244.

Rominger, D.H, Rominger, C.M., Fitzgerald, L.W., Grzanna, R., Largent, B.L. and Zaczek, R. (1998) Characterization of [^{125}I]sauvagine binding to CRH2receptors: membrane homogenate and autoradiographic studies, *J. Pharm. Exp. Therapeutics* **286** (1), 459-468.

Schlaeger, E.J. and Christensen, K. (1999) Transient gene expression in mammalian cells grown in serum free suspension culture, *Cytotech.* **30**, 71-83.

Schlaeger, E.J., Christensen, K. and Kitas, E.A. (1999) Scale-up of transient gene expression in mammalian cells grown in serum free suspension culture, *Abstract Paper 4th gene delivery and cellular protein expression conference,* Oct 17-21, Lake Tahoe, California.

Schorr, J., Moritz, P., Seddon, T. and Schleef, M. (1995) Plasmid DNA for human gene therapy and DNA vaccines, *Annals NY Acad Sc.* **772,** 271-273.

Smerdou, C. and Liljeström, P. (2000) Alphavirus vectors : highly efficient sytems for transient gene expression,, *This book.*

Swartz, J.R., Kim, D.-M., Zawada, J. and Michel-Reydellet, N. (2000) Cell-free protein expression: Designing for efficiency and low cost, *Abstract Paper, 219th ACS national meeting,* San Francisco - March 26-30.

Tairi, A.-P., Hovius, R., Pick, H., Blasey, H., Bernard, A., Surprenant, A., Lundström, K., and Vogel, H. (1998) Ligand binding of the serotonin 5HT3 receptor studied with a novel fluorescent ligand, *Biochemistry* **37**, 15850-15864.

Tolbert, W.R., Hitt, M.M. and Feder, J. (1980) Cell aggregate suspension culture for large-scale production of biomolecules, *In Vitro* **16,** 486-490.

Tyers, M.B. (1991) 5-HT3 receptors and the therapeutic potential of 5-HT3 receptor antagonists, *Therapie* **46,** 431-435.

Weber, M., Möller, K., Welzeck, M. and Schorr, J. (1995) Effects of lipopolysaccharide on transfection efficiency in eukaryotic cells, *Biotechniques* **19,** 930-940.

Wurm, F. and Bernard, A. (1999) Large-scale transient expression in mammalian cells for recombinant protein production, *Curr. Opin. Biotech.* **10,** 156-159.

INDEX PAGE NUMBER

Cell Engineering

1. M. Al-Rubeai (ed.): *Cell Engineering*. 1999 ISBN 0-7923-5790-6
2. M. Al-Rubeai (ed.): *Cell Engineering*. Transient Expression. 2000 ISBN 0-7923-6596-8

Kluwer Academic Publishers – Dordrecht / Boston / London